Outdoor Insulation and Gas Insulated Switchgears

Outdoor Insulation and Gas Insulated Switchgears

Editors

Issouf Fofana
Stephan Brettschneider

MDPI • Basel • Beijing • Wuhan • Barcelona • Belgrade • Manchester • Tokyo • Cluj • Tianjin

Editors

Issouf Fofana
Université du Québec à
Chicoutimi (UQAC)
Canada

Stephan Brettschneider
Université du Québec à
Chicoutimi (UQAC)
Canada

Editorial Office
MDPI
St. Alban-Anlage 66
4052 Basel, Switzerland

This is a reprint of articles from the Special Issue published online in the open access journal *Energies* (ISSN 1996-1073) (available at: https://www.mdpi.com/journal/energies/special_issues/outdoor_insulation).

For citation purposes, cite each article independently as indicated on the article page online and as indicated below:

LastName, A.A.; LastName, B.B.; LastName, C.C. Article Title. *Journal Name* **Year**, *Volume Number*, Page Range.

ISBN 978-3-0365-5941-4 (Hbk)
ISBN 978-3-0365-5942-1 (PDF)

Cover image courtesy of Stephan Brettschneider

Contents

About the Editors

Issouf Fofana

Issouf Fofana, IET Fellow, held the Canada Research Chair on insulating liquids and mixed dielectrics for electrotechnology (ISOLIME) from 2005 to 2015. At his university, he serves as Director of the International Research Centre on Atmospheric Icing and Power Network Engineering (CENGIVRE) and serves as the research chair on the Ageing of Power Network Infrastructure (ViAHT). Prof Fofana is a member of a number of technical/scientific committees of international conferences (including IEEE ICDL, IEEE CEIDP, IEEE ICHVE, CATCON, ISH, etc.) and the IEEE DEIS Administrative Committee. He is also serving the scientific community as Subject Editor in the Transformers, Transmission Lines and Cables category in IET Generation, Transmission & Distribution; is a Guest/Academic Editor for IEEE TDEI and Energies and chair of the IEEE DEIS Technical Committee on Dielectric Liquids. He is also member of a few CIGRE and ASTM working groups. Prof Fofana's research in the area of HV engineering has focused on insulation systems relevant to power equipment. His lifetime publication record includes more than 350 scientific publications and 3 patents.

Stephan Brettschneider

Stephan Brettschneider is a member of the International Research Centre on Atmospheric Icing and Power Network Engineering (CENGIVRE) and the Modeling and Diagnosis of Power Line Equipment Laboratory (MODELE) at the University of Quebec in Chicoutimi (UQAC). In August 2019, he was appointed professor at this university in the fields of power systems and high-voltage engineering. Previously, he worked in the power industry for 19 years. He obtained his PhD at UQAC in 2000, where he conducted research on high-voltage discharges.

Editorial

Outdoor Insulation and Gas-Insulated Switchgears

Issouf Fofana * and Stephan Brettschneider

Modelling and Diagnostic of Electrical Power Network Equipment Laboratory (MODELE), Department of Applied Sciences, Université du Québec à Chicoutimi, Chicoutimi, QC G7H 2B1, Canada
* Correspondence: ifofana@uqac.ca

Citation: Fofana, I.; Brettschneider, S. Outdoor Insulation and Gas-Insulated Switchgears. *Energies* **2022**, *15*, 8141. https://doi.org/10.3390/en15218141

Received: 10 August 2022
Accepted: 12 October 2022
Published: 1 November 2022

1. Introduction

With the growth of the world's population and faster-developing industries, larger amounts of electric energy are needed [1–6]. To reduce Joule losses at longer distances, voltages delivered by generators are increased with step-up power transformers, and electric energy is transported at large voltages [7,8]. Consequently, many ultra/extra AC/DC high-voltage transmission projects have been commissioned or are under construction in many countries: Canada (735 kV), Venezuela (800 kV), China (1100 kV), Japan (1100 kV), and India (1200 kV) [9–18]. For the power to be delivered to end-users, transmission grids, including towers, conductors, insulators, as well as substations, are essential [19–22]. The main equipment in substations includes, but is not limited to: power transformers, circuit breakers, surge arrestors, relays, insulators, disconnector switches, busbars, capacitor banks, batteries, wave trapper, switchyard, as well as protection, control, and metering instruments, etc. When the equipment is installed outside, it is refereed as an "outdoor substation" and an "indoor substation" when set inside a building. One of the main advantages of an indoor substation over an outdoor substation is the independence from meteorological impacts [23–25]. Nowadays, indoor substations are commonly gas-insulated substations (GIS), as they require a much smaller footprint [23].

These transmission lines that carry power to end-users cross various regions, exposing all outdoor equipment to various atmospheric conditions [26–33]. Actually, the more or less complex profiles of insulators are essentially determined to meet a certain number of criteria in relation to the shape of the voltage waves and the conditions of pollution, fog, and rain. Understanding the fundamental aspects of outdoor insulation is, therefore, important for properly designing and monitoring practical HV transmission lines and hardware [26–33].

Listed hereafter are the topics of interest considered in the call for papers:

- Lightning phenomena and related applications;
- Long air gap discharges;
- Switching surges;
- Insulation coordination;
- Breakdown and pre-breakdown phenomena in gases;
- Insulating material efficiency improvement by chemical admixtures;
- Gas-insulated switchgear (GIS);
- Measurement, monitoring and diagnostic techniques.

With fast changes occurring in both current and future grids, developing reliable insulators for outdoor insulation applications, studying the influence of atmospheric conditions on the grid performance, investigating the very fast transient overvoltage behavior of gas-insulated substations, and protecting power equipment against lightning have become essential tasks in maintaining a reliable link for the future power grids [34–39].

This Special Issue, in its final form, focuses on theoretical and practical developments with special emphasis on pollution/icing of the power grid hardware problems and the improvement in silicone rubber performances for applications in the composite insulators. The performance of gas-insulated switchgears (GIS) and lighting protection are also concerned.

2. An Outlook of the Special Issue

"Outdoor Insulation and Gas Insulated Switchgears", a Special Issue of *Energies* has been successfully organized with the support extended from the editorial staff of the journal and the MDPI publishing team. This Special Issue was quite remarkable and successful, with a good geographical distribution of authors and coauthors: Canada (12), Chile (1), China (10), Malaysia (1), Saudi Arabia (5), the UK (10), Romania (3), Italy (1), Finland (2), Germany (3), Algeria (4), and Côte d'Ivoire (2).

A brief status of the Special Issue is as follows:

- Published papers: (13);
- Rejected papers: (3);
- Median Article Processing Time: 42 days;
- Article Type: Article (12); Review (1).

The guest editors would like to thank all the reviewers for their prompt responses during the reviewing and revising of the manuscripts.

The articles published in this issue are discussed in the subsequent sections of this editorial.

3. A Review of the Special Issue

The articles [40–45] address the flashover performance of polluted insulators. In [40], an attempt has been made by Amrani et al. to propose a mathematical model allowing the prediction of the AC surface arc propagation on polluted insulators under divergent electric fields. Several electro-geometric parameters such as the distance between electrodes, pollution conductivity, radius of high-voltage electrodes, and length of the plane electrode were taken into account. The results indicate good agreement between computed and experimental results for various test configurations.

In [41], Chen et al. proposed a dynamic pollution prediction model of insulators based on atmospheric environmental parameters (meteorological data and wind speed). Two insulators' coefficients, c1 and c2 (c1: pollution ratio of U210BP/170 to XP-160; c2: calculated pollution ratio of U210BP/170T to XP-160), were computed as monitoring parameters. It is shown that the NSDD (non-soluble deposit density) of insulators with different structures can be predicted using the insulators' structure coefficient and the reference XP-160 insulator's NSDD. The results were verified against experimental data under natural pollution.

In [42], Nan et al. investigated the pollution flashover characteristics of four types of AC composite crossarm insulators with diameters ranging from 100 mm to 450 mm under different voltage grades (from 66 to 1000 kV). The pollution grade was varied between 0.2 and 1.0 mg/cm^2. The results involving the effects of the surface hydrophobicity state of silicone rubber, core diameter, umbrella structure, arrangement, and insulation distance on the pollution flashover voltage of the composite crossarm insulators can be helpful for the structural design and optimization of the composite crossarm insulators.

In [43], Slama et al. reported a preliminary study of an in situ monitoring of 400 kV SiR textured insulators in a polluted environment. The artificially polluted HTV-textured silicone rubber insulators were electrically and thermally monitored. It was found that the level of pollution, which acted as a parameter, and the voltage affect the discharge activity and its nature. In addition, the magnitude and pulses of the rms leakage current, along with the average dissipated power, depend on the pollution levels and the dry-bands formation. Finally, emphasis is laid on the differentiation and quantification between dry-band discharge onset and dry-band arc inception.

Fofana et al. [44] has shared field experience from post-installation pollution levels assessment. It is shown that the pollution level should not be considered static due to the dynamics of environmental parameters. Tests were performed on some distribution insulators removed from service. To assess the pollution levels of the insulators, various parameters such as equivalent salt deposit density (ESDD) and non-soluble deposit density (NSDD), surface resistance, and leakage current characteristics (density, third harmonic amplitude, and phase) were monitored. It was confirmed that the dynamics of the local environmental parameters should be considered for grid reliability. The authors recommended post-installation investigations be conducted whenever the surrounding insulator's area undergoes changes (construction, habitation, changes in factory processes, etc.).

In addition to pollution concerns, the power grid may be endangered by atmospheric icing in cold regions of the world. This may cause structural damages or insulator flashovers. Over the years, various techniques have been proposed to guarantee the reliability of the grid. This includes active (heating, chemicals, and mechanical methods) and passive (nanotechnology) solutions. A snapshot of some significant developments on this topic over the last four decades is presented by the guest editors [45]. The problems in using these techniques, their applications, and perspectives are discussed.

Liu et al. [46] proposed a method to monitor of the icing degree of an insulator during icing accretion. Two algorithms (direct equalization and median filter methods) have been used to reduce the noise and enhance the contrast of the images obtained in simulated laboratory conditions. From the image, the maximum entropy threshold segmentation algorithm is used to discriminate between the insulator and ice surface. The boundaries of the non-iced insulator and the ice thickness are determined using an improved Canny operator edge detection algorithm. The location of the icicles is accurately determined using a regional growing method. This allows locating the air gap in the image of the iced insulator, the icicle length. The operator may then evaluate the likelihood of ice bridging the insulator strings. The proposed approach allows predicting icing induced flashover.

The next three articles [47–49] focus on aspects of synthetic insulation materials, either specifically silicone rubber insulation housing composites' modification by suitable fillers or more generally with a review on polymeric insulation applications.

Ghunem et al. [47] pinpoint some perceptions and design paradigms about the admixture of inorganic fillers to silicone rubber housing composites for outdoor insulation with the aim of improving resistance to erosion against dry band arcing. From their critical review, the authors recommended two aspects considered in the design: (1) the volume effect of the fillers and (2) the shield effect. They also recommended supplementary characterization tests and measurements, useful in understanding the protective filler actions in enhancing the erosion resistance of silicone rubber such as TGA, SEM, leakage current measurements, and thermal conductivity measurements.

Alqudsi et al. [48] experimentally investigated the effect of ground and fumed silica fillers (incorporated at different loading levels) on suppressing DC+ erosion in silicone rubber. To evaluate the erosion performance of the composites, the inclined plan test (IPT) under DC+ voltage was used. The analytical analyses included the leakage currents, simultaneous thermogravimetric–differential thermal analysis (TGA–DTA) under nitrogen and air atmospheres to understand the thermal decomposition characteristics of the composites, surface morphology by scanning electron microscopy (SEM) and surface roughness analysis by an optical microscope. From the investigations, it is reported that fumed silica and its interaction with silicone rubber were effective in promoting the formation of a coherent shielding residue, which resulted in suppressing the DC+ erosion.

Haque et al. [49] proposed a comprehensive survey including 144 citations, focusing on the application and suitability of polymeric materials as insulators in power equipment such as cables, transformers, insulators, etc. In this survey, the authors emphasized and highlighted the basic physicochemical properties of polymer materials, thermoplastics, and thermosets. Recent studies on their performances are reviewed. The main assessment techniques for HV applications are also discussed. Finally, future research hotspots and notable research topics are discussed for the benefit of researchers.

Two papers pertaining to gas-insulated switchgears were published in this Special Issue. In [50], Alexandru et al. proposed an electromagnetic field theory approach to study the effect of several model configurations of gas-insulated substations under very fast transient overvoltage (VFTO) using suitable computer-aided design models. The partial equivalent element circuit (PEEC) approach embedded into an XGSLab software package was adopted. The results can be seen as benchmark hints for grounding grid designers, particularly for the proper development and implementation of transient ground potential rise (TGPR) mitigation techniques across a gas-insulated substation.

Götz et al. [51] investigated the partial discharge behaviour of a distorted weakly inhomogeneous electrode arrangement in sulphur hexafluoride (SF_6) and synthetic air under high DC voltage stress. The insulation gas pressure, the gas type, the electric field strength, and the voltage polarity were chosen as parameters. Depending on the experimental parameters (discharge impulses, discharge impulses with superimposed pulseless discharges, subsequent smaller discharges and pulseless discharges), the authors identified four different discharge types. The paper ended with some suggestions for reliable partial discharge measurements under DC voltage stresses.

The last contribution in this Special Issue focuses on lightning protection. Pourakbari-Kasmaei et al. [52] proposes an inductor-based filter for the protective performance of surge arresters against indirect lightning strikes. Different surge arresters with different ratings and two different classes of energy were tested under different indirect lighting impulses. The obtained results indicate that equipping an MV transformer with surge arresters with a proper filter size enhances the performance of the surge arrester significantly.

4. Closing Remarks

High-voltage transmission networks requiring outdoor insulation or GIS play a very important role in our modern societies, and they are becoming even more essential as efforts are underway to decarbonize our energy usage through increasing electrification (for example, in the personal transportation sector) [53]. The articles published testify that research in this field is very active and that the stakes of electrical insulation are multiple.

- The insulation performance under pollution conditions will continue to be an actual topic, as urbanization and industrialization will continue to increase around the globe. In addition, recent events in various regions of the world show a tendency that extreme weather conditions will become more frequent (such as severe ice storms or wild fires during heating periods). Furthermore, more transmission lines need to be constructed through environmentally harsh areas to interconnect new renewable energy resources to the existing networks. For example, there exists large solar power potential in desert areas, important wind power potential along coast lines and offshore, and potential for new hydroelectric power stations in remote mountainous areas.
- The durability and long-term performance aspects of synthetic insulation materials also remain an ongoing issue. Among others, the hydrophobic and glaciophobic properties of the insulation surfaces are of interest to increase the performance of exterior insulation systems with respect to the environmental challenges of transmission systems. However, the conservation of their beneficial properties in the long term remains a challenge.
- The role of compact GIS insulation will increase, and the number of such installations will rise due to the urbanization and densification of residential areas. Therefore, the continuation of research efforts and improvements in this field are of interest.

The papers published in this Special Issue report on the progress made in various application areas. The combination of experimental work, modelling efforts, and analysis of utility installations help to confirm and explain the results obtained. The outcomes will contribute to design more reliable and more durable insulation systems in the future.

Author Contributions: Conceptualization, I.F. and S.B.; writing—original draft, I.F. and S.B.; writing—review and editing, I.F. and S.B. All authors have read and agreed to the published version of the manuscript.

Funding: This research received no funding.

Acknowledgments: The guest editors would like to thank the authors for submitting their excellent contributions to this Special Issue. Thanks are also extended to the blind reviewers for evaluating the manuscripts and providing helpful suggestions.

Conflicts of Interest: The authors declare no conflict of interest.

References

1. Keiner, D.; Ram, M.; Barbosa, L.D.S.N.S.; Bogdanov, D.; Breyer, C. Cost optimal self-consumption of PV prosumers with stationary batteries, heat pumps, thermal energy storage and electric vehicles across the world up to 2050. *Sol. Energy* **2019**, *185*, 406–423. [CrossRef]
2. Shan, B.; Jia, D.; Zhang, L.; Cao, F.; Sun, Y. Analysis of energy demand forecasting model in the context of electric power alteration. In Proceedings of the 2017 8th IEEE International Conference on Software Engineering and Service Science (ICSESS), Beijing, China, 24–26 November 2017; pp. 798–801. [CrossRef]
3. Çelik, D.; Meral, M.E.; Waseem, M. The progress, impact analysis, challenges and new perceptions for electric power and energy sectors in the light of the COVID-19 pandemic. *Sustain. Energy Grids Netw.* **2022**, *31*, 100728. [CrossRef]
4. Li, X.; Ge, Y.; Mao, X.; Xue, W.; Xu, N. Improved Energy Structure Prediction Model Based on Energy Demand Forecast. In Proceedings of the 2018 2nd IEEE Conference on Energy Internet and Energy System Integration (EI2), Beijing, China, 20–22 October 2018; pp. 1–5. [CrossRef]
5. Hamed, M.M.; Ali, H.; Abdelal, Q. Forecasting annual electric power consumption using a random parameters model with heterogeneity in means and variances. *Energy* **2022**, *255*, 124510. [CrossRef]
6. Kim, J.-Y.; Cho, S.-B. Explainable prediction of electric energy demand using a deep autoencoder with interpretable latent space. *Expert Syst. Appl.* **2021**, *186*, 115842. [CrossRef]
7. Papailiou, K.O. Overhead Lines. In *CIGRE Green Books*; CIGRE: Paris, France, 2017.
8. EPRI. *EPRI AC Transmission Line Reference Book—200 kV and Above*, 3rd ed.; EPRI: Palo Alto, CA, USA, 2005.
9. Li, Y.; Cui, J.; Yang, R.; Huang, P.; Chen, B.; Wang, J.; Gan, Z. Research on over-voltage characteristics of ultra high-voltage direct current transmission system. *Energy Rep.* **2022**, *8*, 804–811. [CrossRef]
10. Rajamani, P.; Aravind, K.A.; Rao, K.R.; Sandhya, K.; Nirgude, P. Experimental Studies on Radio Interference of 1200 kV Unipolar UHV DC Transmission Line. In Proceedings of the 2019 International Conference on High Voltage Engineering and Technology (ICHVET), Hyderabad, India, 7–8 February 2019; pp. 1–4. [CrossRef]
11. Trinh, N.G.; Maruvada, P.S.; Flamand, J.; Valotaire, J.R. A Study of the Corona Performance of Hydro-Quebec's 735-kV Lines. *IEEE Power Eng. Rev.* **1982**, *PER-2*, 30. [CrossRef]
12. Asplund, G.; Astrom, U.; Lescale, V. 800 kV HVDC for Transmission of Large Amount of Power Over Very Long Distances. In Proceedings of the 2006 International Conference on Power System Technology, Chongqing, China, 22–26 October 2006; pp. 1–10. [CrossRef]
13. Swarup, S.K. Selection of design parameters of equipment for 800 kV transmission systems in India. In Proceedings of the International Conference on AC and DC Power Transmission, London, UK, 17–20 September 1991; pp. 235–240.
14. Åström, U.; Lescale, V.F.; Menzies, D.; Weimin, M.; Zehong, L. The Xiangjiaba-Shanghai 800 kV UHVDC project, status and special aspects. In Proceedings of the 2010 International Conference on Power System Technology, Hangzhou, China, 24–28 October 2010; pp. 1–6. [CrossRef]
15. Zhang, S.; Peng, Z.; Liu, P.; Li, N. Design and dielectric characteristics of the ±1100 kV UHVDC wall bushing in China. *IEEE Trans. Dielectr. Electr. Insul.* **2015**, *22*, 409–419. [CrossRef]
16. Tao, L.; Xinnian, L.; Shaobo, L.; Xiao, L.; Yifu, Z.; Jinhua, Z. Analysis on the Dynamic Characteristics of ±1100 kV Ultra HVDC Integrated AC System. In Proceedings of the 2018 International Conference on Power System Technology (POWERCON), Guangzhou, China, 6–8 November 2018; pp. 2509–2517. [CrossRef]
17. Krylov, S.V.; Timashova, L.V. Experience of live-line maintenance on 500–1200 kV lines in Russia. In Proceedings of the ESMO '93. IEEE 6th International Conference on Transmission and Distribution Construction and Live-Line Maintenance, Las Vegas, NV, USA, 12–17 September 1993; pp. 359–368. [CrossRef]
18. Maruvada, P.a.; Trinh, N.G.; Dallaire, R.D.; Rivest, N. Corona Studies for Biploar HVDC Transmission at Voltages Between ±600 kV and ±1200 kV Part 1: Long-Term Bipolar Line Studies. *IEEE Trans. Power Appar. Syst.* **1981**, *PAS-100*, 1453–1461. [CrossRef]
19. Hammons, T.J.; Lescale, V.F.; Uecker, K.; Haeusler, M.; Retzmann, D.; Staschus, K.; Lepy, S. State of the Art in Ultrahigh-Voltage Transmission. *Proc. IEEE* **2011**, *100*, 360–390. [CrossRef]
20. Allard, S.; Mima, S.; Debusschere, V.; Quoc, T.T.; Criqui, P.; Hadjsaid, N. European transmission grid expansion as a flexibility option in a scenario of large scale variable renewable energies integration. *Energy Econ.* **2020**, *87*, 104733. [CrossRef]
21. Yogi Goswami, D. Transmission infrastructure, interconnection of mini and main grids and energy storage for grid stability. *Sol. Compass* **2022**, *2*, 100026. [CrossRef]
22. Lacerda, J.C.; Freitas, C.; Macau, E.E. Elementary changes in topology and power transmission capacity can induce failures in power grids. *Phys. A Stat. Mech. Appl.* **2021**, *590*, 126704. [CrossRef]
23. Macomber, A. Outdoor vs. indoor substations. *Proc. Am. Inst. Electr. Eng.* **1914**, *33*, 165–171. [CrossRef]

24. Lee, C.-H.; Chang, C.-N. Comparison of 161/69-kV Grounding Grid Design Between Indoor-Type and Outdoor-Type Substations. *IEEE Trans. Power Deliv.* **2005**, *20*, 1385–1393. [CrossRef]
25. Fullerton, H.L. Indoor and outdoor substations in Pennsylvania. *Proc. Am. Inst. Electr. Eng.* **1914**, *33*, 173–189. [CrossRef]
26. CIGRE Working Group 33-04, TF 01. Polluted insulators: A review of current knowledge. *CIGRE Rep.* **2000**, *158*, 128–129.
27. Farokhi, S.; Farzaneh, M.; Fofana, I. Experimental investigation of the process of Arc propagation over an ice surface. *IEEE Trans. Dielectr. Electr. Insul.* **2010**, *17*, 458–464. [CrossRef]
28. Farzaneh, M.; Baker, A.C.; Bernstorf, R.A.; Burnhan, J.T.; Cherney, E.A.; Chisholm, W.A.; Fofana, I.; Gorur, R.S.; Grisham, T.; Gutman, I.; et al. Selection of Line Insulators With Respect to Ice and Snow—Part II: Selection Methods and Mitigation Options. *IEEE Trans. Power Deliv.* **2007**, *22*, 2297–2304. [CrossRef]
29. Jianhui, Z.; Lichun, S.; Caixin, S.; Leguan, G. DC flashover performance of iced insulators under pressure and pollution conditions. In Proceedings of the 3rd International Conference on Properties and Applications of Dielectric Materials, Tokyo, Japan, 8–12 July 1991; Volume 2, pp. 957–960. [CrossRef]
30. Kong, X.; Liu, Y.; Han, T.; Gao, Y.; Du, B.X. Effects of Air Humidity and Hanging Angle on Accumulation Characteristics of Pollution Particles on Anti-icing Polymer Insulators. In Proceedings of the 2019 2nd International Conference on Electrical Materials and Power Equipment (ICEMPE), Guangzhou, China, 7–10 April 2019; pp. 508–511. [CrossRef]
31. Shu, L.; Farzaneh, M.; Li, Y.; Sun, C. AC flashover performance of polluted insulators covered with artificial ice at low atmospheric pressure. In Proceedings of the 2001 Annual Report Conference on Electrical Insulation and Dielectric Phenomena (Cat. No.01CH37225), Kitchener, ON, Canada, 14–17 October 2001; pp. 609–612. [CrossRef]
32. Hussain, M.M.; Farokhi, S.; McMeekin, S.G.; Farzaneh, M. Impact of wind on pollution accumulation rate on outdoor insulators near shoreline. In Proceedings of the 2016 IEEE International Power Modulator and High Voltage Conference (IPMHVC), San Francisco, CA, USA, 6–9 July 2016; pp. 453–456. [CrossRef]
33. Brettschneider, S.; Farzaneh, M.; Srivastava, K. Nanosecond streak photography of discharge initiation on ice surfaces. *IEEE Trans. Dielectr. Electr. Insul.* **2004**, *11*, 250–260. [CrossRef]
34. Zhan, H.; Duan, S.; Li, C.; Yao, L.; Zhao, L. A Novel Arc Model for Very Fast Transient Overvoltage Simulation in a 252-kV Gas-Insulated Switchgear. *IEEE Trans. Plasma Sci.* **2014**, *42*, 3423–3429. [CrossRef]
35. Pathak, N.; Bhatti, T.S. Ibraheem Study of very fast transient overvoltages and mitigation techniques of a gas insulated substation. In Proceedings of the 2015 International Conference on Circuits, Power and Computing Technologies [ICCPCT-2015], Nagercoil, India, 19–20 March 2015; pp. 1–6. [CrossRef]
36. Feng, R.; Che, C.; Zhao, J.; Wang, Q.; Zhao, L.; Yang, W. Research on Influencing Factors of Very Fast Transient Overvoltage in EHV GIS. In Proceedings of the 2020 IEEE International Conference on High Voltage Engineering and Application (ICHVE), Beijing, China, 6–10 September 2020; pp. 1–4. [CrossRef]
37. Martinez, J.A. Statistical assessment of very fast transient overvoltages in gas insulated substations. In Proceedings of the 2000 Power Engineering Society Summer Meeting (Cat. No.00CH37134), Seattle, WA, USA, 16–20 July 2000; Volume 2, pp. 882–883. [CrossRef]
38. Qureshi, S.A.; Shami, T.A. Effects of the GIS parameters on very fast transient overvoltages. In Proceedings of the MELECON '94. Mediterranean Electrotechnical Conference, Antalya, Turkey, 12–14 April 1994; Volume 3, pp. 960–963. [CrossRef]
39. Gao, Y.; Yuan, H.; Wang, E.; Cao, Y. Calculation of very fast transient over-voltage and its distribution in transformer windings in 110 kV GIS. In Proceedings of the ICEMS'2001. Proceedings of the Fifth International Conference on Electrical Machines and Systems (IEEE Cat. No.01EX501), Shenyang, China, 18–20 August 2001; Volume 1, pp. 208–211. [CrossRef]
40. Amrani, M.L.; Bouazabia, S.; Fofana, I.; Meghnefi, F.; Jabbari, M.; Khelil, D.; Boudiaf, A. Modelling Surface Electric Discharge Propagation on Polluted Insulators under AC Voltage. *Energies* **2021**, *14*, 6653. [CrossRef]
41. Chen, S.; Zhang, Z. Dynamic Pollution Prediction Model of Insulators Based on Atmospheric Environmental Parameters. *Energies* **2020**, *13*, 3066. [CrossRef]
42. Nan, J.; Li, H.; Wan, X.; Huo, F.; Lin, F. Pollution Flashover Characteristics of Composite Crossarm Insulator with a Large Diameter. *Energies* **2021**, *14*, 6491. [CrossRef]
43. Slama, M.E.A.; Albano, M.; Haddad, A.M.; Waters, R.T.; Cwikowski, O.; Iddrissu, I.; Knapper, J.; Scopes, O. Monitoring of Dry Bands and Discharge Activities at the Surface of Textured Insulators with AC Clean Fog Test Conditions. *Energies* **2021**, *14*, 2914. [CrossRef]
44. Fofana, I.; N'cho, J.S.; Betie, A.; Hounton, E.; Meghnefi, F.; Yapi, K.M.L. Lessons to Learn from Post-Installation Pollution Levels Assessment of Some Distribution Insulators. *Energies* **2020**, *13*, 4064. [CrossRef]
45. Brettschneider, S.; Fofana, I. Evolution of Countermeasures against Atmospheric Icing of Power Lines over the Past Four Decades and Their Applications into Field Operations. *Energies* **2021**, *14*, 6291. [CrossRef]
46. Liu, Y.; Li, Q.; Farzaneh, M.; Du, B.X. Image Characteristic Extraction of Ice-Covered Outdoor Insulator for Monitoring Icing Degree. *Energies* **2020**, *13*, 5305. [CrossRef]
47. Ghunem, R.A.; Hadjadj, Y.; Parks, H. Common Perceptions about the Use of Fillers in Silicone Rubber Insulation Housing Composites. *Energies* **2021**, *14*, 3655. [CrossRef]
48. Alqudsi, A.Y.; Ghunem, R.A.; David, E. A Novel Framework to Study the Role of Ground and Fumed Silica Fillers in Suppressing DC Erosion of Silicone Rubber Outdoor Insulation. *Energies* **2021**, *14*, 3449. [CrossRef]

49. Haque, S.M.; Ardila-Rey, J.A.; Umar, Y.; Mas'ud, A.A.; Muhammad-Sukki, F.; Jume, B.H.; Rahman, H.; Bani, N.A. Application and Suitability of Polymeric Materials as Insulators in Electrical Equipment. *Energies* **2021**, *14*, 2758. [CrossRef]
50. Alexandru, M.; Czumbil, L.; Andolfato, R.; Nouri, H.; Micu, D.D. Investigating the Effect of Several Model Configurations on the Transient Response of Gas-Insulated Substation during Fault Events Using an Electromagnetic Field Theory Approach. *Energies* **2020**, *13*, 6231. [CrossRef]
51. Götz, T.; Kirchner, H.; Backhaus, K. Partial Discharge Behaviour of a Protrusion in Gas-Insulated Systems under DC Voltage Stress. *Energies* **2020**, *13*, 3102. [CrossRef]
52. Pourakbari-Kasmaei, M.; Lehtonen, M. Enhancing the Protective Performance of Surge Arresters against Indirect Lightning Strikes via an Inductor-Based Filter. *Energies* **2020**, *13*, 4754. [CrossRef]
53. Aslam, A.A.; Majid, A.A.; Muhamad, M.; Marayati, B.M.; Yong, J.Y.; Muhamad, S.; Abd, R.; Nur, A.S. NEPLAN-Based Analysis of Impacts of Electric Vehicle Charging Strategies on Power Distribution System. *IOP Conf. Ser. Mater. Sci. Eng.* **2021**, *1127*, 012033.

MDPI

Article

Modelling Surface Electric Discharge Propagation on Polluted Insulators under AC Voltage

Mohamed Lamine Amrani [1], Slimane Bouazabia [1], Issouf Fofana [2,*], Fethi Meghnefi [2], Marouane Jabbari [2], Djazia Khelil [1] and Amina Boudiaf [1]

[1] Laboratory of Electrical Industrial Systems (LSEI), Department of Electrical Engineering, University of Science and Technology Houari Boumediene, BP 32 El Alia, Bab Ezzouar 16111, Algeria; ml.amrani@yahoo.fr (M.L.A.); sbouazabia@yahoo.fr (S.B.); khelil_djazia@yahoo.fr (D.K.); boudiaf.a2@gmail.com (A.B.)

[2] Modelling and Diagnostic of Electrical Power Network Equipment Laboratory (MODELE), Department of Applied Sciences, Université du Québec à Chicoutimi, Chicoutimi, QC G7H 2B1, Canada; fmeghnef@uqac.ca (F.M.); marouane.jabbari1@uqac.ca (M.J.)

* Correspondence: Issouf_Fofana@uqac.ca; Tel.: +1-418-545-5011 (ext. 2514)

Abstract: In this contribution, a mathematical model allowing for the prediction of the AC surface arc propagation on polluted insulators under non-uniform electric field is proposed. The approach is based on the experimental concept of Claverie and Porcheron. The proposed model, which makes it possible to reproduce the surface electric discharge, includes a condition for arrest of the propagating discharge. The electric field at the tip of the discharge is the key parameter governing its random propagation. A finite element approach allows for mapping of the electric field distribution while the discharge propagation process is simulated in two dimensions. The voltage drop along the arc discharge path at each propagation step is also taken into account. The simulation results are validated against experimental data, taking into account several electro-geometric parameters (distance between electrodes, pollution conductivity, radius of high-voltage electrode, length of the plane electrode). Good agreement between computed and experimental results were obtained for various test configurations.

Keywords: pollution; flashover; inception voltage; arc propagation; finite element method (FEM)

Citation: Amrani, M.L.; Bouazabia, S.; Fofana, I.; Meghnefi, F.; Jabbari, M.; Khelil, D.; Boudiaf, A. Modelling Surface Electric Discharge Propagation on Polluted Insulators under AC Voltage. *Energies* **2021**, *14*, 6653. https://doi.org/10.3390/en14206653

Academic Editor: Abu-Siada Ahmed

Received: 10 September 2021
Accepted: 8 October 2021
Published: 14 October 2021

1. Introduction

For the safe operation of any power system network, it is important that key equipment such as transmission line hardware operate properly for many years. Transmission lines pass through several hundreds of kilometres, being subjected to various climatic conditions (sea salts, domestic pollution, natural pollution, dirt, and chemical residues in the industrial areas, etc.). Pollution of insulators is recognised as a major engineering concern since in polluted areas, overhead lines may see their reliability and performance decline [1–4].

This is because when the contaminated surface is wet, the leakage currents then increase with potential flashover of the insulating surface, which means power outage [5]. Even though many investigations have been reported on the topic, there is still a dearth of information to gather. The physics of the flashover of polluted insulators is still not completely understood due to the complexity of the involved phenomenon resulting from the interaction of the electric field, pollution layer, and environmental conditions [6].

Due to fast-growing cities, industrialisation, and climate change, power grids are poised to experience more pollution. The study of the flashover of polluted insulators is therefore of interest due to potential power-related outages. Significant efforts have been made to better understand the evolution of insulator flashover. A large number of papers and books on the insulator surface flashover are available [7]. It is now well established that the flashover process of polluted insulators includes three steps: (i) arc initiation, (ii) arc propagation, and (iii) flashover. The electric field distribution along the insulator, which depends on several factors, has a great influence on the arc initiation and propagation. The initiation of an arc starts in a region of high local electrical stress.

Theoretical models for simulating electrical discharge may allow for a better understanding of the physical processes involved in the flashover of polluted insulators. From the engineering point of view, the ultimate goal of the simulation/modelling is to predict the behaviour of polluted insulators from purely macroscopic data. Therefore, reliable models may considerably reduce the time factor required for laboratory tests. Thus far, several flashover models, mathematical, numerical, or experimental investigations, e.g., [7–12], have been reported to determine the critical characteristics of the electric arc.

Few mathematical or numerical studies have been conducted to predict the critical flashover voltage of polluted insulators considering that an arc behaves like an equipotential surface in contact with the pollution layer and assumes propagation in one dimension [13–15]. Some models describe the electric arc behaviour using various criteria: electric field (Hampton's criteria), leakage current variation (Hesketh's criteria), applied voltage and energy (Anjana and Lakshminarasimha's criteria), power variation (Wilkins' criteria), and equivalent impedance [16–20]. All these models are based on equations involving the static arc constants (n and N) [12]. A survey of the literature shows that their values vary over a wide range for different types of arcs [21]. These values depend not only on the arc medium but also on the electrolyte used to form the resistive layer. The fundamental shortcoming is related to the large range between these values when comes to the time to select values for a specific application. Slama et al. [2] proved that the parameters n and N are, in reality, not static and depend on the thermal characteristics of the arc along with the equivalent electrical circuit simulating the phenomena.

The present paper aimed at studying the inception and propagation of an electric arc discharge up to flashover under polluted conditions. The main features include the electric field distribution mapping and the fact that the propagation process is superficial (2D). The electric arc discharge model is based on finite element method (FEM). It is generally reported that in the case of uniform pollution, the discharge occurs in the dry bands generated by the circulation of the leakage current [11,22–24]. For non-uniform pollution, the arc develops in unpolluted areas [11,22–24]. These observations allow for confirmation that the electric arc starts with a discharge in the air in contact with the pollution layer. The model established in this paper is based on this result by exploiting Peek's work on electric discharges in gases.

The electrical arc discharges, initiated from the cylindrical electrode, propagate by stepping randomly along the polluted insulator plate once the field exceeds the threshold level computed at each time step. When the electric arc progresses, a new configuration is imposed on the system, represented by an increase of the arc length and changes in its characteristics. This increase in the arc length causes the inception field to change at each step. The electrical discharge is modelled by small branches in contact with the pollution layer. The proposed model is validated against experimental results. The comparison of the computed and measured flashover results indicates the validity of the proposed model for predicting the breakdown voltage. From this contribution, explicit information onto the influence of the electro-geometrical parameters on the electric field distribution that affects the flashover process of the polluted insulators is reported. Three aspects of the electrical discharge were analysed: arc initiation, discharge propagation law, and voltage drop. The influence of the electro-geometric parameters on the characteristics of the electric discharge is also discussed. In particular, the proposed model makes it possible to predict the flashover voltage of a polluted insulator according to the geometry of the electrodes, gap distance, and the conductivity of the pollution.

The rest of the paper is organised as follows. The laboratory model is described in Section 2. The experimental investigations are reported in Section 3. In Section 4, the mathematical model is described. The comparison of the simulated results with experimental ones is reported in Section 5. Conclusions and future directions of research are provided in Section 6.

2. Laboratory Model

The rod–plane arrangement constitutes one of the basic configurations in the investigation of non-uniform electric fields and insulation properties in high-voltage studies [25,26]. The physical laboratory model (Figure 1) is based on the plane model of Claverie and Porcheron [10,27]. It consists of a glass insulating plate on which a uniform pollution layer with specific conductivity (γ) is deposited. A cylindrical electrode with a radius (Rp) served for high voltage (U) application while a rectangular electrode of length (L) and width (a) served as the ground electrode. The cylindrical and rectangular electrodes are separated by a distance (D). It is also assumed that distance between the centre of the circular electrode and the end of the plane electrode is 'XL'.

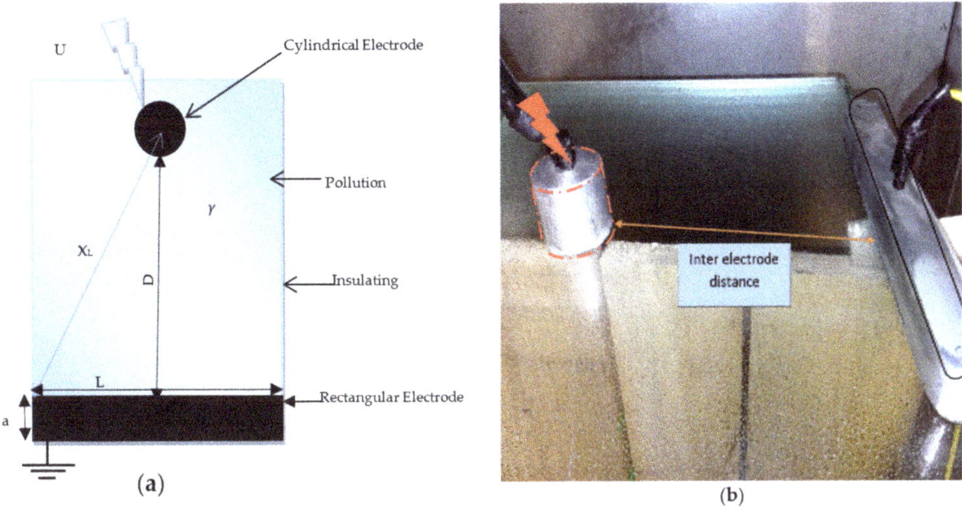

Figure 1. Overview of the laboratory physical model. (**a**) Schematic illustration. (**b**) Photo.

3. Experimental Investigations

The experimental circuit (Figure 2) adopted for the present study consists of a HV transformer (110 kV, 5 kVA) and voltage and leakage current measuring systems (voltage divider and shunt resistor). The arc discharge's light emission intensity was recorded with a R3896 Hamamatsu Photonics photomultiplier (PMT) directed at the HV electrode. The covered wavelength ranges from ultraviolet light to the near-infrared region (185 to 900 nm), allowing the detection of early discharge inception since corona discharges in air emit light mainly in the 230–405 nm range. The most intense emissions are between 300 and 360 nm, whereas those of the arc phase are between 600 and 800 nm [7]. The PMT tube had a rise time of 22 ns. The applied voltage was increased at a rate of 1 kV/s using a High Volt WSM3 control device. Measurements signals (current, voltage, and PMT) were recorded via a National Instrument data acquisition system (NI PCI 6251).

The leakage current was recorded through a 100 Ω power shunt resistor (R_{SH}).

The preparation of the saline solution, used for the pollution of insulators, was based on the salt fog method according to standard IEC 60507. The level of the pollution layer was chosen on the basis of earlier work by Teguar et al. [22–24,28]. These authors considered that the highly conductive layers have a conductivity greater than 400 μS/cm, while the weakly conductive layers have a conductivity lower than 40 µS/cm. The tap water was referred as median conductivity level.

The physical laboratory model was carefully cleaned with distilled water so that all traces of drifts were removed and dried with papers. The conductivity of the pollution layer was adjusted to a desired value using a conductivity meter of Yokogawa type by adding sodium chloride (NaCl) to deionised water. The pollution layer was deposited by spraying the surface of the insulating plate with the salty solution while respecting the number of sprays and distance to obtain a uniform and reproducible pollution layer.

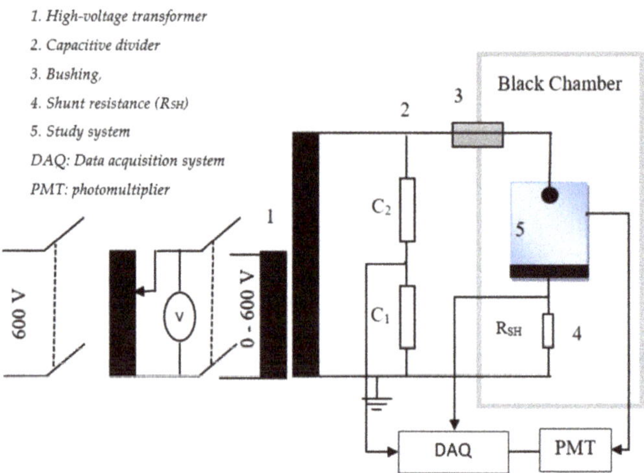

Figure 2. Measuring circuit for AC inception and breakdown test.

4. Mathematical Modelling

4.1. Basic Concept

The backbone of the model originated from the Obenaus concept [29]. This concept was firstly developed to describe the flashover under DC conditions. Furthermore, some researchers such as Claverie [10] and Rizk [12,30,31] provided the essential criteria to adapt the Obenaus concept to AC voltages. The AC flashover on a polluted surface is considered as an arc in series with a residual resistance consisting of a pollution layer that is not bridged by the arc (Figure 3). The circuit equation then reads as follows:

$$U = Varc + Rp\ (l_{arc}).I \quad (1)$$

$$Varc = Nl_{arc}I^{-n} \quad (2)$$

where

U: applied voltage (V);
N and n: arc constants;
l_{arc}: arc length (cm);
I: leakage current (A);
$Varc$: arc voltage (V);
$Rp\ (l_{arc})$: residual resistance of the pollution layer (Ω).

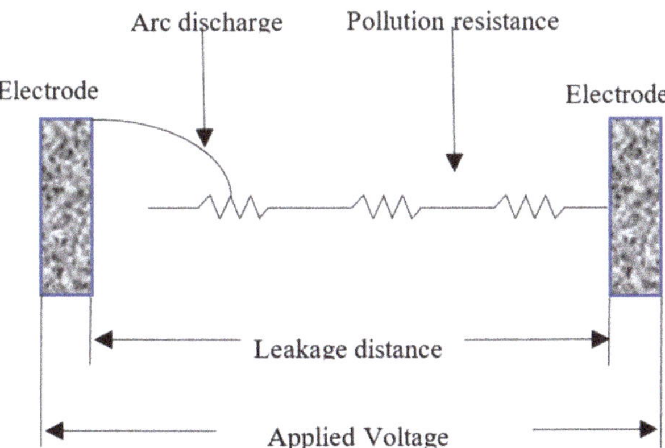

Figure 3. Model of flashover on a polluted insulator [29].

The arc tends to extinguish at every half-cycle as the current passes through zero. The re-ignition can occur when the applied voltage reaches the critical value (depending on leakage current *I* and arc length *larc*) according to Claverie and Porcheron [10,27]:

$$U_{rei} \geq \frac{800\ l_{arc}}{\sqrt{I}} \tag{3}$$

On the basis of some experimental results, Hampton established that the arc cannot propagate unless the voltage gradient in the pollution layer *Ep* exceeds the voltage gradient in the arc column *Earc* [16]. Therefore, the propagation condition yields

$$Ep > Earc \tag{4}$$

On the basis of these equations, we propose a two-dimensional model. The fundamentally new feature of the model is its use of a field mapping criterion to describe the arc propagation. The angle between the arc discharge step and the axis of the gap is random with an assumed form of uniform probability distribution. The initiation and propagation criteria of the electric discharge are estimated by the modified Peek's formula. The main features are described in the next section.

4.2. Simulation Model

Computations of the model were performed using the scientific library MATLAB®. To reproduce the arc discharge progression forms, we established a circular mesh (Figure 4), made up of ($n_{mesh} \times m_{mesh}$) elements with angular ($\Delta\theta$) and radial (Δr) steps. The nodes of this mesh represent the targeted points by the arc discharge.

The development of a superficial model of electrical arc discharge propagation is reproduced under the following assumptions:

- The discharge progresses stepwise on a plane (two-dimensional) where the conductive surface of the polluted insulator is considered constant.
- The electrical arc initiates at the cylindrical electrode subjected to high voltage and progress towards the grounded plane electrode.
- The discharge propagates randomly from the most intense field points, emphasising the predominant effect of the electric field.

In the computer program, the system is reproduced by a matrix of dimension $m_{mesh} \times n_{mesh}$ where the centre of the circular electrode constitutes the element (1, 1) with the coordinates (0, 0). The other points are defined geometrically (x, y) or as a matrix (i, j) abscissas, as defined by Equations (5) and (6).

The difficulty of an electric arc rises as it takes place on a polluted surface. The conductivity and the distribution of pollution layers not only influences the inception/propagation of a surface electric discharge but is itself influenced by this electric discharge. The established model concerns two steps of discharge evolution: inception and propagation.

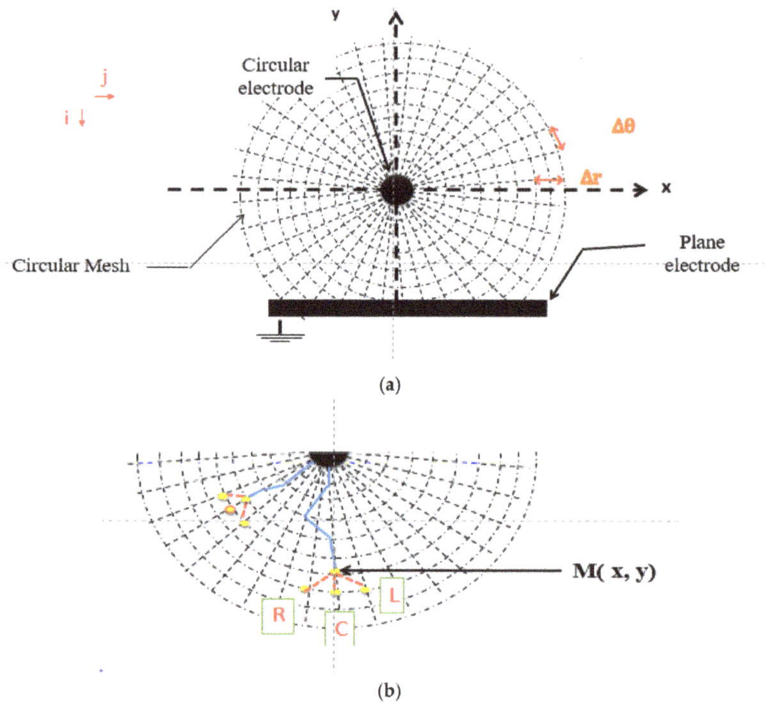

Figure 4. Progression discharge grid. (**a**) Mesh. (**b**) Target points.

The coordinates of the points are calculated by the following relationships:

$$x = i\,\Delta r\,\cos(j\,\Delta\theta) \tag{5}$$

$$y = i\,\Delta r \sin(j\,\Delta\theta) \tag{6}$$

with

$\Delta\theta$, Δr: angular and radial steps, respectively;
j: index varying between 1 and $m_{mesh} = 2\pi/\Delta\theta$;
m_{mesh}: number of angular divisions;
i: index varying between 1 and $n_{mesh} = (X/\Delta r) + 1$;
n_{mesh}: number of radial divisions.

4.2.1. Discharge Inception

When the applied voltage stress (electric field) between two electrodes in an insulation system reaches a threshold value, a first corona discharge of initial length L_0 is initiated. If the conditions for the discharge propagation are satisfied, the discharge channel fed by the arc discharge current lengthens. Otherwise, the channel switches off and disappears unless the applied voltage reaches the threshold re-ignition value [32,33].

The threshold electric field is estimated using empirical Peek's law [33–38]:

$$E_c = E_{ai}\delta\left(1 + \frac{K}{\sqrt{\delta R_p}}\right)m_1.m_2 \quad (7)$$

The threshold voltage of the discharge inception for the considered configuration is [7,37,38]

$$Uc = E_{ai}\delta\left(1 + \frac{K}{\sqrt{\delta Rp}}\right)\frac{\left(In\left(\frac{2D+Rp}{Rp}\right)\right)}{\left(\frac{1}{Rp} + \frac{1}{2D+Rp}\right)}m_1\ m_2 \quad (8)$$

where

E_{ai} = 21.2 kVrms/cm, K = 0.3;

R_p: radius of rod electrode in cm;

m_1: geometric correction coefficient (between 0 and 1) being equal to

$$m_1 = \frac{\log(X_L)}{2\ \pi} \quad (9)$$

X_L: distance between the centre of the circular electrode and the end of the plane electrode;

m_2: correction coefficient of the pollution (0.3 to 0.9). For a clean surface, m_2 = 0.9, whereas for highly polluted one, m_2 = 0.3.

The surrounding condition is also taken into account through the density:

$$\delta = (3.92\ P)/(273 + t) \quad (10)$$

P: air pressure in cm Hg;

t: the surrounding temperature in °C.

The Equations (8)–(10) are used to determine the threshold voltage according to the electro-geometrical parameters and climatic conditions [7,37,38].

4.2.2. Discharge Propagation Criteria

A discharge initiation on a polluted insulator may not lead to a breakdown process unless two conditions are satisfied. First, the required conditions for discharge propagation must be satisfied. Second, the discharge should not stop, and therefore the first condition remains valid during the progression of the arc up to the flashover (otherwise the voltage has to be increased).

A mandatory condition for an electric discharge to be initiated is related to the distribution of electric fields, which greatly depend on geometrical dimensions and electric parameters of pollution.

The propagation of electric discharge being the succession of avalanches of critical size and the last phase is called the final jump, which corresponds to the establishment of the flashover between the electrodes.

In this work, the electric discharge starts from the energised electrode (cylindrical electrode) where the electric field value is high enough to sustain the discharge propagation to the grounded electrodes. Indeed, the electric discharge is considered to evolve randomly when the field at the target point exceeds the threshold field of corona appearance.

The proposed multi-arc FEM model uses the inception field calculated from Equation (7) as condition to initiate the discharge.

At each evolution step, the extreme point of the discharge M, a three-options random probability propagation is offered (Figure 4b): L (left), C (centre), and R (right).

Considering the target point with electric field E, the discharge evolution criterion (Hampton criteria) towards this point is given by

$$E > \vartheta\ E_C \quad (11)$$

where

E_C: the critical field (Equation (7);

ϑ: random variable generated at each step and each branch by a uniform probability distribution in the model.

The uniform law allows for pseudo-random number generation algorithms, to generate random variables with the same probability of occurrence over an interval.

If the likelihood is sufficiently high enough, a new segment connects the extremity M to the targeted point, which becomes itself a new discharge extremity.

After each propagation step, the electric field distribution is modified by the newly added branches. The electric field distribution is therefore computed.

The voltage at the new extremity referred to as V_{arc} is given by the following equation:

$$V_{arc} = U - \Delta v \tag{12}$$

with Δv [V] representing the linear voltage drop along the discharge channel.

When the electric arcs progress, a new configuration is imposed on the system by the arc length and its voltage drop. This new configuration affects the critical field at each step of the evolution. The inception field is recalculated at each step of the electric discharge evolution (Equation (13)).

$$E_c(l_{arc}) = E_{ai}.\delta\left(1 + \frac{K}{\sqrt{\delta.R(l_{arc})}}\right)m_1.m_2 \tag{13}$$

where

$$R(l_{arc}) = R_p + l_{arc} \tag{14}$$

l_{arc}: the maximum radial length of the discharge

$$l_{arc} = \Delta\text{r}.\ N_{be} \tag{15}$$

N_{be}: number of steps;

Δr: radial steps.

The model used to compute the breakdown voltage is based on Obenaus [29] and Claverie-Porcheron [10,27] models. The electric field distribution is obtained by the Finite Element Method Magnetics software (FEMM) [38]. To do this, a rectangular domain defines the calculation limits with a Neumann condition. MATLAB is used to introduce the electro-geometric data in the form of an executable program and the exploitation of the FEMM results.

During each calculation step, if the propagation condition is satisfied, the discharge lengthens. Otherwise, the peak value of the applied voltage (U) has to be increased and the parameters initialised.

The flashover voltage is obtained when the discharge length exceeds the inter electrode gap D. Figure 5 presents the general flowchart of simulation.

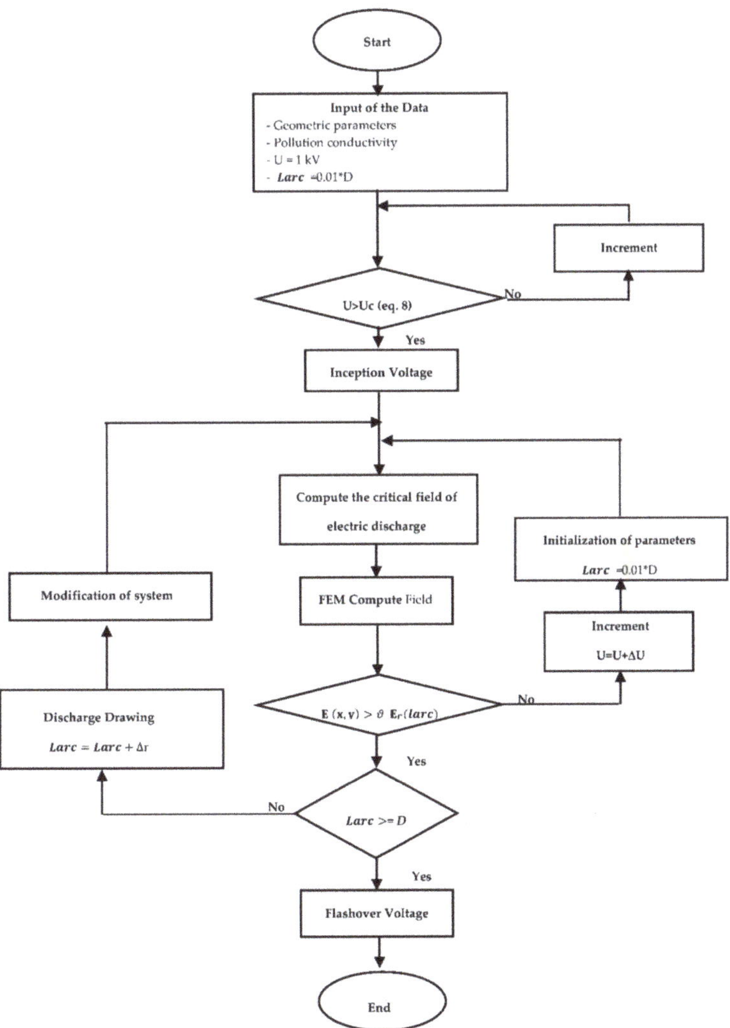

Figure 5. General flowchart of the simulation.

5. Simulation Results and Validation against Experiments

The flashover of the polluted insulator was measured and simulated while considering the effect of various electro-geometric parameters including conductivity, inter-electrode distance, HV electrode radius, and ground electrode width. For this purpose, we considered various electro-geometrical constraints as follows

- Radius of circular electrode RP = 1.5 cm, 2.5 cm.
- The length of the plane electrode L = 15.5 cm, L = 25.5 cm.
- Inter-electrode distance D = 10 cm, 20 cm, 30 cm, 40 cm.
- Four layers of pollution of different conductivity are considered (γ = 10 µS/cm, 80 S/cm, 400 µS/cm, γ = 900 µS/cm).

A comparison between simulated and experimental data to validate the proposed model is reported hereafter. The variation of the flashover voltage as a function of the above-mentioned electro-geometric parameters under AC voltage is given in Figures 6–13.

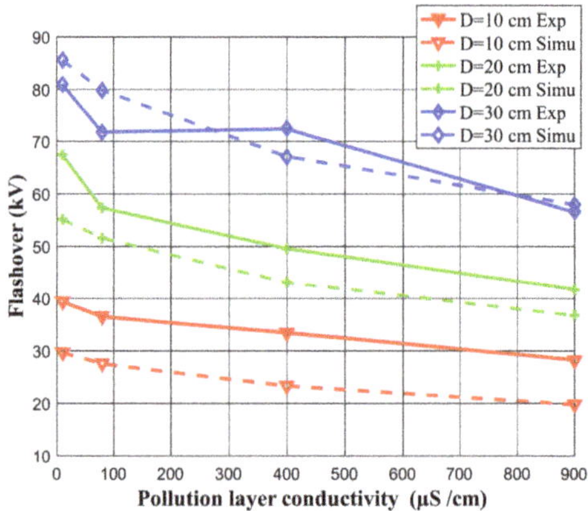

Figure 6. Flashover vs. pollution layer conductivity (Rp = 1.5 cm, L = 25.5 cm).

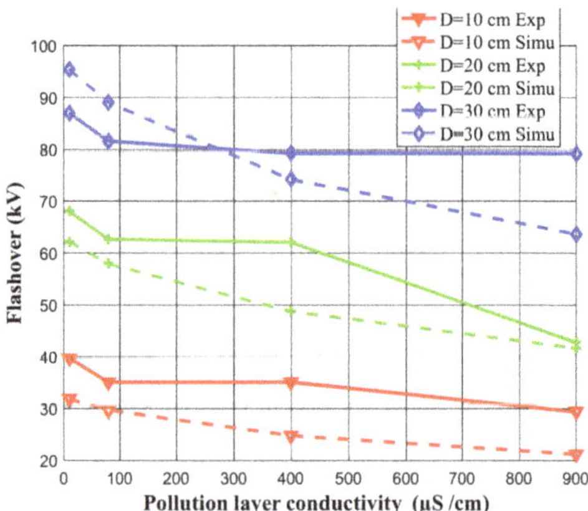

Figure 7. Flashover vs. pollution layer conductivity (Rp = 2.5 cm, L = 25.5 cm).

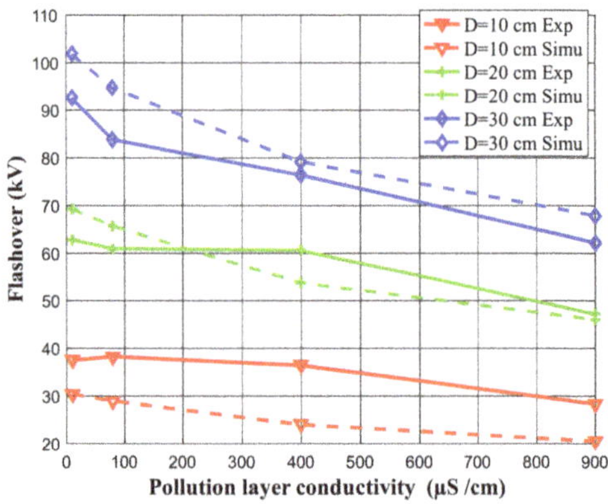

Figure 8. Flashover vs. pollution layer conductivity (Rp = 1.5 cm, L = 15.5 cm).

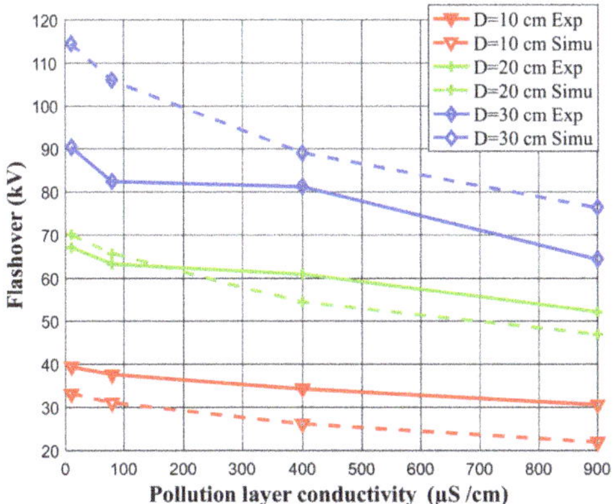

Figure 9. Flashover vs. pollution layer conductivity (Rp = 2.5 cm, L = 15.5 cm).

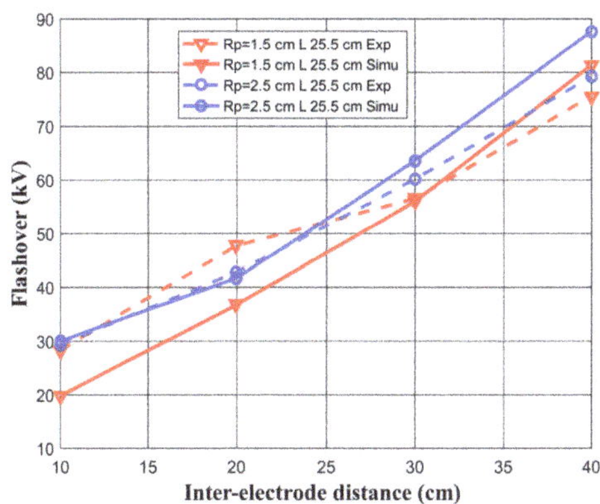

Figure 10. Flashover vs. inter-electrode distance for electrode radius (γ = 900 µS/cm, L = 25.5 cm).

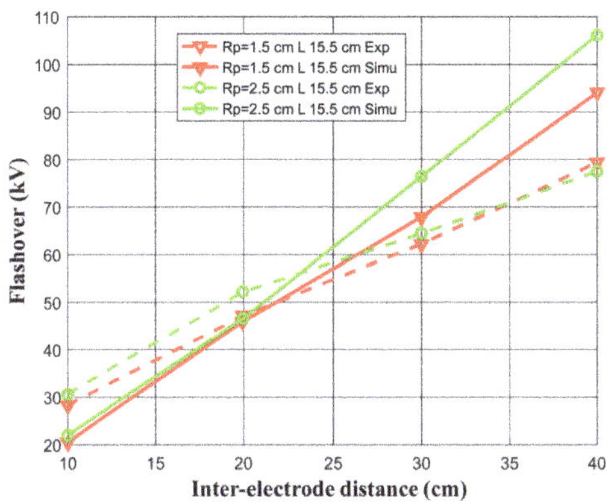

Figure 11. Flashover vs. inter-electrode distance for electrode radius (γ = 900 µS/cm, L = 15.5 cm).

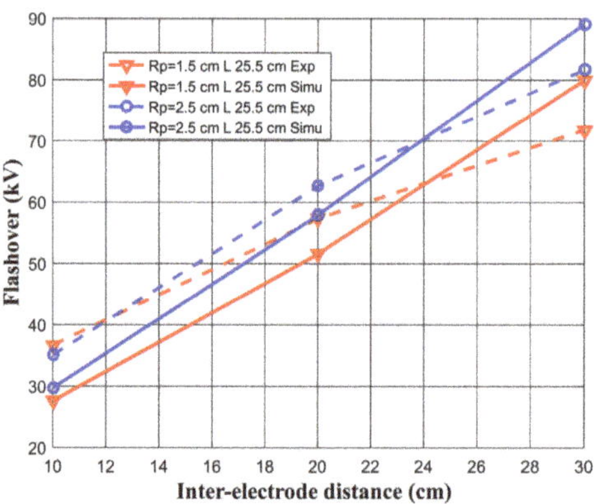

Figure 12. Flashover vs. inter-electrode distance for different electrode radius (γ = 80 μS/cm, L = 25.5 cm).

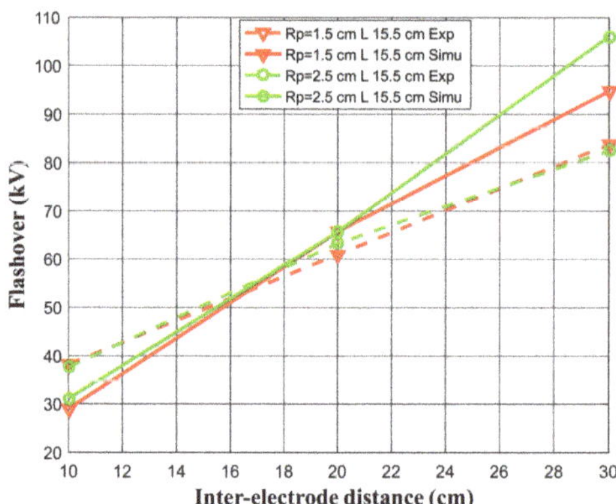

Figure 13. Flashover vs. inter-electrode distance for electrode radius (γ = 80 μS/cm, L = 15.5 cm).

It can be observed from Figures 6–9 that the flashover voltage decreased with the pollution layer's conductivity. In other words, increasing the pollution conductivity caused the electrical performance to worsen. It can also be seen that increasing inter-electrode distance and energised electrode radius caused the disruptive voltage to increase, regardless of the pollution layer's conductivity and the length of the grounded electrode.

The electric field decreased with increasing D or Rp, thus requiring a higher level of applied voltage to generate a conductive current and dry bands that stimulate generation of arcs, leading to flashover.

The breakdown voltage was reduced with increasing pollution conductivity. The presence of pollution favours a conductive current in the regions of high electrical field stress, which create dry bands and facilitate arc generations.

The variation of the geometrical parameters such as the radius and the length of the ground electrode, as reported in Figures 10–13, impacts the flashover by affecting the distribution of the electric field. For a given inter-electrode distance, with any increase in the electrode radius or the length of grounded electrodes, the electric field tends to be uniform. However, the length effect is less important than that of the inter-electrode distance.

For a given radius and inter-electrode distance, it should be noticed that the experimental data have the same order of magnitude as those predicted with the established numerical model (less than 20% error). This difference can be traced to the possible displacement of the wet part of the pollution, the difficulty in applying uniformly the pollution in real life, and/or the residues from the previous discharges.

It can be seen that the simulation results were close to the experimental ones for small radius, large gap electrodes, and large plane electrode. Furthermore, the difference between the simulation and experimental results was reduced when the HV electrode radius decreased and the ground electrode length increased. The random variable generated at each step and each branch by a uniform probability distribution in the model may be seen as a clear limitation of the model in its actual form.

6. Conclusions

In this contribution, experimental investigations on the flashover of a non-uniform field polluted insulators under AC was carried out. Different electro-geometrical parameters (conductivity, radius, inter-electrode distance, and the plane length) were considered.

The established model makes it possible to predict the flashover voltage of a polluted insulator, which is very useful information for electrical engineering. In addition, it is possible to determinate the initial value of the insulator degradation represented by the inception voltage. Knowledge of this data allows designers, builders, and managers of electrical power transmission lines to combat the inconvenience of insulator pollution in a technical manner. Moreover, it was confirmed that the flashover phenomenon depends on the interaction between the electric field and the conductivity of pollution layers.

The established model is in essence based on the electric field and the randomness of the evolution of the electric discharge on the surface of the polluted insulator. The choice of the progression criterion, which considers the modified Peek's law, confirms that the discharge bypasses the dry bands and evolves in the air in contact with the pollution layer.

From these investigations, we report that the flashover voltage increased with increasing inter-electrode distance and high-voltage electrode radius. It decreased with increasing conductivity. The arc development was particularly dependent on the pollution layer's conductivity and electrical field stress, which plays a primordial role on the flashover process. These observations were confirmed by the model.

The prediction of the electric arc initiation and development on a polluted surface is of significant interest for designing an HV system suitable for heavily polluted areas. It is hoped that the results presented can be viewed as a benchmark and a challenge for further research. It is also expected to have an impact on further modelling/simulation of discharge inception and propagation on polluted surface. Work is now in progress to improve the propagation concept and extend the model to non-uniformly polluted surfaces and real-life insulators. This should be of great help in outdoor insulation design.

Author Contributions: This research was conducted by M.L.A. under the supervision of S.B., I.F. and F.M.; M.L.A., F.M. and M.J. conducted the experimental investigations. Writing—original draft preparation, M.L.A.; writing—review and editing, S.B., I.F., D.K. and A.B. All authors have read and agreed to the published version of the manuscript.

Funding: This work was partially sponsored by the Natural Sciences and Engineering Research Council of Canada (NSERC) under grant no. RGPIN-201504403.

Institutional Review Board Statement: Not applicable.

Informed Consent Statement: Not applicable.

Data Availability Statement: Not applicable.

Acknowledgments: M.L.A. is especially thankful to the Directorate General for Scientific Research and Technological Development—Algerian Ministry of Higher Education and Scientific Research for supporting his research stay in Canada.

Conflicts of Interest: The authors declare no conflict of interest.

Abbreviations

Exp	Experimental results
FEM	Finite element method
Exp	Experimental results
FEM	Finite element method
FEMM	The Finite Element Method Magnetics software
HV	High voltage
IEC	International Electrotechnical standards
PMT	Photonics photomultiplier simulation results
Nomenclature	
A	Width of the rectangular electrode
D	Inter-electrode distance
E	Electric field
Earc	The voltage gradient in the arc column
E_c	The threshold electric field
Ep	The voltage gradient in the pollution layer
I	Leakage current
L	Length of the rectangular electrode
l_{arc}	Arc length, the maximum radial length of the discharge
m_1	Geometric correction coefficient
m_2	Correction coefficient of the pollution
m_{mesh}	Number of angular divisions
N and n	Arc constants
n_{mesh}	Number of radial divisions
Nbe	Number of steps
P	Air pressure
Rp	Radius of the cylindrical electrode
$Rp\ (l_{arc})$	Residual resistance of the pollution layer
t	Temperature
U_c	The threshold voltage
U_{rei}	Critical voltage of the re-ignition condition
$Varc$	Arc voltage
U	Applied voltage
XL	Distance between the centre of the circular electrode and the end of the plane electrode
γ	Conductivity of pollution layer
$\Delta\theta$	Angular steps
Δr	Radial steps
Δv	The linear voltage drop
δ	The air density
ϑ	Random variable

References

1. Venkataraman, S.; Gorur, R.S. Prediction of flashover Voltage of non-ceramic Insulators under Contaminated Conditions. *IEEE Trans. Dielectr. Electr. Insul.* **2006**, *13*, 862–869. [CrossRef]
2. Slama, M.E.-A.; Beroual, A.; Hadi, H. Analytical Computation of Discharge Characteristic Constants and Critical Parameters of Flashover of Polluted Insulators. *IEEE Trans. Dielectr. Electr. Insul.* **2006**, *17*, 1764–1771. [CrossRef]
3. Hussain, M.M.; Farokhi, S.; McMeekin, S.G.; Farzaneh, M. Risk Assessment of Failure of Outdoor High Voltage Polluted Insulators under Combined Stresses Near Shoreline. *Energies* **2017**, *10*, 1661. [CrossRef]
4. Hussain, M.M.; Farokhi, S.; McMeekin, S.G.; Farzaneh, M. Mechanism of saline deposition and surface flashover on outdoor insulators near coastal areas part II: Impact of various environment stresses. *IEEE Trans. Dielectr. Electr. Insul.* **2017**, *24*, 1068–1076. [CrossRef]
5. Fofana, I.; N'cho, J.S.; Betie, A.; Hounton, E.; Meghnefi, F.; Yapi, K.M.L. Lessons to Learn from Post-Installation Pollution Levels Assessment of Some Distribution Insulators. *Energies* **2020**, *13*, 4064. [CrossRef]
6. Farzaneh, M.; Chisholm, W.A. *Insulators for Icing and Polluted Environments*; Wiley: Hoboken, NJ, USA, 2009.
7. Amrani, M.L. Etude de L'évolution en Bidimensionnel de L'arc Electrique sur une Surface Diélectrique Polluée. Ph.D. Thesis, Université des Sciences et de la Technologie Houari Boumediene, Alger, Algerie, 2019.

8. Rumeli, A.; Hızal, M.; Demir, Y. Analytical Estimation of Flashover Performances of Polluted Insulators. *Madras* **1981**, *1*, 1–6.
9. Aydogmus, Z.; Cebeci, M. A New Flashover Dynamic Model of Polluted HV Insulators. IEEE Trans. *Dielectr. Electr. Insul.* **2004**, *11*, 577–584. [CrossRef]
10. Claverie, P. Predetermination of the Behaviour of Polluted Insulators. *IEEE Trans.* **1971**, *90*, 1902–1908. [CrossRef]
11. Mekhaldi, A. Etude des Phénomènes de Conduction et de Décharge Electrique sur des Surfaces Isolantes Polluées Sous Tension Alternative 50 Hz. Ph.D. Thesis, Ecole Nationale Polytechnique, Alger, Algerie, June 1999.
12. Rizk, F.A.M. Mathematical Models for Pollution Flashover. *Electra* **1981**, *78*, 71–103.
13. Volat, C.; Farzaneh, M.; Mhaguen, N. Improved FEM models of one- and two-arcs to predict AC critical flashover voltage of ice-covered insulators. *IEEE Trans. Dielectr. Electr. Insul.* **2011**, *18*, 393–400. [CrossRef]
14. Jabbari, M.; Volat, C.; Farzaneh, M. A New Single-arc AC Dynamic FEM Model of Arc Propagation on Ice Surfaces. In Proceedings of the 31st Electrical Insulation Conference (EIC), Ottawa, ON, Canada, 2–5 June 2013; pp. 360–364.
15. Jabbari, M.; Volat, C.; Fofana, I. Application of a New Dynamic Numerical Model to Predict Polluted Insulator Flashover Voltage. In Proceedings of the Electrical Insulation Conference, Philadelphia, PA, USA, 8–11 June 2014; pp. 102–106.
16. Hampton, B.F. Flashover Méchanism of Polluted Insulation. *Proc. PIEE.* **1964**, *111*, 985–990.
17. Hesketh, S. General criterion for the prédiction of pollution flashover. *Proc. IEE.* **1967**, *114*, 531–532. [CrossRef]
18. Anjana, S.; Lakshminarasmha, C.S. Computed of Flashover Voltages of Polluted Insulators using Dynamic Arc Model. In Proceedings of the 6th International Symposium on high-voltage Engineering, New Orleans, LA, USA, 28 August–1 September 1989.
19. Wilkins, R. Flashover Voltage of High Voltage Insulators with Uniform Surface Pollution Films. *Proc. IEE.* **1969**, *116*, 457–465. [CrossRef]
20. Dhahbi-Megriche, N.; Beroual, A.; Krahenbuhl, L. A New Proposal Model for Polluted Insulators Flashover. *J. Phys. D Appl. Phys* **1997**, *30*, 889–894. [CrossRef]
21. Tavakoli, C.; Farzaneh, M.; Fofana, I.; Beroual, A. Dynamics and Modeling of AC Arc on Surface of Ice. *IEEE Trans. Dielectr. Electr. Insul.* **2006**, *13*, 1278–1285. [CrossRef]
22. Teguar, M.; Mekhaldi, A.; Boubakeur, A.; Harid, N. Flashover performance of non-uniformly polluted glass insulating surface under hvac stress. *Int. J. Power Energy Syst.* **2017**, *37*. [CrossRef]
23. Mekhaldi, A.; Namane, D.; Bouazabia, S.; Beroual, A. Flashover of Discontinuous Pollution Layer on HV Insulators. *IEEE Trans. Dielectr. Electr. Insul.* **1999**, *6*, 900–906. [CrossRef]
24. Mekhaldi, A.; Bouazabia, S. Conduction Phenomena on Polluted Insulating Surfaces under AC High Voltage. In Proceedings of the 9th International Symposium on HV Engineering, Graz, Austria, 19–23 October 1995; pp. 31–74.
25. Fofana, I.; Farzaneh, M.; Hemmatjou, H.; Volat, C. Study of Discharge in Air from The Tip of an Icicle. *IEEE Trans. Dielectr. Electr. Insul.* **2008**, *15*, 730–740. [CrossRef]
26. Farokhi, S.; Farzaneh, M.; Fofana, I. Experimental Investigation of the Process of Arc Propagation over an Ice Surface. *IEEE Trans. Dielectr. Electr. Insul.* **2010**, *92*, 464–485. [CrossRef]
27. Claverie, P.; Porcheron, Y. How to Choose Insulators for Polluted Areas. *IEEE Trans. Power Appar. Syst.* **1973**, *92*, 1121–1131. [CrossRef]
28. Teguar, M. Modélisations D'isolateurs Pollués Soumis à Divers Paramètres Electrogéométriques. Ph.D. Thesis, Ecole Nationale Polytechnique, Alger, Algerie, 2003.Obenaus, F. Fremdschichtüberschlag und Kriechweglänge. *Dtsch. Elektrotech.* **1958**, *4*, 135–136.
29. Rizk, F.A.M. Analysis of Dielectric Recovery with Reference to Dry—Zone Arcs on Polluted Insulators. In Proceedings of the IEEE PES Winter Power Meeting, New York, NY, USA, 28 January–2 February 1971.
30. Rizk, F.A.M. A Criterion for AC Flashover of Polluted Insulators. In Proceedings of the IEEE PES Winter Power Meeting, New York, NY, USA, 28 January–2 February 1971.
31. Dhahbi, N.; Beroual, A. Flashover dynamic model of polluted insulators under ac voltage. *IEEE Trans. Dielectr. Electr. Insul.* **2000**, *7*, 283–289. [CrossRef]
32. Pelissier, R. *L'effet Couronne sur les Lignes Aériennes*; Indice D-160; Technique de L'ingénieur: Paris, France, 1951; Volume 1.
33. Rakotonandrasana, J.H.; Beroual, A.; Fofana, I. Modelling of the negative discharge in long air gaps under impulse voltages. *J. Phys. D Appl. Phys.* **2008**, *41*, 105210. [CrossRef]
34. Peek, F.W. *Dielectric Phenomena in High-Voltage Engineering*; McGraw-Hill: New York, NY, USA, 1929.
35. Warner, E.H. *Corona Discharge*; Forgotten Books: London, UK, April 2018; 146p.
36. Amrani, M.L.; Bouazabia, S.; Fofana, I.; Meghnefi, F.; Jabbari, M. Modified Peek formula for calculating positive DC corona inception voltage on polluted insulator. *Electr. Eng.* **2019**, *101*, 489. [CrossRef]
37. Amrani, M.L.; Bouazabia, S.; Fofana, I.; Meghnefi, F.; Jabbari, M. On Discharge Inception Voltage for Insulators under Non-uniform Field with AC Voltage. In Proceedings of the International on Advanced Electrical Engineering 2019 (ICAEE 2019), El Hamma, Algiers, 19–21 November 2019. [CrossRef]
38. Meeker, D. *Finite Element Method Magnetics*; Version 4.2, User's Manual; University of Virginia: Charlottesville, VA, USA, 2004.

Article

Dynamic Pollution Prediction Model of Insulators Based on Atmospheric Environmental Parameters

Siyi Chen * and Zhijin Zhang

State Key Laboratory of Power Transmission Equipment & System Security and New Technology, Chongqing University, Chongqing 400044, China; zhangzhijing@cqu.edu.cn
* Correspondence: 20163155@cqu.edu.cn; Tel.: +86-152-2300-5117

Received: 8 May 2020; Accepted: 12 June 2020; Published: 13 June 2020

Abstract: Pollution-induced flashover is one of the most serious power accidents, and the pollution degree of insulators depends on atmospheric environmental parameters. The pollution models used in the power system research are usually static, but the environmental parameters are dynamic. Therefore, the study on the dynamic pollution prediction model is of great importance. In this paper, the dynamic pollution prediction model of insulators based on atmospheric environmental parameters was built, and insulators' structure coefficients were proposed based on the model. Firstly, the insulator dynamic pollution model based on meteorological data (PM2.5, PM10, TSP (total suspended particulate), and wind speed) was proposed, and natural pollution tests were also conducted as verification tests. Furthermore, insulator structure coefficients *c*1, *c*2 (*c*1: pollution ratio of U210BP/170 to XP-160; *c*2: calculated pollution ratio of U210BP/170T to XP-160) were then obtained, and their influence factors were discussed. At last, insulator structure coefficients were calculated, and it can be seen that the calculated error of insulator structure coefficients was acceptable, with the average *re* (relative errors) at 9.0% (*c*1) and 13.5% (*c*2), which verifies the feasibility of the model. Based on the results in this paper, the NSDD (non-soluble deposit density) of insulators with different structures can be obtained using the insulators' structure coefficient and the reference XP-160 insulator's NSDD.

Keywords: dynamic pollution model; reference insulators; insulator structure coefficient; natural pollution tests; finite element method

---

1. Introduction

Pollution-induced flashover is among the most serious power accidents, which seriously threatens the safety and stability of the power system [1–5]. In the past three decades, China has suffered from air pollution due to rapid economic growth, industry-led urbanization, and a lack of environmental protection [6]. Severe air pollution aggravates the possibility of pollution flashover [7]. Therefore, much literature has focused on the issue of insulator contamination [8–21].

It will cost a lot of workforces and material resources to test the pollution degree of insulators operating in transmission lines. Therefore, some scholars have first studied insulator contamination in a wind tunnel and other pollution accumulation systems [8–12]. For example, insulator contamination characteristics were studied by wind tunnel simulation in the literature [8], and the results suggest that insulator structures, wind speed and RH (relative humidity) have obvious impacts on the contamination degree of insulators. Research on the contamination characteristics in the winter environment [9,10] also show that the wind speed has a greater effect on insulator contamination and NSDD. Y. Liu et al. [11] set up a natural pollution accumulation system and uses it to analyze the contamination characteristics and the micro-shape features of the insulator surface. The results show that the DC electric field has a significant effect on agglomeration characteristics.

In addition, the analysis methods based on finite element, grey theory, etc. are also applied to the insulator pollution model [13–18]. For example, the coupling physics model of a three-unit XP-160 insulator string was established in the literature [13]. Moreover, the contamination deposition process was simulated using the multiphysics simulation software Comsol. Based on the grey theory, X. Qiao et al. [14] established the insulator pollution model, but the result shows that there are still some errors. The error is the error between the calculated ESDD (equivalent salt deposit density) and the actual ESDD. The error mainly comes from the defect of the algorithm and the calculation accuracy fluctuates with the sample selection. Z. Zhang et al. [15] presented the contamination results using a volume fraction which was obtained by a Eulerian two-phase flow model and further proved the feasibility of this method. What is more, little literature has studied insulator contamination in the natural environment [19–21]. However, Z. Zhang et al. [19] pointed out that it will take long time and high expense to get reliable results.

In addition, the pollution model used in the power system is usually static, but the environmental parameters are dynamic. There is less study on the quantitative relationship between the pollution degree and the dynamic environmental parameters. Moreover, now in the power system, the actual insulator's NSDD (non-soluble deposit density) is usually determined by measuring the reference XP-160 insulator's NSDD. However, even in the same pollution condition, the pollution levels of insulators with different structures are various.

Therefore, the dynamic pollution prediction model of insulators based on atmospheric environmental parameters was built, and insulator structure coefficients were proposed based on the model in this paper.

2. Dynamic Pollution Prediction Model of Insulators

2.1. Numerical Simulation Based on Eulerian Two-Phase Model

The practicability of the method based on the Eulerian two-phase model in engineering has been verified in the literature [15,22–24]. In the Eulerian simulation model, the different phases are treated mathematically as interpenetrating continua. In the simulation model, the standard k-ε model is used to describe the effects of turbulent fluctuations of velocities. The basic equations of the k-ε model have been given in our previous research [15]. Ti (turbulence intensity) can be calculated according to Re (Reynolds number). More specifically, $Ti = 0.16Re^{-1/8}$.

In this paper, seven-unit 3D insulator string models are established by the Solidworks. The insulators' structure and structural parameters are shown in Table 1. The seven-unit 3D insulator string models are imported into Ansys, and a wind tunnel computational domain (9 m × 2.5 m × 7 m) is created according to the studies [13,25,26], as is shown in Figure 1a. One side of the computational domain is the air and particles inlet, and the other side is the flow outlet.

Table 1. Profile parameters of insulators.

Samples	Parameters			Structure
	H (mm)	*D* (mm)	*L* (mm)	
U210BP/170	155	300	450	

Table 1. *Cont.*

Samples	Parameters			Structure
	H (mm)	*D* (mm)	*L* (mm)	
U210BP/170T	170	330	525	
XP-160	155	255	305	

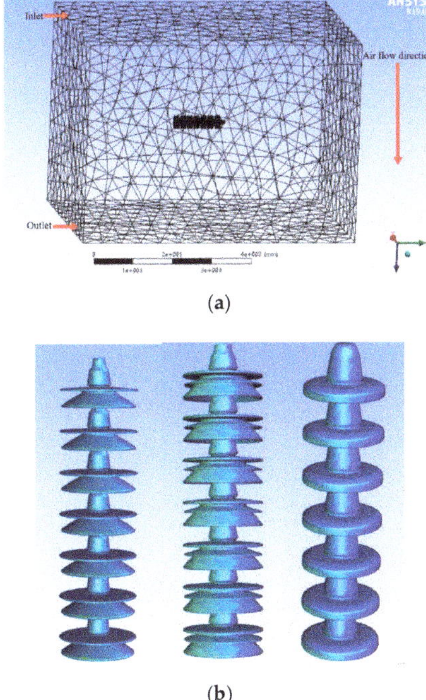

(**a**)

(**b**)

Figure 1. Calculation model and simulation results: (**a**) the calculation model in Ansys; (**b**) the simulation results of U210BP/170, U210BP/170 and XP-160.

Furthermore, a size function is attached to the four regions around the insulator to control the size of the grid cells, and these regions mesh with tetrahedral cells. The outer regions are meshed with hexahedron and prism cells to reduce the number of grids. Practical experience shows [15] that this grid meshing technique improves the calculation accuracy and reduces the cost of calculation time. Then, the boundary condition setting of the calculation domain is processed. The inlet of the domain is set to the "velocity-inlet" boundary type, and the outlet of the domain is set to the "out-flow" boundary type.

The calculated results in this paper are shown in Figure 1b. When the initial concentration settings are the same (0.06), the pollution performance of the insulator was mainly affected by the particle diameter and wind speeds. Specifically, when the initial concentration is 0.06, the simulation results relationship between the volume fraction of the three insulators and the different particle diameters and different wind speeds are shown in Figure 2. It can be seen that the wind speed (W_i) and particle size (d_p) have a great influence on the pollution performance, but the influence on each insulator are not the same. Besides, the Euler two-phase flow simulation led to steady-state results; therefore, it is necessary to establish a connection between the simulated accumulation results and the actual accumulated pollution results. Thus, the simulation results with the same environmental parameters are compared with the wind tunnel experiments. Finally, the NSDDs of each insulator string are obtained by simulation and comparison.

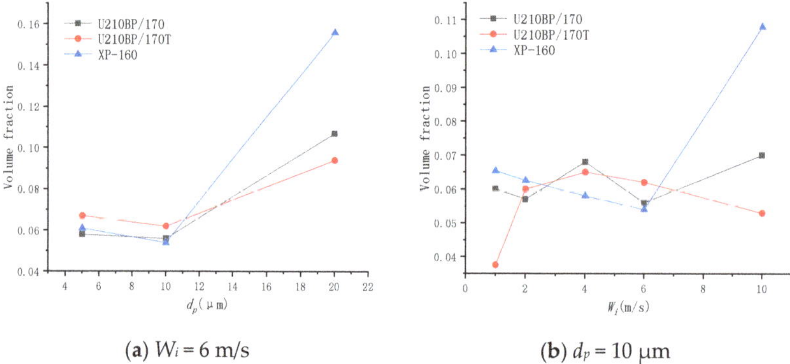

(a) W_i = 6 m/s (b) d_p = 10 μm

Figure 2. Simulation results: (**a**) the volume fraction with a different particle diameter; and (**b**) the volume fraction with a different wind speed.

2.2. Dynamic XP-160 Pollution Model Based on Meteorological Data

The pollution accumulation degree depends on meteorological conditions. For example, the number of pollution particles adhered to the surface of insulator increases with the increase in pollution concentration. According to the corresponding meteorological data, the pollution amount of the insulator surface area under the condition of the pollution concentration can be obtained. In consideration of the time-varying dynamic change of atmospheric environmental parameters, the pollution amount of insulator surface area should be superposed by the accumulated pollution amount of each period, namely:

$$\begin{array}{c} \Delta\phi_{mi} = \int_0^{d_{pM}} \frac{c_{pi}(d_p)}{c_{p0}} \cdot t_i \cdot \rho_m\left(V_i, d_p\right) dd_p \\ \phi_m(H) = \sum\limits_{i=1}^{N} \Delta\phi_{mi}, H = \sum\limits_{i=1}^{N} t_i \end{array} \quad (1)$$

where, c_{p0} is the reference concentration, 15 mg/m^3. $c_{pi}(d_p)$ is the concentration corresponding to the polluted particles with the particle size of d_p in the i time period, mg/m^3. V_i is the wind speed in this time period, m/s; t_i is the time length in the i time period. $\rho_m(V_i,d_p)$ is the pollution accumulation per unit time on the insulator surface. d_{pM} is the maximum particle size of the atmospheric particles in the i time period, μm. $\Delta\Phi_{mi}$ is the pollution increment of the insulator surface area in the i time period, mg/cm^2. H is the total time in each time period. $\Phi_m(H)$ is the final accumulated pollution amount of the polluted particles on the insulator surface, mg/cm^2.

The $c_{pi}(d_p)$ cannot be directly measured, but it needs to be obtained through the relationship between the particle size and its mass concentration, but the relationship is difficult to measure and obtain. Therefore, the approximate method is adopted, and it is considered that the mass fraction and particle size of the polluted particles meet the rosin rammer distribution:

$$\lambda_i(d_p) = 1 - \exp\left(-n_2 \cdot {d_p}^{n_1}\right) \tag{2}$$

where $\lambda_i(d_p)$ is the mass fraction of the polluted particles, whose particle size is less than d_p in the i time period. n_1 is the distribution characteristic index. n_2 is the distribution characteristic coefficient.

The meteorological department usually classifies the polluted particles according to the air quality index standard of real-time monitoring: PM2.5 (d_p < 2.5 μm), PM10 (d_p < 10 μm), TSP (total suspended particulate) (d_p < 100 μm). The units of these three parameters are μg/m^3. According to the data measured in the i time period, the PM2.5, PM10 and TSP can be obtained:

$$\begin{aligned} &\mathrm{PM2.5/TSP} = 1 - \exp(-n_2 \cdot 2.5^{n_1}) \\ &\mathrm{PM10/TSP} = 1 - \exp(-n_2 \cdot 10^{n_1}) \\ &1 = 1 - \exp(-n_2 \cdot 100^{n_1}) \end{aligned} \tag{3}$$

By fitting Equation (3), n_1, n_2 can be obtained, and then the mass fraction particle size distribution function of the i time period can be obtained. Taking the air pollution monitoring data of an area as an example, the TSP, PM10 and PM2.5 measured in the period i are about 200 μg/m^3, 120 μg/m^3, 24 μg/m^3, then the values of n_1, n_2 are 1.42 and 0.03, respectively:

$$\lambda_i(d_p) = 1 - \exp\left(-0.03 \cdot {d_p}^{1.42}\right) \tag{4}$$

The fitting degree R^2 is 0.99, which shows that the fitting result is reasonable. After the mass fraction-particle size relationship is obtained, Equation (2) can be discretized to approximate the mass fraction size corresponding to each particle size, and then the concentration $c_{pi}(d_p)$ can be obtained:

$$c_{pi}(d_p) \approx \mathrm{TSP} \cdot \left[\lambda(d_p) - \lambda(d_p - \Delta d_p)\right] \tag{5}$$

In order to improve the calculation efficiency and take into account the accuracy, setting Δd_p as 1 μm. In general, the probability of an air pollution particle size less than 50 μm is 99%. Therefore, only the influence of a pollution particle size less than 50 μm needs to be considered in the prediction of pollution accumulation. Based on the discretization of Equation (5), the following results are obtained:

$$\begin{aligned} &c_{pi}(1) = \mathrm{TSP} \cdot [1 - \exp(-n_2 \cdot 1^{n_1})] \\ &c_{pi}(2) = \mathrm{TSP} \cdot [-\exp(-n_2 \cdot 2^{n_1}) + \exp(-n_2 \cdot 1^{n_1})] \\ &\ldots \\ &c_{pi}(50) = \mathrm{TSP} \cdot [-\exp(-n_2 \cdot 50^{n_1}) + \exp(-n_2 \cdot 49^{n_1})] \end{aligned} \tag{6}$$

Taking the particle mass fraction–particle size distribution function obtained in Equation (4) as an example, and using the method of Equation (6) to discretize, the pollution particle concentration $c_{pi}(d_p)$ corresponding to each d_p value can be obtained, as shown in Figure 3:

Taking Equation (6) into Equation (1) to get the final amount of air pollution particles on the insulator surface after H time of pollution accumulation:

$$\phi_m(H) = \sum_{i=1}^{N} \Delta\phi_{mi} = \sum_{i=1}^{N}\left(\sum_{n=1}^{50} \frac{c_{pi}(d_p)}{c_{p0}} \cdot t_i \cdot \rho_m(V_i, d_p)\right) \tag{7}$$

As mentioned before, $\rho_m(V_i,d_p)$ is the pollution accumulation per unit time on the insulator surface, which can be obtained through several ways, such as the numerical simulation of pollution deposition, wind tunnel tests and nature tests. Furthermore, the pollution degree and insulator structure pollution coefficients (the coefficients will be discussed in the discussion section) can be calculated according to the flow chart based on the proposed model in this paper, as is shown in Figure 4:

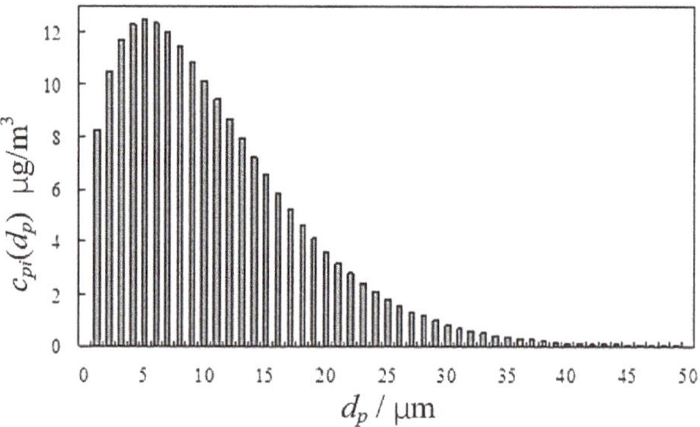

Figure 3. An example of the calculation result of the particle concentration–diameter relationship.

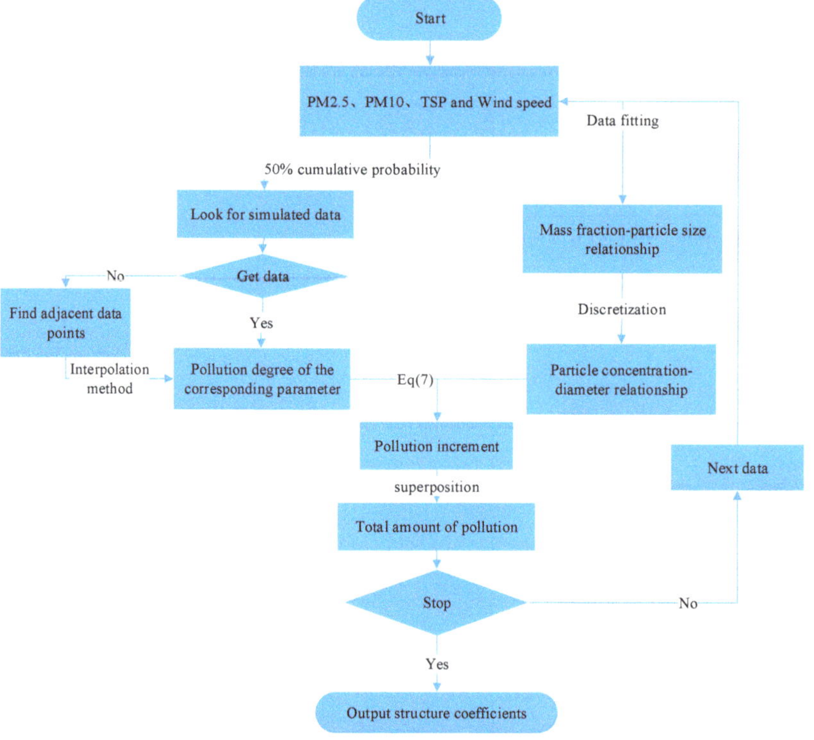

Figure 4. Flow chart of the pollution degree and the coefficients calculation.

2.3. Model Validation

2.3.1. The Experiment Procedures

The contamination samples were three types of insulators arranged in different towers. The test samples arrangement are shown in Figure 5. Firstly, seven units of three types of insulator strings (U210BP/170, U210BP/170T and XP-160) were suspended in the test towers. When the tests were completed, the samples were carefully taken out of the towers. The measuring method of contamination in our previous research [8] is shown in Figure 6. The measuring process is referred to as the IEC Standard 60,507 [8]. It should be noted that the experimental results were the average values of insulator pieces results. When one experiment ended, the next experiment cycle continued.

Figure 5. Field pollution tests.

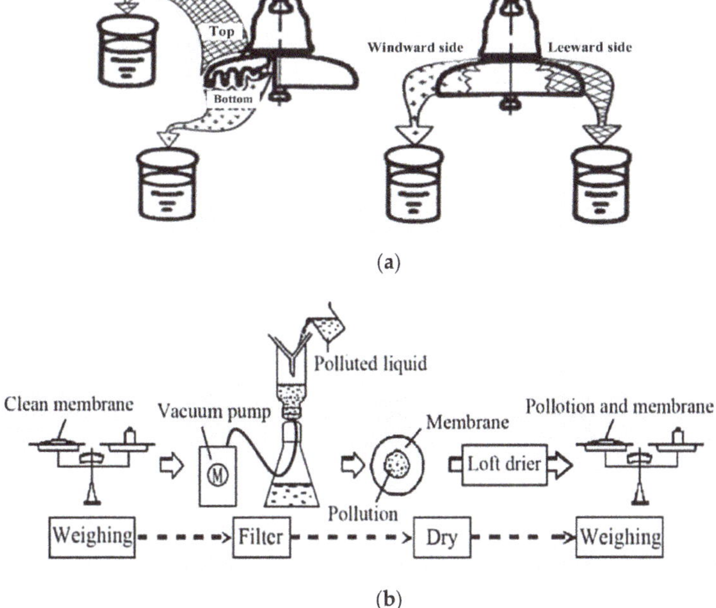

Figure 6. The measuring method of contamination [8]: (**a**) the collection of contamination; and (**b**) the contamination weighing process.

2.3.2. The Experiment Results and Calculated *re*

The experimental results are shown in Table 2 (corresponding meteorological parameters of the first period are shown in Figure 7). Since the meteorological parameters are constantly changing, it is necessary to process the meteorological data before applying the model calculation. In this paper, 50% cumulative probability density is used to express the meteorological parameters in the *H* period. The cumulative probability distribution, also known as the cumulative distribution function, the distribution function and so on, is used to describe the probability of random variables falling on any interval, which is often regarded as the characteristics of the data. If the variable is continuous, the cumulative probability distribution is the function obtained by integrating the probability density function; if the variable is discrete, the cumulative probability distribution is a function obtained by adding the distribution law.

Table 2. Tests result and calculated *re*.

Samples	U210BP/170		U210BP/170T		XP-160	
($\times10^{-3}$ mg/cm^2)	NSDD	*re*	NSDD	*re*	NSDD	*re*
T1	4.3	12.6%	6.2	9.8%	9.8	13.5%
T2	3.1	10.6%	3.4	11.3%	5.6	14.0%
T3	3.5	12.6%	4.1	12.4%	6.5	7.5%
T4	2	8.5%	2.7	11.8%	4.2	10.9%
T5	3.5	7.5%	4.2	9.6%	7.6	7.9%
T6	6.8	15.9%	6.9	17.5%	10.2	14.9%
T7	5.1	20.5%	5.5	18.9%	7.3	21.5%

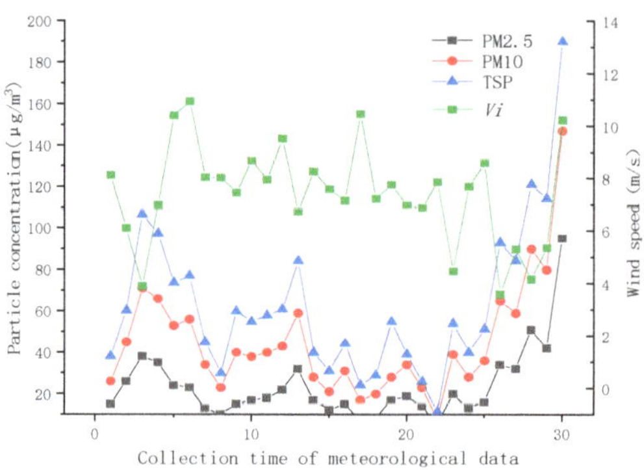

Figure 7. Thirty sets of meteorological data.

According to the results in Table 2, the relative errors between the predicted outcome and the measured results are basically within 20%. Therefore, a new method was proposed to predict insulator contamination by using meteorological monitoring data, which provides a new idea for insulator contamination in a natural environment prediction. At present, the prediction method in this paper can only be used for predicting NSDD. There are complex chemical reactions involved in the ESDD (formation of salt density) and the calculation is more complicated, since ESDD is closely related to the composition of sulfur-containing and nitrogen-containing gases in the atmosphere.

3. Discussion

3.1. Model Error Analysis

According to the above calculation results, this section further discusses the causes of errors. Figure 8 is obtained from the calculation results. It can be seen from Figure 8 that the error of the dynamic pollution model based on atmospheric parameters is different. The *re* of T1–T5 are relatively small, with its value varying from 7.5% to 14%. However, the *re* of T6–T7 are relatively larger, with its value standing at about 20%.

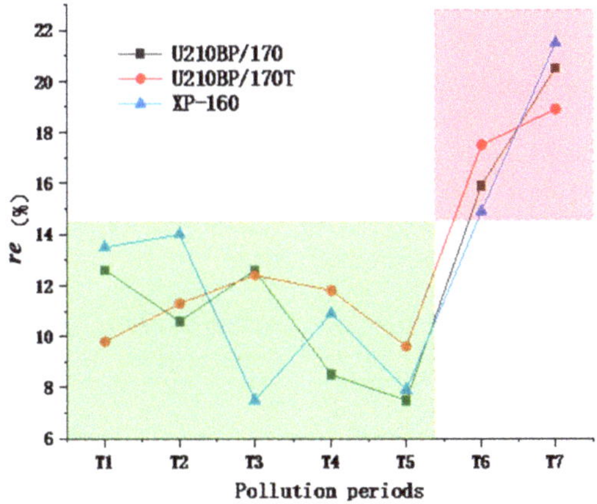

Figure 8. Calculated *re*.

According to the differences between the different *re*, the characteristics of meteorological parameters are further investigated. It was found that, there are five days of rainfall in the sixth stage; while in the seventh stage, there are eight days of rainfall, and the rainfalls are heavy. Therefore, it can be concluded that rainfall mainly causes the errors in the dynamic accumulation model [27].

3.2. Insulator Structure Coefficients

Now in the power system, the actual insulator's NSDD is usually determined by measuring the NSDD of the reference insulator XP-160. However, even under the same pollution condition, the pollution degrees of insulators with different structures varies. Therefore, insulator structure coefficients *c*1 and *c*2 are further calculated. Table 3 and Figure 9 can be obtained.

Table 3. Insulator structure coefficients.

	T1	T2	T3	T4	T5	T6	T7
*c*1	0.439	0.554	0.538	0.476	0.461	0.667	0.699
*c*2	0.633	0.607	0.631	0.643	0.553	0.676	0.753

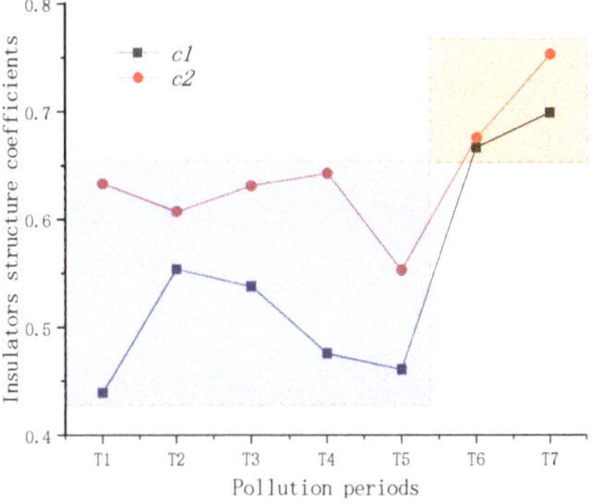

Figure 9. Insulator structure coefficients.

As shown in Figure 9, the value of $c1$ ($c2$) witnessed fluctuation, and the change was evident with its figure varying between 0.439 and 0.699 (0.553 and 0.753). In addition, rainfall had a significant influence on the insulators' structure coefficients, since $c1$ and $c2$ increased to a certain extent during the rainfall period. It can be discovered by further calculation that insulator structure coefficient varies due to the change of wind speed, particle size and other factors, with no obvious changing patterns. Hence, this indicates that insulator structure coefficients are dynamic.

Therefore, the original method for obtaining the pollution level of transmission line insulators only by reference insulator is unreasonable. Insulator structure coefficients should be taken into consideration to get the pollution degree. However, it is time consuming and labor consuming to obtain insulator structure coefficients from natural experiments. Therefore, the insulator structure pollution coefficients can be calculated according to the flow chart based on the proposed model in this paper. And the calculated results are show in Table 4:

Table 4. Calculated insulator structure coefficients.

	T1	T2	T3	T4	T5	T6	T7	Average *re*
$c1'$	0.523	0.612	0.596	0.651	0.423	0.536	0.642	
$c2'$	0.694	0.745	0.712	0.731	0.501	0.545	0.684	
$re1$	0.192	0.106	0.107	0.077	0.081	0.021	0.081	9.0%
$re2$	0.097	0.227	0.129	0.137	0.093	0.125	0.092	13.5%

It can be seen that the calculated error of insulator structure coefficients is acceptable, with the maximum value at 19.2% ($re1$) and 22.7% ($re2$), and the average value at 9.0% ($re1$) and 13.5% ($re2$), which verifies the feasibility of the model.

4. Conclusions

In this paper, the dynamic pollution prediction model of insulators based on atmospheric environmental parameters was built, and insulator structure coefficients were proposed based on the model.

Firstly, the insulator dynamic pollution model based on meteorological data (PM2.5, PM10, TSP, and wind speed) was proposed, and natural pollution tests were also conducted as a verification test. Furthermore, insulator structure coefficients $c1$ and $c2$ were then obtained, and its influence factors were discussed. Rainfall has a significant influence on the insulator structure coefficient; plus, $c1$ and $c2$ increase to a certain extent during the rainfall period. Further calculation of insulator structure coefficients show that it varied with the change of wind speed, particle size and other factors, and there is no obvious changing rule. This result indicates that insulator structure coefficients are dynamic.

At last, insulator structure coefficients $c1'$, $c2'$ were calculated; besides, it can be seen that the calculated error of insulator structure coefficients is acceptable, with the average *re* at 9.0% ($re1$) and 13.5% ($re2$), which verifies the feasibility of the model.

Author Contributions: Methodology, S.C.; Software, S.C.; Data curation, S.C.; Writing—original draft preparation, S.C.; Writing—review and editing, S.C. and Z.Z.; Visualization, S.C.; Supervision, Z.Z. All authors have read and agreed to the published version of the manuscript.

Funding: This research received no external funding.

Acknowledgments: The authors thank all members of external insulation research team (Especially Xinhan Qiao, Dongdong Zhang, Wei Zhang, Jinwei You and others) in Chongqing University for their hard work to obtain the experimental data in this paper.

Conflicts of Interest: The authors declare no conflict of interest.

References

1. Sima, W.; Yuan, T.; Yang, Q.; Xu, K.; Sun, C. Effect of non-uniform pollution on the withstand characteristics of extra high voltage (EHV) suspension ceramic insulator string. *IET Gener. Transm. Distrib.* **2010**, *4*, 445–455. [CrossRef]
2. Samakosh, J.D.; Mirzaie, M. Investigation and analysis of AC flashover voltage of SiR insulators under longitudinal and fan-Shaped non-Uniform pollutions. *Int. J. Electr. Power Energy Syst.* **2019**, *108*, 382–391. [CrossRef]
3. Samakosh, J.D.; Mirzaie, M. Flash-Over voltage prediction of silicone rubber insulators under longitudinal and fan-shaped non-Uniform pollution conditions. *Comput. Electr. Eng.* **2019**, *78*, 50–62. [CrossRef]
4. Qiao, X.; Zhang, Z.; Jiang, X.; Sundararajan, R.; Ma, X.; Li, X. AC failure voltage of iced and contaminated composite insulators in different natural environments. *Int. J. Electr. Power Energy Syst.* **2020**, *120*, 105993. [CrossRef]
5. Zhang, Z.; Qiao, X.; Zhang, Y.; Tian, L.; Zhang, D.; Jiang, X. AC flashover performance of different shed configurations of composite insulators under fan-shaped non-uniform pollution. *IET High Voltage* **2018**, *3*, 199–206. [CrossRef]
6. Zhao, D.; Chen, H.; Li, X.; Ma, X. Air pollution and its influential factors in China's hot spots. *J. Clean. Prod.* **2018**, *185*, 619–627. [CrossRef]
7. Ahmadi-veshki, M.; Mirzaie, M.; Sobhani, R. Reliability assessment of aged SiR insulators under humidity and pollution conditions. *Int. J. Electr. Power Energy Syst.* **2020**, *117*, 105679. [CrossRef]
8. Qiao, X.; Zhang, Z.; Jiang, X.; Zhang, D. Contamination Characteristics of Typical Transmission Line Insulators by Wind Tunnel Simulation. *Electr. Power Syst. Res.* **2020**, *184*, 106288. [CrossRef]
9. Ravelomanantsoa, N.; Farzaneh, M.; Chisholm, W.A. Effects of Wind Velocity on Contamination of HV Insulators in Winter Conditions. In Proceedings of the Annual Report Conference on Electrical Insulation and Dielectric Phenomena, Quebec, QC, Canada, 26–29 October 2008; pp. 240–244.
10. Ravelomanantsoa, N.; Farzaneh, M.; Chisholm, W.A. Simulation method for winter pollution contamination of HV insulators. In Proceedings of the 2011 Electrical Insulation Conference (EIC), Annapolis, MD, USA, 5–8 June 2011; pp. 373–376.
11. Liu, Y.; Wu, G.; Guo, Y.; Zhang, X.; Liu, K.; Kang, Y.; Shi, C. Pollution agglomeration characteristics on insulator and its effect mechanism in DC electric field. *Int. J. Electr. Power Energy Syst* **2020**, *115*, 105447. [CrossRef]
12. Gu, C.; Li, J.; Li, M.; Leizhou, Z.; Wei, Z.; Qiaogen, Z.; Zhiyin, Q.; Rong, X.; Chenzhao, F. Study on contamination deposition characteristics of transmission line insulators. In Proceedings of the IEEE 2011 Asia-Pacific Power and Energy Engineering Conference (APPEEC), Wuhan, China, 25–28 March 2011; pp. 1–4.
13. Zhang, D.; Zhang, Z.; Jiang, X.; Shu, L.; Wu, B. Simulation Study on the Effects of DC Electric Field on Insulator Surface Pollution Deposit. *Energies* **2018**, *11*, 626. [CrossRef]
14. Qiao, X.; Zhang, Z.; Jiang, X.; Yushen, H.; Xun, L. Application of grey theory in pollution prediction on insulator surface in power systems. *Eng. Fail. Anal.* **2019**, *106*, 104153. [CrossRef]
15. Zhang, Z.; Zhang, W.; You, J.; Wu, B.; Maoqiang, B.; Xingliang, J.; Jian, W.; Wei, Z. Influence factors in contamination process of XP-160 insulators based on computational fluid mechanics. *IET Gener. Transm. Distrib.* **2016**, *10*, 4140–4148. [CrossRef]
16. Lv, Y.; Li, J.; Zhan, X.; Pang, G. A simulation study on pollution accumulation characteristics of XP13-160 porcelain suspension disc insulator. *IEEE Trans. Dielectr. Electric Insul.* **2016**, *23*, 2196–2206. [CrossRef]
17. Wu, X.; Zhang, X.; Wu, G. Simulation and Analysis of Contamination Depositing Characteristic of Roof Insulator. In Proceedings of the 2016 IEEE International Conference on High Voltage Engineering and Application (ICHVE), Chengdu, China, 19–22 September 2016; pp. 1–4.
18. Gu, C.; Li, J.; Zhang, L.; Li, M.; Xiao, R.; Qian, Z. Experimental and numerical study of the pollution accumulation on outdoor insulators. *High Volt. Eng.* **2011**, *11*, 2822–2829.
19. Zhang, Z.; Qiao, X.; Yang, S.; Jiang, X. Non-Uniform Distribution of Contamination on Composite Insulators in HVDC Transmission Lines. *Appl. Sci.* **2018**, *8*, 1962. [CrossRef]
20. Zhang, Z.; Zhang, D.; Jiang, X.; Liu, X. Study on natural contamination performance of typical types of insulators. *IEEE Trans. Dielectr. Electr. Insul.* **2014**, *21*, 1901–1909. [CrossRef]

21. Chao, Y.; Huang, F.; Zhao, S.; Wang, C.; Wang, F.; Yue, Y. Study on natural pollution accumulating characteristics of cap and pin suspension ceramic insulator with composite shed of DC 500 kV transmission line in central China. In Proceedings of the 2016 IEEE International Conference on High Voltage Engineering and Application (ICHVE), Chengdu, China, 19–22 September 2016; pp. 1–4.
22. Fu, P.; Farzaneh, M. A CFD approach for modeling the rime-ice accretion process on a horizontal-axis wind turbine. *J. Wind Eng. Ind. Aerodyn.* **2010**, *98*, 181–188. [CrossRef]
23. Drew, D.A.; Lahey, R.T. *In Particulate Two-Phase Flow*; Butterworth-Heinemann: Boston, MA, USA, 1993; pp. 509–566.
24. Launder, B.E.; Spalding, D.B. *Lectures in Mathematical Models of Turbulence*; Academic Press: London, UK, 1972.
25. Xiaohuan, L. Study on the Pollution Accumulation and AC Pollution Flashover Performance of Insulators with Typical Sheds. Master's Thesis, Chongqing University, Chongqing, China, 2013.
26. Haibo, L. Hydrodynamics-Based Contamination Depositing Characteristics Research of Suspension Insulator String. Masters's Thesis, Chongqing University, Chongqing, China, 2010.
27. Jiang, Z.; Jiang, X.; Zhang, Z.; Guo, Y.; Li, Y. Investigating the Effect of Rainfall Parameters on the Self-Cleaning of Polluted Suspension Insulators: Insight from Southern China. *Energies* **2017**, *10*, 601. [CrossRef]

Article

Pollution Flashover Characteristics of Composite Crossarm Insulator with a Large Diameter

Jing Nan [1,2,*], Hua Li [1], Xiaodong Wan [2], Feng Huo [2] and Fuchang Lin [1]

[1] State Key Laboratory of Advanced Electromagnetic Engineering and Technology, Huazhong University of Science and Technology, Wuhan 430074, China; leehua@mail.hust.edu.cn (H.L.); fclin@hust.edu.cn (F.L.)

[2] State Key Laboratory of Power Grid Environmental Protection, China Electric Power Research Institute, Wuhan 430074, China; wanxiaodong@epri.sgcc.com.cn (X.W.); huofeng@epri.sgcc.com.cn (F.H.)

* Correspondence: nanjing@epri.sgcc.com.cn

Abstract: The composite crossarm insulator differs greatly from the suspension insulator in structure and arrangement. This study aims to determine the pollution flashover characteristics of composite crossarm insulators under different voltage grades. Four types of AC composite crossarm insulators with diameters ranging from 100 mm to 450 mm are subjected to artificial pollution test, and then the effects of the surface hydrophobicity state of silicone rubber, core diameter, umbrella structure, arrangement, and insulation distance on the pollution flashover voltage of the composite crossarm insulators are analyzed. Under the pollution grade 0.2/1.0 mg/cm^2 and voltage grade from 66 kV to 1000 kV, if the silicone rubber surface changes from HC5 to HC6, the pollution flashover voltage of the composite crossarm insulator will increase by 13.5% to 21.0% compared with the hydrophilic surface. If the core diameter changes from 100 mm to 300 mm, the pollution flashover voltage gradient decreases with the increase in core diameter; if the core diameter changes from 300 mm to 450 mm, the pollution flashover voltage gradient increases with core diameter. Under the same insulation height and core diameter, the umbrella structure will have a certain impact on pollution flashover voltage by up to 1.7% to 5.4%. Under the horizontal arrangement, the pollution flashover voltage can increase by 10.5% to 12.1% compared with that under the vertical arrangement. Under the hydrophilic surface and weak hydrophobicity state, the pollution flashover voltage has a linear relationship with the insulation distance. The above results can provide a reference for the structural design and optimization of the composite crossarm insulator.

Keywords: composite crossarm; pollution flashover characteristics; core diameter; hydrophobicity; umbrella structure; voltage gradient

Citation: Nan, J.; Li, H.; Wan, X.; Huo, F.; Lin, F. Pollution Flashover Characteristics of Composite Crossarm Insulator with a Large Diameter. *Energies* **2021**, *14*, 6491. https://doi.org/10.3390/en14206491

Academic Editor: Stephan Brettschneider

Received: 29 August 2021
Accepted: 28 September 2021
Published: 11 October 2021

1. Introduction

The composite crossarm insulator features good pollution, lightning, and wind deviation protection effects, and is lightweight and easy to install. In addition, it can save power transmission corridors. When being used for electric transmission lines, they can save project costs, improve operational reliability, and produce numerous economic and social benefits [1,2].

Many countries have studied composite crossarms. Since the 1960s, Japan has studied the use of fiber-reinforced polymers (FRPs) for crossarms of electric transmission lines, which could greatly solve flashover accidents due to wind deviation [3–5]. Multiple companies in the USA used FRPs for practical production. Shakespeare Composite Structures™ was the first to develop electric poles with composite materials. These electric poles were installed in Hawaii, where high salt spray corrosion and hurricanes occur frequently, and had been used there for over 40 years. At present, they still have good service conditions. In addition, many FRP manufacturers, such as Newmark, Strongwell, and Ebert, have developed their own FRP power transmission towers [6–8]. In Europe, the 3D electric field computation of a composite crossarm was carried out, and a trial site has been developed within a substation on the Northeast coast of Scotland for the electrical testing of high voltage composite crossarms [9–11]. Although studies on composite crossarms started late in China, they have developed rapidly. In recent years, many scientific institutions and universities have been devoted to studying composite crossarms. China Electric Power Research Institute developed electrical tests, mechanical tests, and construction technology research for composite crossarms under multiple voltage grades [12,13]. The Northwest Electric Power Design Institute cooperated with Xi'an Jiaotong University to study the electric field distribution, electrical test, and mechanical structure for a 750 kV composite crossarm tower, obtain the interstitial discharge characteristics of tower head and the voltage-resistant characteristics of composite crossarm insulators reflecting pollution conditions in the northwest region and use a 750 kV composite crossarm for the electric transmission line of Hami Nan-Shazhou project [14,15]. Tsinghua University cooperated with China Southern Power Grid to conduct a pollution flashover test for composite crossarms under a 500 kV high altitude and observed arc development [16]. The composite crossarm manufacturers NARI and SHEMAR studied the internal materials of large-size composite outdoor insulators and ensured the internal insulation strength of composite insulators with a large diameter through air inflation and filling in polymer materials [17,18].

In sum, numerous studies have focused mainly on the mechanical performance of composite crossarm structures, the simulation of electric field distribution, the insulating property of materials in crossarm, and the electrical property of a single voltage grade. However, the pollution flashover characteristics of composite crossarm insulators have not been systematically studied. In addition, composite crossarm insulators largely differ from suspension insulators in diameter, arrangement form, and umbrella structure [19]. Thus, determining the pollution flashover characteristics of composite crossover insulators with a large diameter is important. In this study, an artificial pollution test is performed for AC composite crossarm insulators with voltage grades 66 kV to 1000 kV to determine their pollution flashover characteristics. Then, the effects of surface hydrophobicity state, core diameter, umbrella structure, arrangement form, and insulation distance on pollution flashover voltage are analyzed. This study may serve as technical support for designing and optimizing the electrical structure of composite crossarm insulators.

2. Test Equipment, Sample, and Method

2.1. Test Equipment

The artificial pollution test for composite crossarm insulators under power frequency voltage hereof is conducted in the large-scale environmental climate lab at the extra-high voltage AC test base. With a clear height of 25 m and a diameter of 20 m, the equipment tank is provided with a TYDZ-4800 kVA/10.5 kV voltage regulator and a YDTCW-6000 kVA/3 × 500 kV test transformer with a rated voltage of 10 kV and a rated current of 600 A at the primary side, and a rated voltage of 500 kV/1000 kV/1500 kV and a rated current of 10 A/6 A/1 A at the secondary side. All above test equipment follows power requirements for AC pollution test under IEC 60507-2013 [20].

2.2. Sample Parameters and Arrangement

The samples hereof include four types of composite crossarm insulators with core diameters of 100, 200, 300, and 450 mm, which correspond to voltage grades 66, 330, 500, and 1000 kV, respectively. Figure 1 shows the sample structure. In Table 1, the creepage factor (CF) is the ratio of the total creepage distance to the insulation distance of the insulator. Table 2 shows the structural parameters.

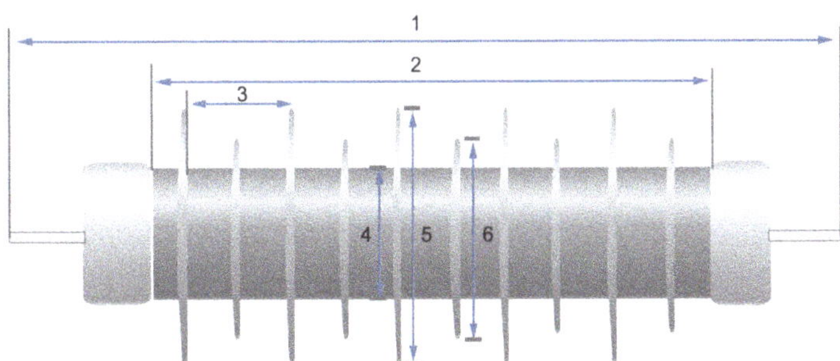

Figure 1. Schematic of sample structure. 1—structure height; 2—insulation distance; 3—shed spacing; 4—core diameter; 5—big shed distance; 6—small shed distance.

Table 1. Composite crossarm insulator parameters.

Voltage Grade (kV)	Core Diameter (mm)	Insulation Height (mm)	Creepage Distance (mm)	Umbrella Skirt Diameter (mm)	Umbrella Distance (mm)	CF
66	100	1 005	3 728	208/152	98/32	3.71
330	200	2 514	9 194	328/296	72/36	3.66
500	300	5 115	19 800	442/408	72/36	3.87
1 000	450	8 633	32 160	571/539	72/36	3.73

Table 2. Parameters of pollution flashover voltage curve.

Voltage Grade/kV	A	α	Degree of Fitting R^2
66	43.88	0.492	0.998
330	124.72	0.391	0.993
500	266.22	0.292	0.996
1 000	575.45	0.202	0.999

The sample of the artificial pollution test for the composite crossarm insulators is arranged horizontally, which is consistent with the practical operation state. Figure 2 shows the arrangement of a typical sample.

Figure 2. Sample arrangement.

2.3. Test Method

The artificial pollution test for the composite crossarm insulators with a large diameter is conducted with the solid chromatography recommended under IEC 60507-2013—"artificial pollution tests on high-voltage ceramic and glass insulators to be used on AC systems" [20]. In the test, commercially available NaCl with a purity of 99.5% and kaolin serve as the soluble salt and inert deposit, respectively, in the simulated deposit.

On the composite crossarm insulators hereof, the umbrella skirt and sheath are made of silicone rubber with good surface hydrophobicity. If the composite insulators' complete loss of surface hydrophobicity under extreme conditions is simulated on-site, kaolin can be applied on the surface to decrease its hydrophobicity, and then pollution coating is applied after complete loss of hydrophobicity. The test is conducted immediately after the pollution layer on the sample has dried [21,22]. If the surface state of the composite crossarm insulators from Grade HC5 to HC6 with weak hydrophobicity is simulated, kaolin is used to reduce surface hydrophobicity, and then pollution coating is applied. Based on temperature and humidity differences in the static environment of the samples, the test is often conducted after the sample has been drying for 4 h to 12 h. In this period, surface hydrophobicity is tested every 1 h. If the surface state satisfies the requirements, the test can be conducted.

In the artificial pollution test, boosting mode includes the constant-voltage lifting and lowering method and the uniform boosting method, where the constant-voltage lifting and lowering method is recommended as the national standard because it is closer to practical operation conditions and shows small data dispersity. The test time of the uniform boosting method is short, but the method is quite different from the actual working conditions and the data is scattered, so it is no longer recommended as a standard method. Therefore, this study uses the IEC 60507-2013-recommended constant-voltage lifting and lowering method to calculate 50% pollution flashover voltage (i.e., at least ten valid tests shall be conducted under standard pollution grade when test conditions are satisfied) and uses the uniform boosting method to quickly find the initial value of the voltage. Then, the 50% pollution flashover voltage U_{50} is calculated under the given pollution grade with the data of the above ten tests. The 50% pollution flashover voltage U_{50} and standard deviation σ are calculated as follows:

$$U_{50} = \sum_{i=1}^{N} U_i / N, \tag{1}$$

$$\sigma = \sqrt{\frac{\sum_{i=1}^{N} (U_i - U_{50})^2}{N}} \times \frac{100\%}{U_{50}}, \tag{2}$$

where U_{50} is the 50% pollution flashover voltage of the composite crossarm insulators; U_i is the applied voltage level, kV; N is the frequency of valid tests; and σ is the relative standard deviation of test results.

3. Test Results and Analysis

3.1. Influence of Pollution Grade on Pollution Flashover Voltage

To determine the flashover characteristics of composite crossarm insulators with a large diameter under different pollution grades, this study conducts an artificial pollution test for composite crossarm insulators with four different core diameters. Considering the pollution grade in the applicable region of composite crossarm insulators, the non-soluble salt density deposit (NSDD) is determined to be 1.0 mg/cm^2, and the equivalent salt density deposition (ESDD) is determined to be 0.1, 0.2, and 0.25 mg/cm^2. Figure 3 shows the test results.

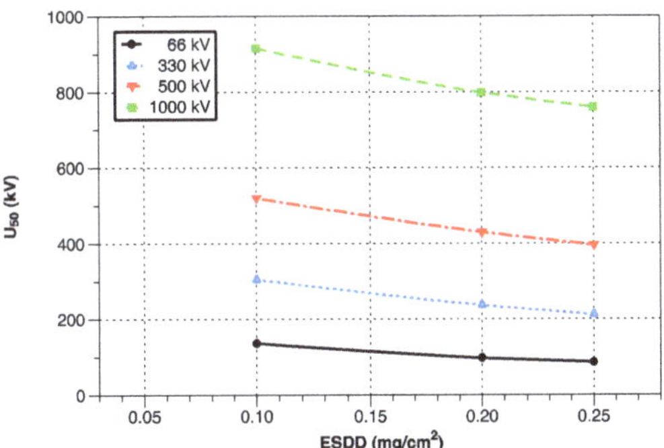

Figure 3. Pollution flashover test results of composite crossarm insulators with different voltage grades.

Test results show that the pollution flashover voltage of the composite crossarm insulators with four core diameters decreases in a nonlinear way with the increase in pollution grade. The relationship between pollution flashover voltage and pollution grade is expressed as below:

$$U_f = A\rho_{ESDD}^{-\alpha}, \tag{3}$$

where U_f is the pollution flashover voltage of the insulator, which is U_{50} hereof; A is the coefficient related to the insulator material and structure; ESDD is the equivalent salt deposit density; and α is the character index that represents the influence of ESDD on pollution flashover voltage.

Table 2 shows the parameters that are acquired through the fitting of the pollution flashover curve.

3.2. Influence of Surface Silicone Rubber State on Pollution Flashover Voltage

The surface state of a composite crossarm insulator largely influences its pollution flashover voltage [23,24]. To understand the influence of surface hydrophobicity state on pollution flashover voltage, this study conducts a pollution flashover test for composite crossarm insulators with a large diameter under surface hydrophilicity and weak hydrophobicity states. Figure 4a,b show the test results of the surface hydrophobicity state for a typical sample. Figure 4a shows the typical hydrophilicity state on the surface of silicone rubber. In accordance with STRI Guide 92/1 Hydrophobicity Classification Guide, the water film covers the whole test area, reaching hydrophobicity grade HC7. Figure 4b shows the typical weak hydrophobicity state on the surface of silicone rubber. The wet area is larger than 2 cm^2. The specific value between the water film area and the test area is close to 90%, and a small dry area can be observed. Thus, the hydrophobicity grade varies from HC5 to HC6.

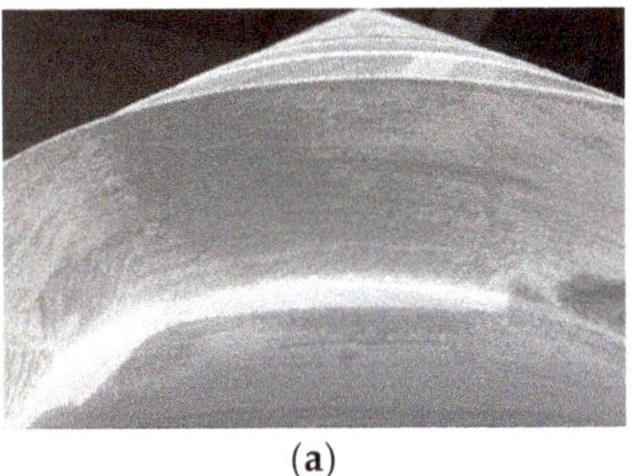

(**a**)

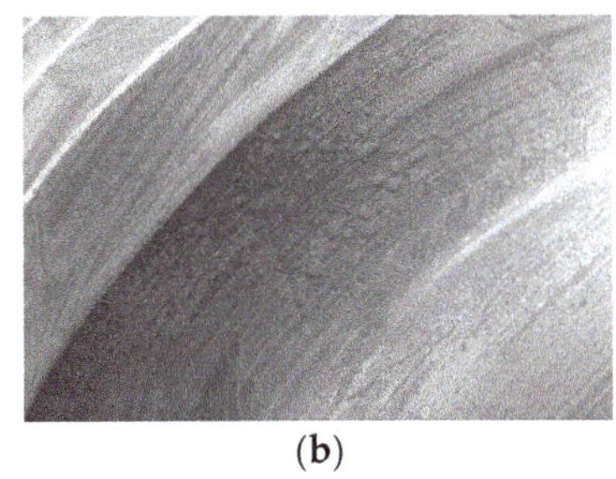

(**b**)

Figure 4. Hydrophobicity state of the sample surface: (**a**) HC7; (**b**) HC5~6.

Considering the pollution level of the area where the composite crossarm is applicable, 0.2/1.0 mg/cm^2 is selected as the test pollution degree, and the following test pollution degree is consistent with this. Table 3 and Figure 5 show the comparison of pollution flashover results under the above two states. Figure 6 shows the trends of a typical leakage current of 500 kV composite crossarm insulators with a diameter of 300 mm under hydrophilicity and weak hydrophobicity states on the surface of silicone rubber.

Table 3. Pollution flashover test results of composite crossarm insulators under hydrophilicity and weak hydrophobicity states.

Voltage Grade (kV)	Hydrophilicity (HC7)		Weak Hydrophobicity (HC5–HC6)	
	U_{50} (kV)	σ(%)	U_{50} (kV)	σ(%)
66	97.2	4.9	112.5	5.7
330	238.0	4.8	288.6	6.2
500	430.6	5.2	508.8	5.6
1000	798.6	5.5	906.3	6.1

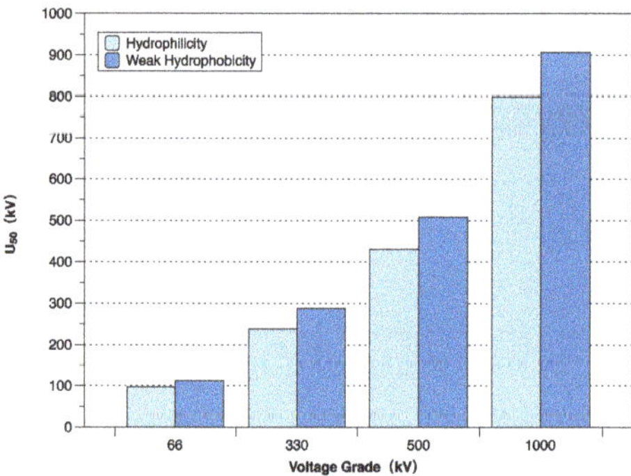

Figure 5. Comparison of pollution flashover test results of composite crossarm insulators with different hydrophobic states.

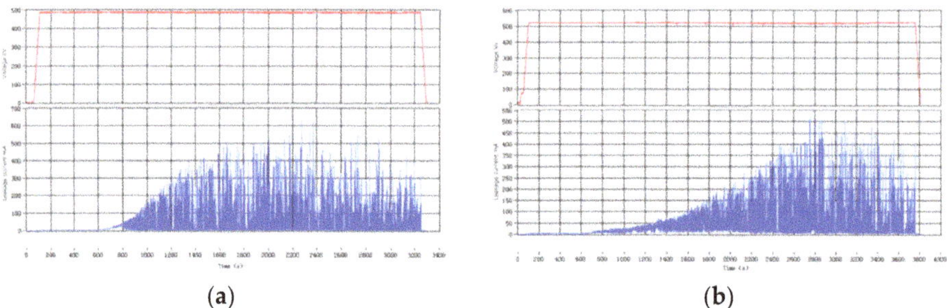

Figure 6. Leakage current trend: (**a**) Hydrophilicity state; (**b**) Weak hydrophobicity state.

As shown in Table 3 and Figure 5, the pollution flashover voltage of the composite crossarm insulator under a weak hydrophobicity state on the surface of silicone rubber is significantly higher than that of the insulator under a hydrophilicity state. In addition, the pollution flashover voltage of the four types of composite crossarm insulators under a weak hydrophobicity state has increased by 13.5% to 21.0% compared with those under a hydrophilicity state.

Leakage current can directly reflect arc size [25–27]. According to the leakage current trends in Figure 6, the surface state of composite crossarm insulators can influence the occurrence time, amplitude, and frequency of the arc. Under a hydrophilicity state, significant discharge can be observed within 10 min because surface pollution coating is easily wet by steam fog. Under a weak hydrophobicity state, the occurrence of significant discharge is several minutes later than under the hydrophilicity state because surface pollution coating cannot be wet by steam fog easily. In addition, the leakage current under the hydrophilicity state in the test is significantly higher than that under the weak hydrophobicity state. Combined with the trends of leakage current in the test, this paper analyzes the difference in pollution flashover performance under hydrophilicity and weak hydrophobicity states. The main reason is that the difference in the hydrophobic performance of pollution coating influences the wetting and arc development of samples [28,29]. In the wetting process, continuous water film can be formed easily on the hydrophilic surface, whereas scattered water drops can form on a strong hydrophobic surface. The water drop and water film can form on weak hydrophobicity surfaces with intermediate hydrophobicity performance. In the arc development process, the dry area on the lower surface with hydrophilicity is concentrated, indicating regional concentration for arc discharge. Due to the distortion effect of water drops on the electric field on the lower surface with hydrophobicity, multi-point discharge may be generated in the flashover process, forming a reticular dry area. In addition, surface discharge is scattered. Thus, multiple discharge branches consume certain energy. A single main arc is difficult to form because of the scattered effect of the discharge arc. As a result, the flashover voltage on the hydrophobicity surface is significantly higher than that on the hydrophilicity surface. For the weak hydrophobicity performance between complete hydrophilicity and strong hydrophobicity states, discharge on the hydrophilic and hydrophobic surfaces can occur in the flashover process, forming regionally concentrated discharge and scattered water drop discharge. That is to say, if the surface hydrophobicity of the composite crossarm insulator is worse, the weak discharge ratio decreases slightly, and the discharge of intermittent arc is transformed into continuous arc discharge, resulting in increased discharge ratio for continuous arc and then flashover. Therefore, the anti-pollution flashover performance of composite crossarm insulators with weak hydrophobic surfaces is significantly higher than those with hydrophilic surfaces.

3.3. Influence of Core Diameter on Pollution Flashover Voltage

The largest difference among selected composite crossarm insulators under different voltage grades hereof is in the core diameter. The gradient of pollution flashover voltage and creepage distance of the samples are calculated to determine the direct influence of core diameter on pollution flashover voltage. Table 4 shows the test results. Figure 7 shows the comparison results of voltage gradient and creepage ratio.

Table 4. Pollution flashover test results of composite crossarm insulators with different core diameters.

Voltage Grade (kV)	Core Diameter (mm)	U_{50} (kV)	σ (%)	Voltage Gradient (kV/m)	Creepage Ratio (mm/ kV)
66	100	97.2	4.9	96.7	38.6
330	200	238.0	4.8	94.7	39.6
500	300	430.6	5.2	84.0	46.0
1 000	450	798.6	5.5	92.5	40.3

Table 4 and Figure 7 show the results of the pollution flashover voltage test for composite crossarm insulators under different voltage grades. As shown in Figure 6a, under the same pollution grade, a 66 kV composite crossarm insulator with a core diameter of 100 mm has a pollution flashover voltage gradient of 96.7 kV/m and shows the best anti-pollution flashover performance, whereas a 500 kV equivalent composite crossarm insulator with a core diameter of 300 mm has a pollution flashover voltage gradient of

84.0 kV/m and shows the worst anti-pollution flashover performance. In the range of core diameter from 100 mm to 300 mm, the voltage gradient decreases with an increase in core diameter. If the core diameter is larger than 300 mm, the voltage gradient increases with the core diameter. The test results are converted into creepage ratio to eliminate the influence of umbrella structure and creepage ratio on pollution flashover results. Figure 6b shows the relationship between creepage ratio and core diameter. If the core diameter varies from 100 mm to 300 mm, the creepage ratio increases with core diameter. If the core diameter is larger than 300 mm, the creepage ratio decreases with the increase in core diameter.

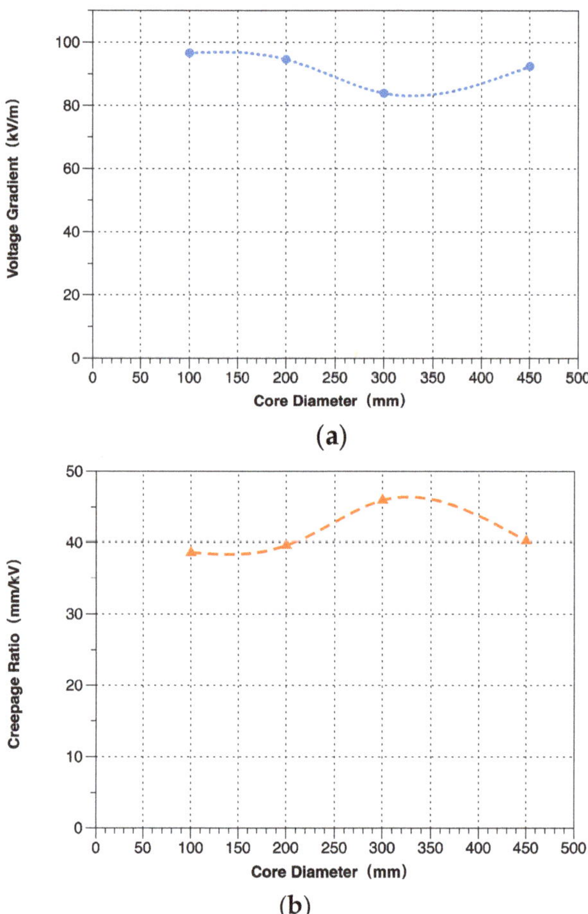

Figure 7. Results of pollution flashover tests: (**a**) Trends of voltage gradient and core diameter; (**b**) Trends of creepage ratio and core diameter.

The rod diameter influences the resistance performance of pollution coating. If other parameters are the same, the resistance of pollution coating increases with leakage distance but decreases with increasing diameter [26,30]. However, the pollution flashover voltage gradient of composite crossarm insulators under different voltage grades has no similar relationship to diameter. Therefore, other influencing factors shall be further studied.

3.4. Influence of Umbrella Structure on Pollution Flashover Voltage

Four groups of samples with the same insulation height and core diameter but different creepage distances, umbrella structures, and umbrella distances are customized to study the influence of umbrella structure on pollution flashover voltage. Samples #1 to #3 are designed with one large umbrella and one small umbrella structure, of which the difference is the diameter of the umbrella skirt and umbrella distance. Sample #4 is designed with one large umbrella and two small umbrella structures. Table 5 shows the structure parameters.

Table 5. Structure parameters.

No.	Insulation Height (mm)	Core Diameter (mm)	Creepage Distance (mm)	Diameter of Umbrella Skirt (mm)	Umbrella Distance (mm)	CF
#1	1 060	220	3 983	356/322	72/36	3.76
#2	1 060	220	3 961	334/294	60/30	3.74
#3	1 060	220	3 971	394/354	96/48	3.75
#4	1 060	220	4 051	378/318	108/36	3.82

Table 6 shows the results of flashover tests for the samples. As listed in Table 5, sample #1 is featured with optimal anti-pollution flashover performance, with a 50% pollution flashover voltage of 110.5 kV. Under the same insulation height and diameter and CF value range from 3.74 to 3.82, the difference in pollution flashover voltage of the four types of crossarm insulators with umbrella structure varies from 1.7% to 5.4%, and the umbrella structure has a minimal influence on the pollution flashover voltage of composite crossarm insulators.

Table 6. Influence of umbrella structure on pollution flashover test results of composite crossarm insulators.

No.	U_{50} (kV)	σ (%)	Voltage Gradient (kV/m)	Creepage Ratio (mm/kV)
#1	110.5	5.1	103.8	36.0
#2	106.6	4.6	100.5	37.1
#3	104.8	4.3	98.9	37.9
#4	108.2	5.3	102.1	37.4

3.5. Influence of Arrangement on Pollution Flashover Voltage

Another significant difference with common composite insulators is the arrangement form. Composite crossarm insulators are mainly arranged in horizontal form, whereas composite insulators are usually arranged in a vertical way. To study the influence of arrangement form on the pollution flashover voltage of composite crossarm insulators with a large diameter, this study conducts a flashover comparison test under a vertical arrangement for samples #1 and #2 in Section 3.3. Table 7 shows the test results.

Table 7. Influence of arrangement on pollution flashover test results of composite crossarm insulators.

No.	Horizontal Arrangement		Vertical Arrangement	
	U_{50} (kV)	σ (%)	U_{50} (kV)	σ (%)
#1	110.5	5.1	98.6	4.5
#2	106.6	4.6	96.5	4.9

Under the same test conditions, the pollution flashover voltage of the composite crossarm insulators is higher under horizontal arrangement than under vertical arrangement. For samples #1 and #2, the pollution flashover voltage under horizontal arrangement increases by 12.1% and 10.5%, respectively, compared with that under the vertical arrangement. The main reason is the difference in pollution loss degree. In the flashover test, the surface pollution on silicone rubber is wet by hot fog to form a high-conductivity water film. Due to gravitation, the water film on the top and bottom surfaces of pollution coating on the composite crossarm insulator under horizontal arrangement can clean the surface of the umbrella sheath and then flow away directly along the sheath, resulting in a serious loss of deposits. It is different for samples under the vertical arrangement. The surface deposit flows away and along the umbrella sheath in the wetting process. The water film cannot play a better cleaning role, and the angle of inclination at the lower surface of the umbrella sheath is small. As a result, the water film cannot be formed on the surface, and deposits cannot flow away easily. Therefore, the loss degree of deposits is lower than the one under the vertical arrangement. These factors decrease the pollution flashover voltage of composite crossarm insulators under the vertical arrangement.

3.6. Relationship between Insulation Distance and Pollution Flashover Voltage

Domestic and foreign scientific research institutions have focused on the relationship between the string length of suspension insulators and pollution flashover voltage. Results show that the pollution flashover voltage of suspension insulators has an approximately linear relationship with string length. However, considering different size structures and arrangement forms, this study conducts an artificial pollution test for 1/4, 1/2, and 3/4 short circuits of EHV composite crossarm insulators to study the relationship between the insulation distance and pollution flashover voltage of composite crossarm insulators with a large diameter. In the test, the surface states of the samples include hydrophilicity and weak hydrophobicity. Figure 8 shows the test results.

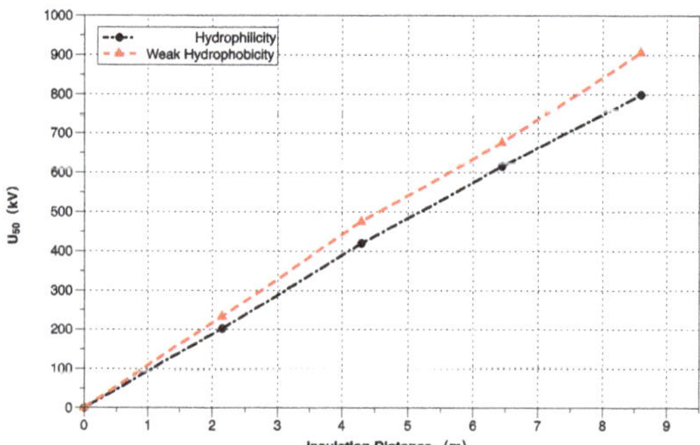

Figure 8. Pollution flashover test results of composite crossarm insulators with different insulation distance.

In the range of insulation distance from 0 m to 8.6 m, the pollution flashover voltage of the composite crossarm insulators with a large diameter under hydrophilicity and weak hydrophobicity states has an approximately linear correlation with insulation distance. The relevance of linear fitting is R12 = 0.999 0 and R22 = 0.999 2, respectively.

In sum, the surface hydrophobicity state of silicone rubber, rod diameter, creepage distance, umbrella structure, and arrangement form can influence the pollution flashover voltage of composite crossarm insulators with a large diameter, where the surface hydrophobicity state, core diameter, creepage distance, and arrangement form exert a large influence on pollution flashover voltage.

4. Conclusions

The presented paper illustrated the influence of composite crossarm insulator core diameter, surface hydrophobicity, umbrella structure, arrangement, and insulation distance on the pollution flashover voltage of large diameter composite crossarm insulators through artificial pollution tests. In the range of core diameter from 100 mm to 450 mm, the pollution flashover voltage gradient decreases with the increase of core diameter and then increases with the increase of core diameter. The hydrophobicity of the surface can significantly increase the pollution flashover voltage, and the influence of the hydrophobicity of the surface should be considered in the external insulation configuration of the project. When the CF value is close, the umbrella structure has little effect on the pollution flashover voltage. Under the same test conditions, the pollution flashover voltage of horizontally arranged composite crossarm insulators is higher than that of vertical ones. The pollution flashover voltage of composite crossarm insulators has an approximately linear relationship with the insulation distance. The above factors need to be considered comprehensively in the structural design of composite crossarm insulators. The pollution flashover of composite crossarm insulators is a complex process that is influenced by multiple factors. Thus, tests are performed to study the pollution flashover voltage. In the future, methods of pollution flashover tests, arc development process, and trends of leakage current of composite crossarm insulators with a large diameter shall be further studied to assess the mechanism underlying their flashover performance.

Author Contributions: Conceptualization, J.N. and F.L.; methodology, J.N., X.W.; formal analysis, J.N.; investigation, F.H.; data curation, X.W.; writing—original draft preparation, J.N., X.W.; writing—review and editing, H.L.; visualization, J.N.; supervision, F.H. All authors have read and agreed to the published version of the manuscript.

Funding: This research was funded by Science & Technology Project of State Grid Corporation of China, grant numbers GY71-17-034 and GCB17201700238-01.

Institutional Review Board Statement: Not applicable.

Informed Consent Statement: Not applicable.

Data Availability Statement: Not applicable.

Acknowledgments: The authors thank all members of UHVAC External Insulation Electrical Characteristics Laboratory of State Grid Corporation of China for their hard work to obtain the experimental data in this paper.

Conflicts of Interest: The authors declare no conflict of interest.

References

1. Zhang, Z.; You, J.; Wei, D.; Jiang, X.; Zhang, D.; Bi, M. Investigations on AC pollution flashover performance of insulator string under different non-uniform pollution conditions. *IET Gener. Transm. Distrib.* **2016**, *10*, 437–443. [CrossRef]
2. Liang, X.; Li, S.; Gao, Y.; Su, Z.; Zhou, J. Improving the outdoor insulation performance of Chinese EHV and UHV AC and DC overhead transmission lines. *IEEE Electr. Insul. Mag.* **2020**, *36*, 7–25. [CrossRef]
3. Fang, D.; Han, J.; Cao, C. Advance in application of composite poles for power transmission. *Fiber Glass* **2008**, *6*, 31–35.
4. Abbasi, A.; Shayegani, A.; Niayesh, K. Pollution performance of HVDC SiR insulators at extra heavy pollution conditions. *IEEE Trans. Dielectr. Electr. Insul.* **2014**, *21*, 721–728. [CrossRef]
5. Guo, Y.; Jiang, X.; Liu, Y.; Meng, Z.; Li, Z. AC flashover characteristics of insulators under haze-fog environment. *IET Gener. Transm. Distrib.* **2016**, *10*, 3563–3569. [CrossRef]
6. Qiao, X.; Zhang, Z.; Jiang, X.; Sundararajan, R.; You, J. DC pollution flashover performance of HVDC composite insulator under different non-uniform pollution conditions. *Int. J. Electr. Power Energy Syst.* **2020**, *185*, 106351. [CrossRef]
7. Zhu, A.; Yang, D.; Liu, Y.; Ma, H.; Wang, Z.; Guo, H. Application and analysis of cross arms made from FRP composite materials in transmission towers. *Power Syst. Clean Energy* **2015**, *31*, 76–82.
8. Holtzhausen, J.P.; Vosloo, W.L. The pollution flashover of ac energized post type insulators. *IEEE Trans. Dielectr. Electr. Insul.* **2001**, *8*, 191–194. [CrossRef]
9. Peesapati, V.; Zachariades, C.; Li, Q.; Rowland, SM.; Cotton, I.; Allison, F.; Chambers, D.; Rhodes, P. Electric field computation for a 400kV composite cross-arm. In Proceedings of the 2012 IEEE Conference on Electrical Insulation and Dielectric Phenomena, Montreal, QC, Canada, 14–17 October 2012; pp. 790–793.
10. Peesapati, V.; Zachariades, C.; Li, Q.; Rowland, SM.; Chambers, D. 3D electric field computation of a composite cross-arm. In Proceedings of the 2012 IEEE Conference on Electrical Insulation, San Juan, WA, USA, 10–13 June 2012; pp. 464–468.

11. Zachariades, C.; Cotton, I.; Rowland, S.M.; Peesapati, V.; Green, P.R.; Chambers, D.; Queen, M. A coastal trial facility for high voltage composite cross-arms. In Proceedings of the 2012 IEEE Conference on Electrical Insulation, San Juan, WA, USA, 10–13 June 2012; pp. 78–82.
12. Xing, H.; Nan, J.; Xing, Z.; Liu, X.; Wu, X. *Research on Application of Key Technology of Composite Insulation Cross Arm for 35kV~750kV Transmission Lines*; Technical Report; China Electric Power Research Institute: Beijing, China, 2020.
13. Nan, J.; Wan, X. *Research on Insulation Characteristics and Surface Insulation Configuration of UHV AC Composite Cross-Arm Insulator under Polluted and Snow-Covered Conditions*; Technical Report; China Electric Power Research Institute: Beijing, China, 2020.
14. Yang, X.; Yu, X.; Shang, Y.; Zhu, A.; Du, J.; Shi, R.; Fan, C.; Gong, X.; Peng, Z. Voltage-sharing characteristics of composite cross-arm in 750 kV AC transmission lines. *Power Syst. Technol.* **2013**, *37*, 1625–1631.
15. Gu, Y.; Yang, L.; Zhang, F.; Xue, Y.; Li, L. DC pollution flashover performance of ultra high voltage convert stations large-size composite post insulators at high altitude areas. *Trans. China Electrotech. Soc.* **2016**, *31*, 93–101.
16. Shen, Y.; Liang, X.; Wang, J.; Wu, C.; Wang, G.; Gao, C. Pollution characteristics of AC 500 kV composite cross-arm in high altitude area. *High Volt. Eng.* **2017**, *43*, 2760–2768.
17. Yang, L.; Wang, H.; Zhao, X. Study on application of composite material in transmission line. *Electr. Power* **2014**, *47*, 53–56.
18. Zhang, C.; Zhang, F.; Chen, C.; Wang, G.; Li, R.; Ma, Y.; Wang, L.; Guan, Z. Research on pollution flashover characteristics of large-size composite outdoor insulation for UHV DC in high altitude area. *Trans. China Electrotech. Soc.* **2012**, *27*, 20–28.
19. Gutman, I.; Dernfalk, A. Pollution tests for polymeric insulators made of hydrophobicity transfer materials. *IEEE Trans. Dielectr. Electr. Insul.* **2010**, *17*, 384–393. [CrossRef]
20. Commission Électrotechnique Internationale. *Artificial Pollution Tests on High-voltage Ceramic and Glass Insulators to be Used on Ac Systems*; International Electrotechnical Commission: London, UK, 2013.
21. Jiang, X.; Yuan, J.; Zhang, Z.; Hu, J.; Sun, C. Study on ac artificial-contaminated flashover performance of various types of insulators. *IEEE Trans. Power Deliv.* **2007**, *22*, 2567–2574. [CrossRef]
22. Gorur, R.S.; Schneider, H.M.; Cartwright, J. Surface resistance measurements on nonceramic insulators. *IEEE Trans. Power Deliv.* **2001**, *16*, 801–805. [CrossRef]
23. Dai, H.; Mei, H.; Wang, L.; Zhao, C.; Jia, Z.; Guan, Z. Volatility of hydrophobicity on the polluted surface during hydrophobicity transfer. In Proceedings of the Chinese Society for Electronical Engineering; Chinese Society of Electrical Engineering: Shanghai, China, 2014; Volume 34, pp. 971–981.
24. Dai, H.; Zhao, C.; Liang, J.; Zhou, J.; Wang, L. Flashover characteristic of polluted silicone rubber with different hydrophobicity. *Trans. China Electrotech. Soc.* **2016**, *31*, 102–111.
25. Li, J.; Sima, W.; Sun, C.; Sebo, S.A. Use of leakage currents of insulators to determine the stage characteristics of the flashover process and contamination level prediction. *IEEE Trans. Dielectr. Electr. Insul.* **2010**, *17*, 490–501. [CrossRef]
26. Savadkoohi, E.M.; Mirzaie, M.; Seyyedbarzegar, S.; Mohammadi, M.; Khodsuz, M.; Pashakolae, M.G.; Ghadikolaei, M.B. Experimental investigation on composite insulators AC flashover performance with fan-shaped non-uniform pollution under electro-thermal stress. *Int. J. Electr. Power Energy Syst.* **2020**, *121*, 106142. [CrossRef]
27. Douar, M.A.; Mekhaldi, A.; Bouzidi, M.C. Flashover Process and Frequency Analysis of the Leakage Current on Insulator Model under non-Uniform Pollution Conditions. *IEEE Trans. Dielectr. Electr. Insul.* **2010**, *17*, 1284–1297. [CrossRef]
28. Zhang, Z.; Jiang, X.; Huang, H.; Sun, C.; Hu, J. Study on the Wetting Process and Its Influencing Factors of Pollution Deposited on Different Insulators Based on Leakage Current. *IEEE Trans. Power Deliv.* **2013**, *28*, 678–685. [CrossRef]
29. Wardman, J.; Wilson, T.; Hardie, S.; Bodger, P. Influence of volcanic ash contamination on the flashover voltage of HVAC outdoor suspension insulators. *IEEE Trans. Dielectr. Electr. Insul.* **2014**, *21*, 1189–1197. [CrossRef]
30. Chandrasekar, S.; Kalaivanan, C.; Montanari, G.; Cavallini, A. Partial discharge detection as a tool to infer pollution severity of polymeric insulators. *IEEE Trans. Dielectr. Electr. Insul.* **2010**, *17*, 181–188. [CrossRef]

Article

Monitoring of Dry Bands and Discharge Activities at the Surface of Textured Insulators with AC Clean Fog Test Conditions

Mohammed El Amine Slama [1,*], Maurizio Albano [1], Abderrahmane Manu Haddad [1], Ronald T. Waters [1], Oliver Cwikowski [2], Ibrahim Iddrissu [2], Jon Knapper [3] and Oliver Scopes [3]

1 Advanced High Voltage Engineering Centre, School of Engineering, Cardiff University, Queen's Buildings, The Parade, Cardiff CF24 3AA, UK; albanoM@cardiff.ac.uk (M.A.); haddad@cardiff.ac.uk (A.M.H.); watersrt@cardiff.ac.uk (R.T.W.)

2 National Grid plc, Warwick CV34 6DA, UK; oliver.cwikowski@nationalgrid.com (O.C.); Ibrahim.Iddrissu@nationalgrid.com (I.I.)

3 Allied Insulators Ltd., Staffordshire ST6 4HN, UK; jon.knapper@alliedinsulators.com (J.K.); oliver.scopes@alliedinsulators.com (O.S.)

* Correspondence: slamame@cardiff.ac.uk

Abstract: The aim of this study is the presentation of the results of an in-lab comparative study of electrical and thermal monitoring of artificially polluted, HTV-textured silicone rubber insulators, with different pollution levels. This work is a preliminary study of an in-situ monitoring of 400 kV SiR textured in a polluted environment. The results showed that the rms leakage current magnitude and pulses, and the average dissipated power depended on the pollution levels and the dry-bands formation. The discharge activity and their nature are governed by the pollution level and the voltage. A differentiation and a quantification between dry-band discharge onset and dry-band arc inception is highlighted.

Keywords: textured insulator; artificial clean fog test; dry bands; discharges; partial arcs; monitoring

Citation: Slama, M.E.A.; Albano, M.; Haddad, A.M.; Waters, R.T.; Cwikowski, O.; Iddrissu, I.; Knapper, J.; Scopes, O. Monitoring of Dry Bands and Discharge Activities at the Surface of Textured Insulators with AC Clean Fog Test Conditions. *Energies* **2021**, *14*, 2914. https://doi.org/10.3390/en14102914

Academic Editors: Issouf Fofana and Stephan Brettschneider

Received: 24 March 2021
Accepted: 14 May 2021
Published: 18 May 2021

Publisher's Note: MDPI stays neutral with regard to jurisdictional claims in published maps and institutional affiliations.

1. Introduction

Optimization of overhead line insulators is one of the key parameters for a high reliability of high voltage (HV) electrical grids. Flashover (FOV) of polluted transmission line insulators is one of the main problems faced by HV overhead line engineers. The performances of insulators under polluted conditions constitute the key factors in the design and dimensioning of HV insulators. The pollutant deposits covering the insulating surfaces can engender a considerable decrease of the dielectric strength of the systems, which can lead to FOVs. According to IEC 60815-1 [1], the FOV process is divided into six phases described separately below. In nature, these phases are not distinct but tend to merge. Depending on the material properties used (ceramic/glass or polymeric), the FOV process of insulators is greatly affected by the insulator's surface properties (hydrophilic or hydrophobic). In the case of glass and ceramic insulators, the surface is hydrophilic, while it is hydrophobic for polymeric insulators. The process of pollution FOV on hydrophilic surfaces begins with the deposition of a contaminant layer (containing salt, insolubles) at the insulator surface. If the pollution is nonconductive (high resistance) when dry, some wetting process is necessary before FOV occurs. In the case of coastal areas, the contaminant is already salted and wetted (salt fog). The second phase of the process is the wetting of the contamination layer according to different mechanisms (fog, light rain, condensation). The salts present in the pollution layer are dissolved and a leakage current (LC) appears and flows at the insulator surface. As the LC flows, a heating effect starts, which dries out parts of the pollution layer. This occurs where the current density is the highest, i.e., where the insulator is at its narrowest. These result in the formation of what are known as dry bands (DBs), and the current flow interrupts. When the DBs appear, discharges are incepted at their board and are electrically in series with the resistance (impedance) of the wetted pollution layer. The kind and the intensity of the discharge depend on the resistance (impedance) of the undried pollution layer. At this instant, the LC flows again and new DBs appears, accompanied with new discharges. The temperature of the Dry Band Discharges (DbDs) becomes higher and evolve to Dry Band Partial Arcs (DBPA). This implies a raising of LC magnitudes and pulses. Depending on the balance between the gradient or the resistance (impedance) of the pollution layer and the DBPAs, the arcs bridging the DBs are sustained and might continue to propagate along the insulator surface [2]. This in turn decreases the resistance (impedance) in series with the arcs, increasing the current until it reaches a critical magnitude, and then, the insulator is completely bridged, and a line-to-earth fault (FOV) is established.

The mechanism of FOV of polymeric hydrophobic insulators initiated by water-droplets rather than dry-bands, was described by Karady and Shah [3]. The authors observed that the electric field changed the shape and elongated the droplets—they become flatter and more elongated. During this process, the coalescence of the droplets forms conductive filaments with a highly resistive layer around them. The elongated filaments shorten the inter-electrode distance and the electric field between them and other neighboring filaments becomes higher. DbDs are initiated between the filaments that lead to a local loss of hydrophobicity, which makes it easier to form new wet regions by the expansion of droplet filaments. In the end, a continuous polluted conductive layer is constituted and the DbDs extends and becomes DBPAs. Then, the situation is similar to the FOV of the hydrophilic surface.

Development of DBs formation and DbDs/DBPAs were studied in [4] for conventional non-textured insulators. They described dynamic DBs and DbDs/DBPAs by IR and visual cameras recording. According to their results, a low LC of only a few mA is enough to create the first DB. They also found that the surface temperature of the pollution layer in the DBs is only a few degrees Celsius above ambient, and is the main cause to prevent rewetting. On the other hand, they observed that DdDs are the probable cause of the delayed re-humidifying of the bands and DBPAs that might result in FOV developing from and across the DbDs.

Many studies were carried out regarding the performances and the optimization of polymeric insulators. Material characteristics as well as geometry and profiles were investigated [5,6]. However, like all polymers, they are vulnerable to long-term degradation due to aging and weathering [7–10]. Field and laboratory experiences showed that the shank regions of polluted polymeric insulators are vulnerable to partial arcs thermal damages [11]. To alleviate this problem, a textured polymeric insulator is one of the promising solution. The main advantage of this kind of insulators is the reduction of both LC density and electrical field strength in such regions, and by the same means to achieve an increase in creepage length [12–14].

In a previous work [15], DBs formation and DbDs/DBPAs at the surface of textured insulator flat samples during an AC inclined plane test was discussed. In this study, dynamic of DBs formation and DbDs/DBPAs at the surface of industrial textured insulator prototype fabricated by our partners Allied Insulators Ltd. was investigated and discussed.

2. Experimental Setup and Procedure

The fog chamber (Figure 1) is the same as described in [12–14]. The test circuit diagram is presented in Figure 2. A Hipotronics AC Dielectrics Test Set supplied the test voltage through a 150 kVA, 50 Hz transformer that could supply a 2 A load at 50 kV (1). The maximum voltage output was 75 kV. The voltage was measured with an embedded digital display on the control panel by tapping on the transformer and externally with a North Star VD-100 RC compensated division high voltage probe (2) of a standard divider ratio 10,000:1 and a <1% error specification. The leakage current was measured through a shunt resistance box (3). An optical link system was used for the connection to the data acquisition system (4).

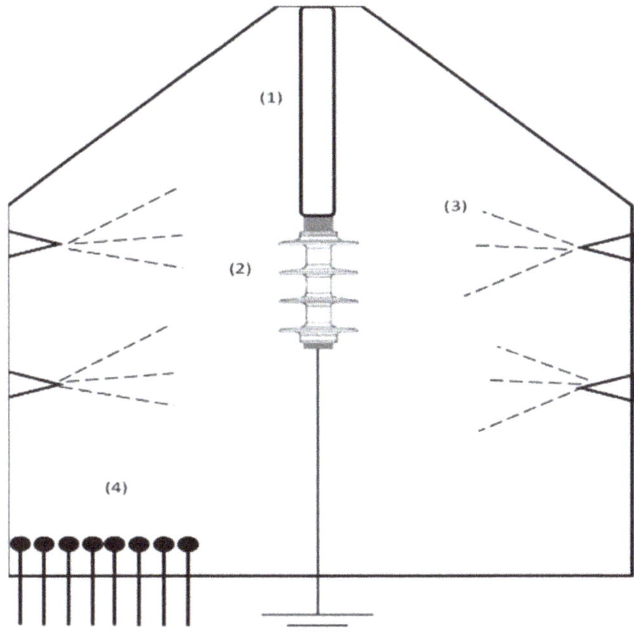

Figure 1. Schematic view of the fog chamber. (**1**) Voltage supply conductor, (**2**) Test insulator, (**3**) Nozzles, and (**4**) Fog control panel.

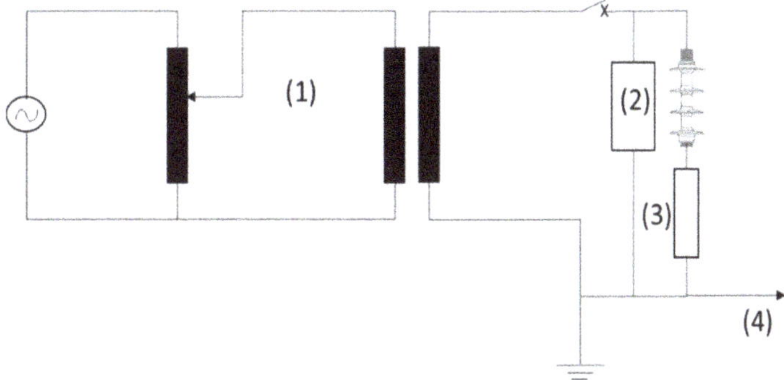

Figure 2. Diagram of the test circuit. (**1**) Voltage supply conductor, (**2**) Test insulator, (**3**) Shunt resistor, and (**4**) to DAQ system.

Applied voltage and leakage current waveforms and signals were acquired, monitored, and stored using a data acquisition (DAQ) system program, based on the NI-LabVIEW platform, as described in [12–14]. Each test facility has its own DAQ system program, as illustrated in Figure 3 and is well-described in [12–14]. As these stored samples were segments of leakage current and voltage waveforms, an appropriate post-processing software program was used to analyze and assess the data recorded during the test [12–14] and was able to calculate the following electrical characteristics.

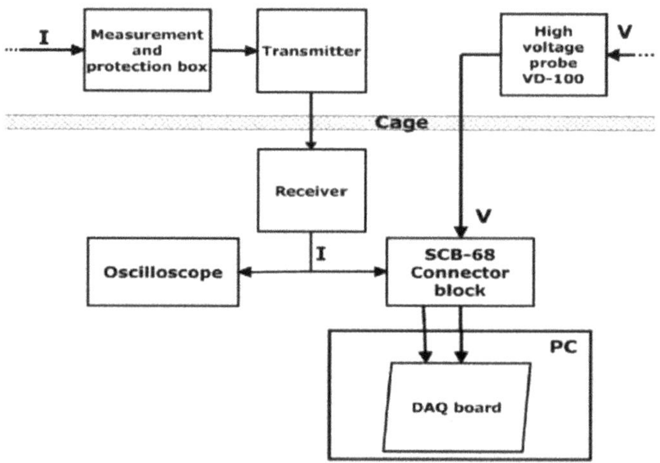

Figure 3. Data acquisition diagram. (I) current, (V) voltage and (PC) Personal Computer.

- The *r.m.s.* values of leakage current i_{rms} and voltage v_{rms} for a cycle was calculated numerically, according to Equations (1) and (2):

$$i_{rms} = \sqrt{\frac{1}{N} \cdot \sum_{j=1}^{N} i_j^2} \tag{1}$$

$$v_{rms} = \sqrt{\frac{1}{N} \cdot \sum_{j=1}^{N} v_j^2} \tag{2}$$

where i_j is the current at sample point j and N is the number of samples per cycle, which in the present work was 200.
- The surface conductance was deduced from the ratio of the current and the voltage.
- The average power was calculated for one cycle by multiplying each voltage element by its counterpart of leakage current, according to the formula:

$$p_{av} = \sqrt{\frac{1}{N} \cdot \sum_{j=1}^{N} v_j \cdot i_j} \tag{3}$$

The second recording device was an infrared (IR) camera FLIR for mapping the thermal stress on the specimen surfaces during the fog test. The standard camera calibration range was maintained, so the measurement interval was 5 °C to 150 °C for spectral thermography. An Ethernet link with a PC, established the communication with the FLIR ResearchIR Max software that processed and stored the captured IR records, and allowed control operation of the camera. A Sony camera was used to have a visual recording along the tests. During tests, heating caused by discharge activity and DB formation in the infrared spectrum were monitored. The goal of the IR imaging was to identify hot-spot formation on the insulator surface, and to correlate with electrical measurements.

The pollution constitution was sodium chloride, 40 g/L of kaolin, wetting agent (Triton), and deionized water. The conductivity of the mixture depended on the quantity of the added sodium chloride. Table 1 shows the used conductivities and their corresponding pollution level, according to IEC 60815.

Table 1. Used pollution levels.

Conductivity (mS/cm)	Pollution Level	ESDD (mg/cm^2)
1.40	Light	0.03
3.50	Medium	0.07
8.60	Heavy	0.19
12.5	Severe	0.28

Before the application of the contaminant, the insulator was pre-conditioned with brushed dry kaolin, according to the Cigré WG C4.303 [16] recommendation. This step was used for reducing the hydrophobicity of the insulator surface. Figure 4 shows an illustration of the effect of pre-conditioning on the HC index for non-textured insulators (conventional CONV). After one hour, 150 mL of contaminant was sprayed (uniform as soon as possible) along the insulator surface. Next, the insulators were left to dry vertically at ambient temperatures, during at least 24 h. The insulators were visually inspected after drying to confirm that the solid pollution layer was more-or-less uniformly formed. The geometrical parameters of the insulators are given in Table 2. Details about texture pattern and how to choose it are well-described in [11–13]. The fog air pressure and flow rate were controlled by the external fog generation control panel, as described previously. We applied a light fog rate 3 L/h (light fog) that simulated the low humidification process.

No pre-conditioned insulators

HC2 to HC3

Pre-conditioned insulators

HC6 to HC7

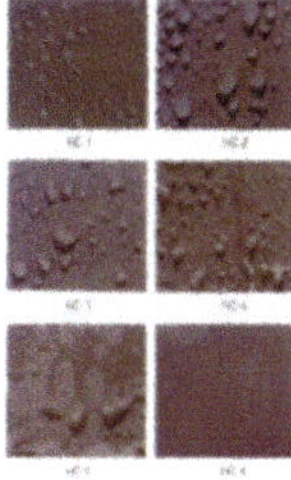

Figure 4. Examples of the effect of pre-conditioning on the HC index.

Table 2. Characteristics of the used insulators.

Arc Length (mm)	Large Shed Diameter (mm)	Small Shed Diameter (mm)	Texture Pattern (mm)	Total Surface (cm^2)
346	160	130	4	4168.6

The testing position of the polluted SiR insulator was vertical, in order to be close to the suspension insulators in the overhead lines. The fog was applied simultaneously with the test voltage. The DAQ system monitored the LC, as the pollution layer absorbed moisture and became gradually conductive. The fog input rate should be sufficient to achieve the maximum value of layer conductance within 10–40 min of the fog application, according to the international standard IEC 60507 [17]. Low-voltage tests were performed to define the wetting duration, corresponding to the maximum SC of the pollution layer for each contamination degree. The test voltage was constant at 0.75 kV to 0.87 kV, which was the minimum voltage output of the power supply, and the retained fog rate was 3 L/h. The insulator was constantly monitored with the IR camera to detect any surface heating, in order to be sure that no drying of the pollution layer due to power dissipation, occurred during the low voltage test.

Figure 5 presents an example of used insulators for the present tests. The insulators are subdivided into 7 sheds and 8 trunk sections.

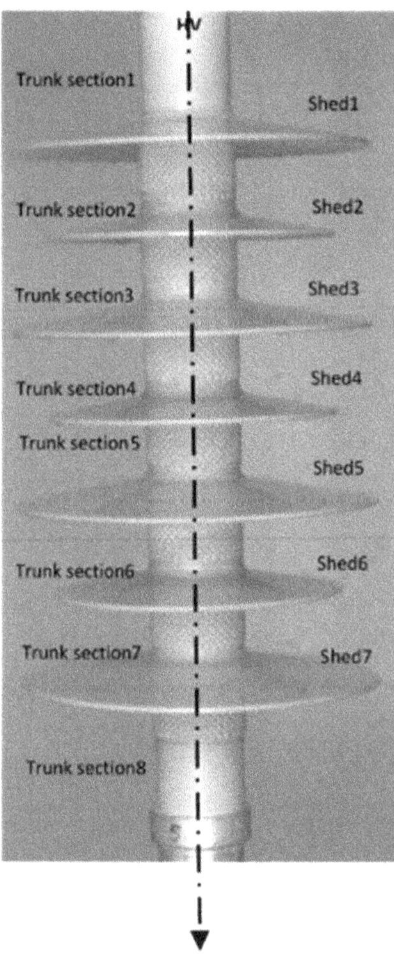

Figure 5. Photo of the used insulator.

The results of the LV tests showed that the conductance of the contaminated insulator was proportional to the pollution level, and the maximum was reached in a time interval of between 10 min and 30 min (depending on the pollution conductivity). For light pollution, the maximum conductance was reached within approximately 30 min and 25 min for medium pollution. In the case of heavy pollution, the maximum conductance was reached at 15 min, while for severe pollution the time was 10 min.

3. Results and Discussion

In this section, the voltage level and the associated SC and LC corresponding to DB onset and DbD/DBPA activity at the insulator surface were investigated.

3.1. Light Pollution

Figure 6 illustrates all steps of a voltage ramp test (from 0.87 kV to 70 kV) that were applied to a fully TXT wetted insulator and the associated LC and SC. At the first voltage steps just before 10 kV (noted (1) in Figure 6), the SC was maintained constant (2.5 μS) while the LC rose with the voltage, until approximatively 2 mA. In this step, only heating was observed without DB formation, as presented in Figure 7a. At 8 kV, the SC and the LC collapsed, resulting in a local DB formation that dramatically increased the insulator surface resistance (noted (2) in Figure 6).

According to Figure 7b, the first dry-band was localized at the beginning of the trunk Section 2 (Figure 7b). This behavior continued with voltage increase, until step (3), corresponding to the first DbD activity. The DbD activity was more intense, and other DB appeared at the other trunk sections. Those DB were larger and more active, as shown in Figure 7c,d. The maximum recorded LC at this final step was 5 mA rms. The maximum temperature at the insulator surface corresponding to DB inception was close to 22.9 °C, while it was around 32 °C when the DBs extended.

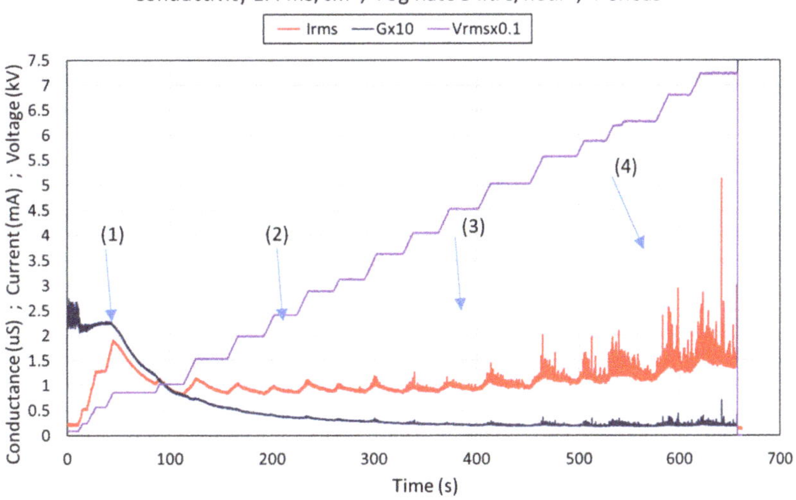

Figure 6. Surface conductance, leakage current, and applied free ramp voltage for light pollution. (1), (2) (3) and (4) are areas where LC activity and magnitude vary.

Figure 8 illustrates the DbD at the insulator surface contaminated with light pollution at the final step voltage (70 kV rms). DbDs were corona/streamers and spark, discharged with ramifications at their heads and purple/white color. Those discharges appeared at the junction between the trunk and the shed. Some discharges also appeared between the sheds due to the formation of the suspended droplets that could be assimilated as a conical electrode. The immediate consequence was that the air broke down between the droplet and the shed.

3.2. Medium Pollution Level

Figure 9 illustrates the records of LC and SC for all steps of a voltage ramp test, as shown for the light pollution level. At the first voltage steps just before 10 kV (noted (1) in Figure 9), the SC was maintained constant, while the LC increased with voltage, until approximatively 7 mA. At 8 kV, the SC and the LC collapsed, resulting in a local DB formation that dramatically increased the insulator surface resistance. Figure 10a,b illustrates the visual and the thermal records of the formation of the first DBs and their extension at the junction of the trunk and at the top of the fourth shed.

At 18 kV, the first DbDs appeared, corresponding to the first current pulses (noted (2) in Figure 9). With an increase in applied voltage, the LC activity was more intense (noted (3) in Figure 9), with a maximum value of 12 mA rms at 27 kV rms. The DbDs were spark discharged with purple/white coloration, as presented in Figure 11a,b. Above 30 kV, the DB activity was more intense, and the LC increased with the voltage. According to Figure 9, the LC in area (4) and (5) had typical values of dry-band arcs (DBPA) and a maximum LC growth up to 40 mA rms.

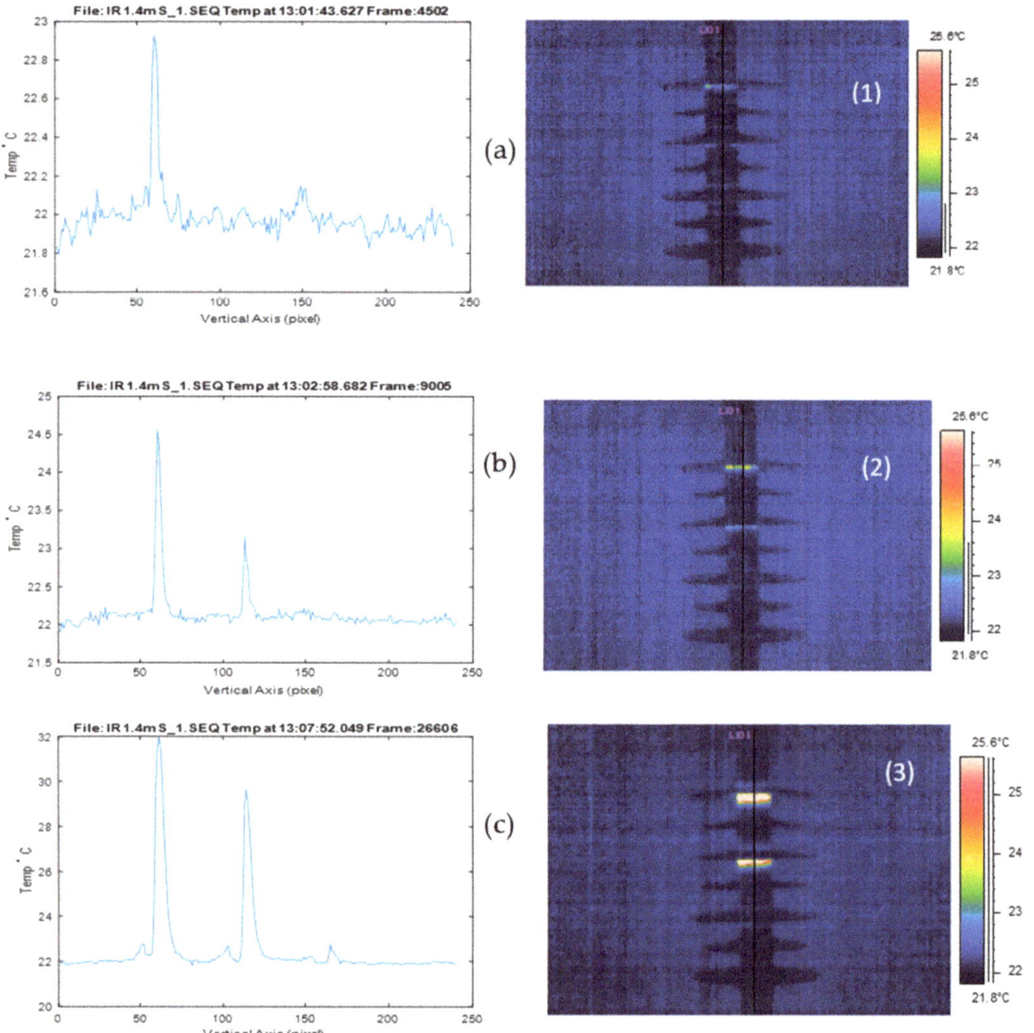

Figure 7. Thermal picture and temperature profile variation for light pollution. (**a–c**) are measured temperatures at the insulator surface. (1), (2) and (3) are the corresponding IR images.

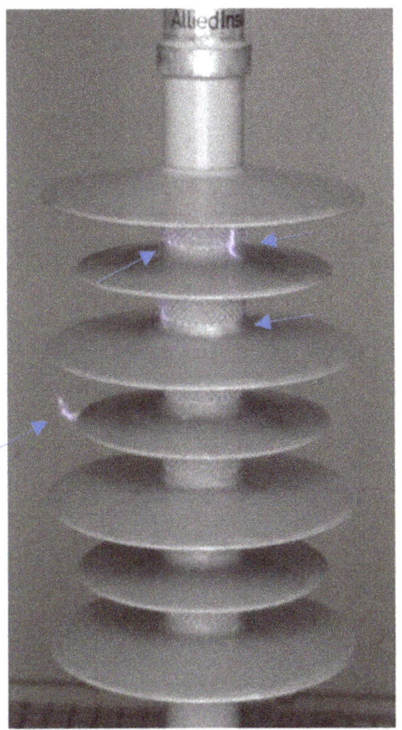

Figure 8. Visual image of Dry Band Discharges (DbD) on the insulator surface with light pollution at 70 kV rms.

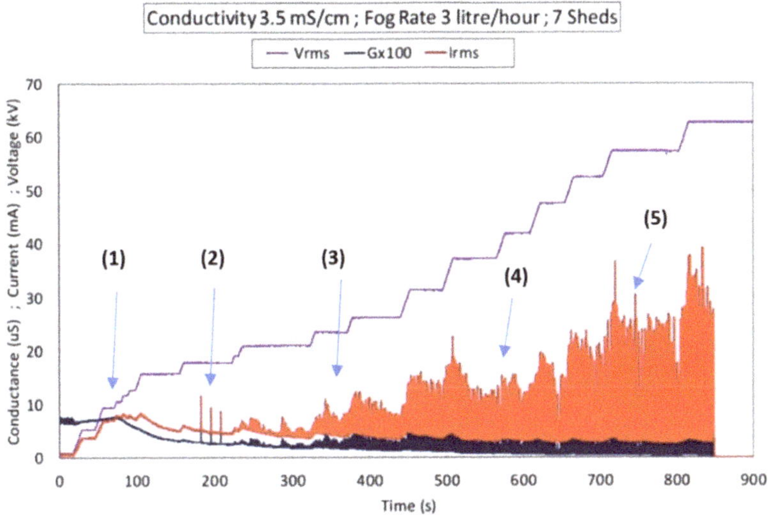

Figure 9. Surface conductance, leakage current, and applied free ramp voltage for medium pollution. (1), (2) (3), (4) and (5) are areas where LC activity and magnitude vary.

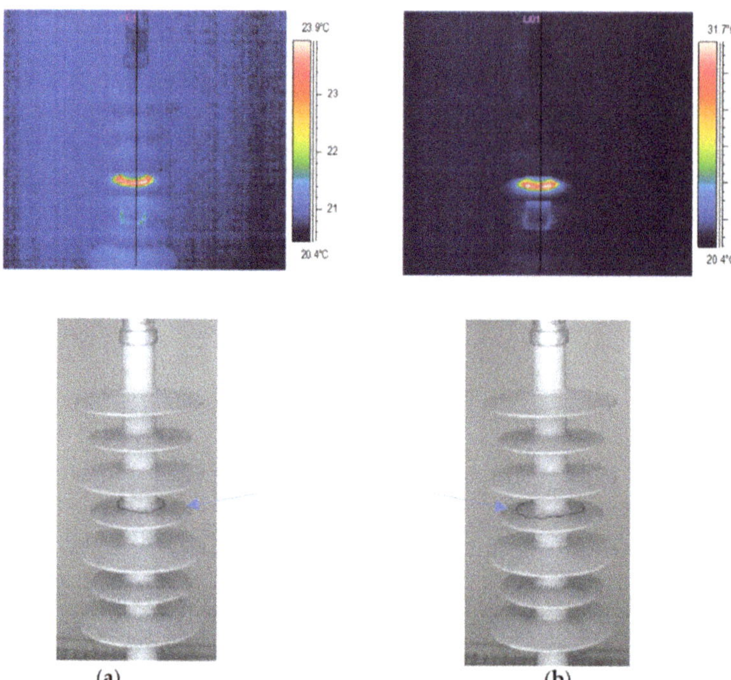

Figure 10. Dry-band formation and extension for medium pollution. (**a**,**b**) represent DB extension.

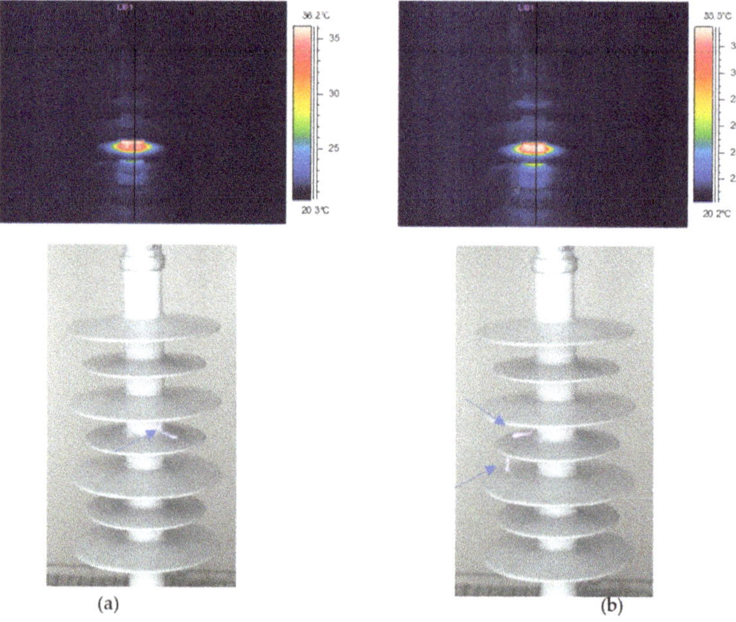

Figure 11. Dry Band Discharges (DbD) development for medium pollution. (**a**,**b**) represent DbDs activity and their corresponding IR images.

3.3. Heavy Pollution Level

Figure 12 illustrates the recording of LC and SC for all steps of a voltage ramp test, as before. DB onset began at 5 kV rms (noted (1) in Figure 12). The SC and LC were likely constant (respectively, 0.9 μS and 5 mA). Figure 13a shows the localization and the temperature profile of the first DB onset corresponding to the area (1) of Figure 12. We remark that DB inception were localized at trunk sections 1, 4, 5, and 6. At 7 kV rms, the SC and the LC collapsed, resulting in a local DB extension, which drastically increased the insulator surface resistance. Figure 12 illustrates visual and thermal images of the DBs and their extension at the trunks of the insulator, especially trunk Section 6.

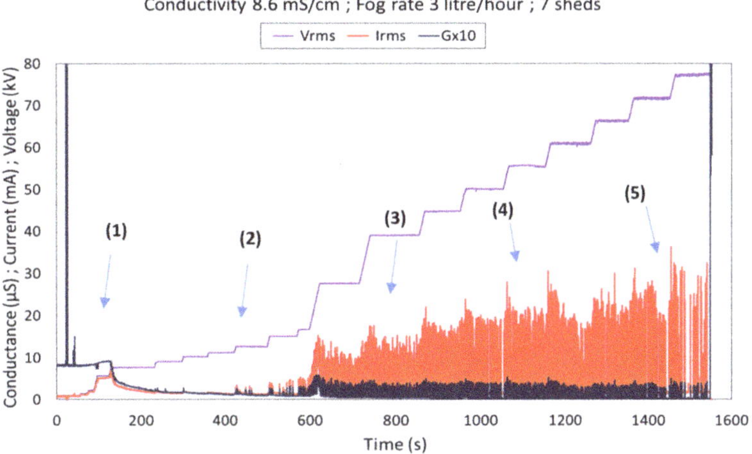

Figure 12. Surface conductance, leakage current, and applied free ramp voltage for heavy pollution. (1), (2) (3), (4) and (5) are areas where Leakage Current (LC) activity and magnitude vary.

At 12 kV, the first DbD appeared corresponding to the first current pulses (noted (2) in Figure 10). With an increase in the applied voltage, the LC activity was more intense (noted (3) in Figure 10). The LC increased drastically at 17 kV rms with a maximum of 20 mA rms. Figures 13 and 14 illustrate the thermal profile and the localization of the DbD activities and DB extension at the trunk sections of the insulator. We remark that the DB of trunk section number 6 extended and the temperature rose.

The LC increased with the applied voltage at each ramp step (noted (4) in Figure 12), with a maximum value of 36 mA rms. Figure 13 illustrates the temperature profile of the insulator, corresponding to area 4 of Figure 12. We remarked that the temperature of the trunk Section 6 was about 40 °C. Over 30 kV, the DB activity was more intense, and the LC increased with the voltage. According to Figure 12, the LC of area (5) had typical values of dry-band arcs (DBPA) and a maximum LC growth up to 40 mA rms. Figure 14 shows the DBPA activity at the insulator surface and infrared camera records of the DB and DBPA evolution. According to Figure 14, we remarked that the DBPA activity increased and many DBPAs were localized at the trunk section and between sheds. In Figure 15, the DBs extended along the trunk section and between the sheds, with voltage rising.

3.4. Severe Pollution Level

Figure 16 presents the recording of LC and SC for all steps of the voltage ramp test. The DB onset begins less than 4 kV rms (noted (1) in Figure 16). Figure 17 shows the DB onset before any discharge activity. We remarked that the DBs were initiated at trunk Sections 2, 3, and 4. The SC and LC decreased until constant values (respectively, 1.13 μS and 1.5 mA). At 16 kV, the first DBPA appeared as indicated in Figure 16 region (2). The LC activity and intensity increased with the voltage ramp steps (region (3) in Figure 16).

This increase was a consequence of the formation/extension of DB (Figure 18) and the inception and development of DBPAs (Figure 19). The LC amplitude increased with the voltage ramp steps (regions (4) and (5) of Figure 16), which indicated a high DBPA activity at the insulator surface, as presented in Figure 19. The maximum LC was around 60 mA rms at 70 kV rms.

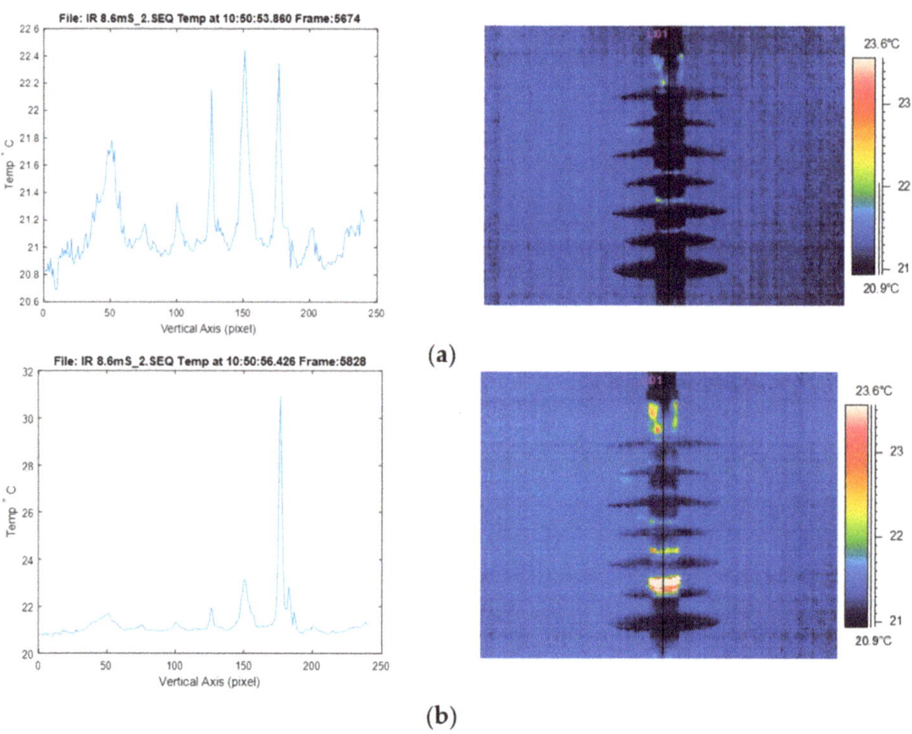

Figure 13. Dry-band onset and extension for heavy pollution (**a**,**b**) represent the (Dry-band Arcs) DBAs activity and corresponding temperature.

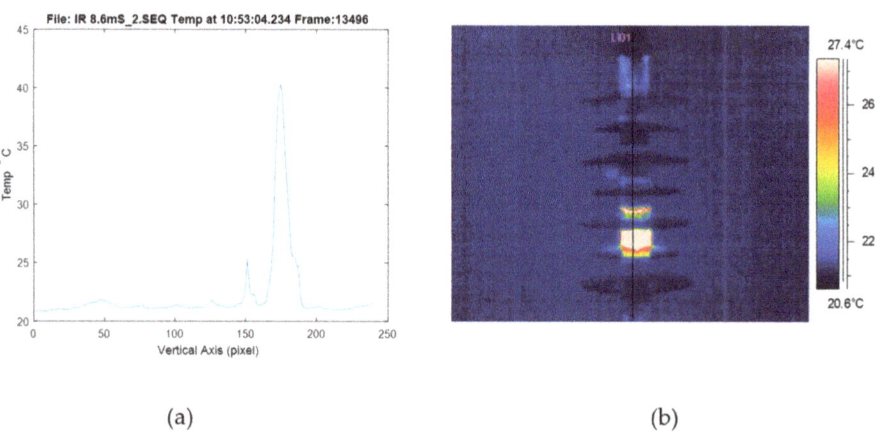

Figure 14. Temperature profile and DB evolution for heavy pollution at 30 kV rms (**a**) temperature profile (**b**) IR image.

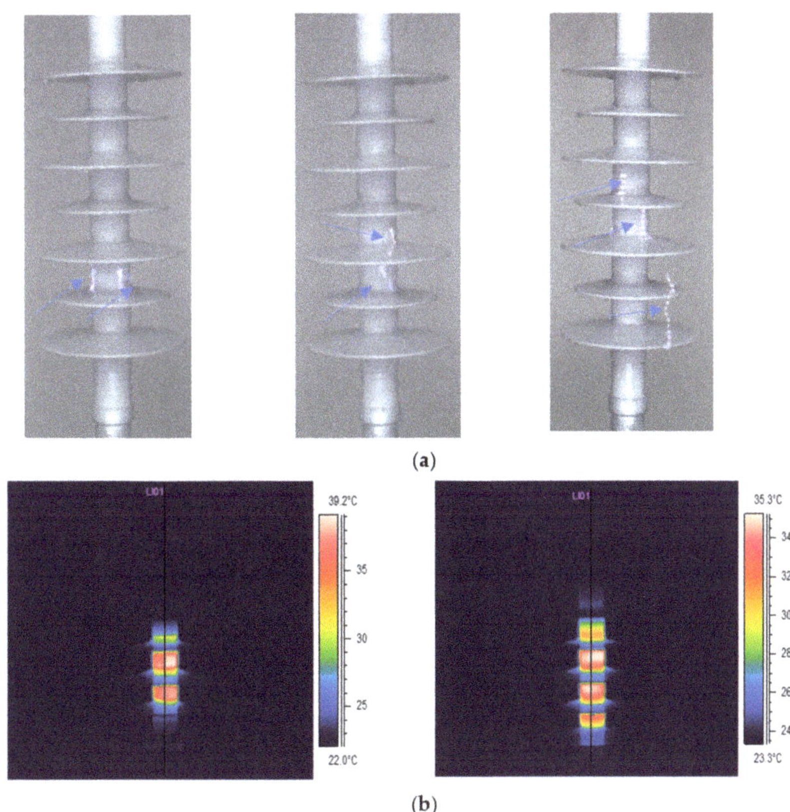

Figure 15. DBPA activity and IR images for heavy pollution over 40 kV rms (**a**) visual image of DBAs activity and (**b**) IR image of DB extension.

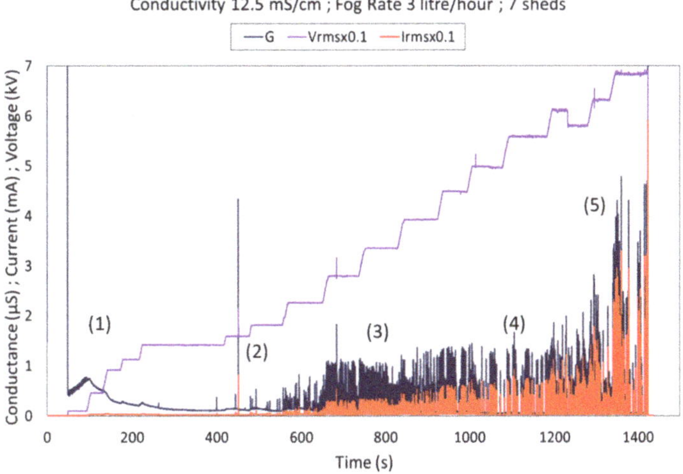

Figure 16. Surface conductance, leakage current, and applied free ramp voltage for severe pollution. Both voltage and current were divided by 10. (1), (2) (3), (4) and (5) are areas where LC activity and magnitude vary.

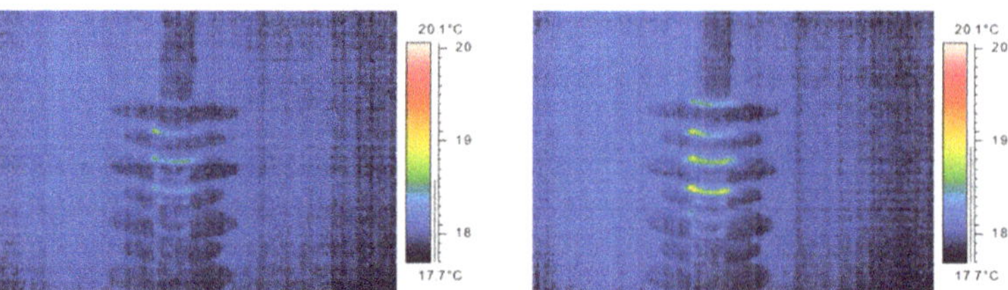

Figure 17. Dry-bands onset for heavy pollution.

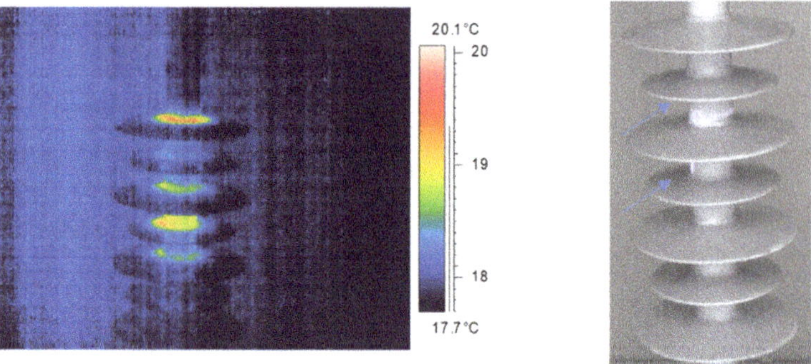

Figure 18. Dry-bands and firsts discharges inception/development for heavy pollution.

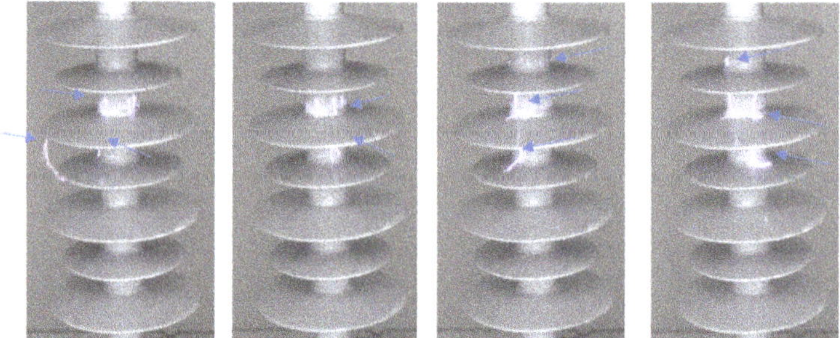

Figure 19. Discharges development with the voltage step increase for heavy pollution.

According to Figure 18, we remarked that the discharges jumped from the trunk section to the shed because of the extension of DB at the junction between these two elements.

3.5. Dry Band Voltage Onset, Dry Band Discharge Voltage Inception, and Average Power

3.5.1. DB Voltage Onset and DbD Voltage Inception

Based on previous measurements, the DB voltage onset and the DbD voltage inception were estimated and plotted as a function of ESDD. Figure 20 shows that both DB voltage onset and the DbD voltage inception decreased with the pollution level. This was attributed to the thermal dissipation of moisture from the pollution layer, which depended on its conductivity. When the conductivity increased, the current on the polluted surface was higher, and consequently, the Joule's dissipation was higher. On the other hand, the DB voltage onset appeared to be linear with the pollution level. This was due to the ohmic behavior of the equivalent electrical circuit, which behaved as a linear resistor with joules effect and heating. The DbD voltage inception presented a different trend and was more likely to be a non-linear function. The main reasons were that the discharges inception depended on many factors, such as ionization process, DBs size, rewetting, and the evaporation process [18] and the non-linearity of the discharge resistance [19].

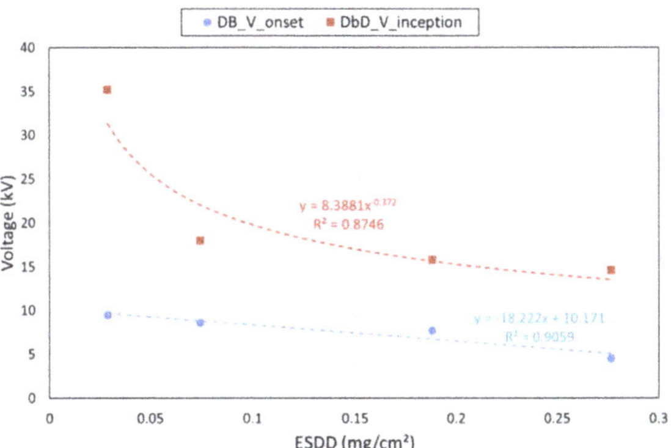

Figure 20. Dry Band (DB) voltage onset and Dry Band Partial Arcs (DBPA) voltage inception vs. pollution level.

3.5.2. Average Power

According to Figure 21a–d, the average power (AP) increased with the pollution layer conductivity. For example, in the case of light pollution, the AP corresponding to DB onset was approximately 16 W, whilst at the DbD inception, it was around 34 W. The maximum AP was 240 W at 70 kV rms. In the case of severe pollution, the AP corresponding to DB onset was approximately 4.5 W, while it was approximately 20 W at the DbD inception. The maximum AP was close to 1 kW for 70 kV rms. According to these results, the energy dissipated for the formation of DBs was less than that needed for the DbDs inception. However, the dissipated energy increased with the voltage and the pollution level.

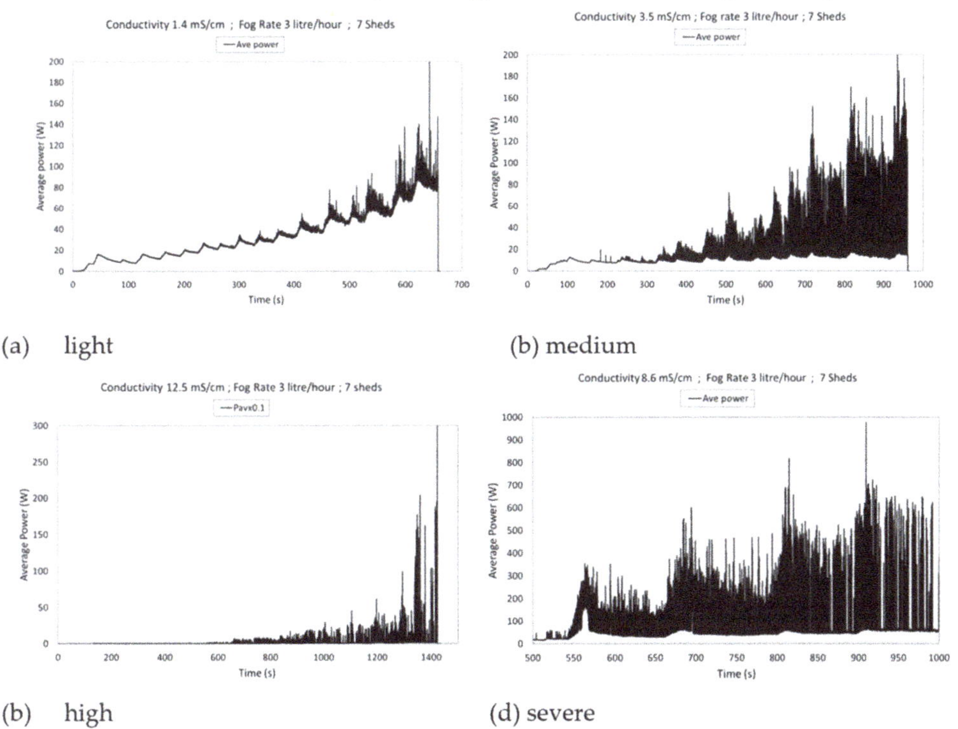

Figure 21. Calculated average power for various pollution levels. (**a**–**d**) are the pollution levels.

4. Conclusions

This study described the dynamic of LC, DB, DbDs, and DBAs of SiR textured insulators under clean fog test conditions. The results showed that the LC rms magnitude and the average power dissipation depended on the pollution level. Additionally, the following results were deduced:

- DB onset began at a relative low voltage (less than 10 kV), depending on the pollution layer conductivity. At this stage, the SC and LC were likely constant.
- After this stage, with increasing voltage ramp step, the SC and LC collapsed, resulting in local DB extension, which drastically increased the insulator surface resistance. Visual and thermal images showed that the DBs were situated at the trunks of the insulator.
- The DbD appeared at a voltage greater than the DB onset. With the increase of the applied voltage, the LC activity was more intense.
- The DbD were spark discharged with purple/white coloration and partial arcs with orange color. The nature of the discharges depended on the voltage level and the pollution layer conductivity.
- The discharge activity was related to the temperature distribution at the insulator surface and the DB dynamic. The nature of those discharges' changed with the pollution level and the applied voltage magnitude.
- DBA inception voltage was higher than DB onset voltage and depended on the applied voltage and the temperature distribution, as well as the pollution level.

Author Contributions: Conceptualization, M.E.A.S.; methodology, M.E.A.S.; validation, M.E.A.S., A.M.H., M.A., and R.T.W.; formal analysis, M.E.A.S., A.M.H., M.A., and R.T.W.; investigation, M.E.A.S.; resources, A.M.H., M.A., O.C., O.S. and J.K.; data curation, M.E.A.S.; writing—original draft preparation, M.E.A.S.; writing—review and editing, M.E.A.S. and A.M.H.; supervision, M.E.A.S., A.M.H., M.A., and R.T.W.; project administration, A.M.H., O.C., and I.I.; funding acquisition, A.M.H. All authors have read and agreed to the published version of the manuscript.

Funding: This research was funded by National Grid UK.

Institutional Review Board Statement: Not applicable.

Informed Consent Statement: Not applicable.

Data Availability Statement: Not applicable.

Conflicts of Interest: The authors declare no conflict of interest.

Abbreviations

ADE	Accumulated Dissipated Energy
AP	Average Power
DB	Dry-Band
DBA	Dry-band Arc
DBPA	Dry Band Partial Arc
ESDD	Equivalent Salt Deposit Density
FOV	Flashover
HC	Hydrophobicity Class
LC	Leakage Current
SC	Surface Conductance
TXT	Textured

References

1. International Electrotechnical Commission. *Selection and Dimensioning of High-Voltage Insulators Intended for Use in Polluted Conditions—Part 1: Definitions, Information and General Principles*; IEC Technical Specification IEC/TS 60815-1, Edition 1.0, 2008-10-29; IEC: London, UK, 2008.
2. Slama, M.E.A.; Beroual, A. Flashover discharges dynamic with continuous and discontinuous pollution layer under lightning impulse stress. *Electr. Eng.* **2021**. [CrossRef]

3. Karady, G.; Shah, M.; Brown, R. Flashover mechanism of silicone rubber insulators used for outdoor insulation-I. *IEEE Trans. Power Deliv.* **1995**, *10*, 1965–1971. [CrossRef]
4. Albano, M.; Waters, R.; Charalampidis, P.; Griffiths, H.; Haddad, A. Infrared analysis of dry-band flashover of silicone rubber insulators. *IEEE Trans. Dielectr. Electr. Insul.* **2016**, *23*, 304–310. [CrossRef]
5. An, Z.; Shen, R.; Gao, W.; Gu, X.; Chen, W.; Yang, L.; Yang, W.; Zhang, Z. Improved flashover performance and tracking resistance of silicone rubber by direct fluorination. *J. Appl. Polym. Sci.* **2020**. [CrossRef]
6. Arshad, G.; Momen, M.; Farzaneh, A.; Nekahi, A. Properties and Applications of Superhydrophobic Coatings in High Voltage Outdoor Insulation: A Review. *IEEE Trans. Dielectr. Electr. Insul.* **2017**, *24*, 3630–3646. [CrossRef]
7. Gubanski, S. Modern outdoor insulation—concerns and challenges. *IEEE Electr. Insul. Mag.* **2005**, *21*, 5–11. [CrossRef]
8. Slama, M.E.A.; Beroual, A. Behaviour of AC high voltage polyamide insulators: Evolution of leakage current in different surface conditions. *Power Eng. Electr. Eng.* **2015**, *13*, 74–80.
9. Elombo, A.I.; Holtzhausen, J.P.; Vermeulen, H.; Pieterse, P.J.; Vosloo, W.L. Comparative evaluation of the leakage current and aging performance of HTV SR insulators of different creepage lengths when energized by AC, DC+ or DC− in a severe marine environment. *IEEE Trans. Dielectr. Electr. Insul.* **2013**, *20*, 421–428. [CrossRef]
10. Abed, M.E.A.; Hadi, H.; Belarbi, A.W.; Slama, M.E.A. Experimental Characterization of Electric Power Insulator Subjected to an Accelerated Environmental Ageing. *Elektrotehniski Vestn.* **2020**, *87*, 183–192.
11. Haddad, A.; Waters, R.T. Insulating Structures. U.S Patent 764268, 2011.
12. Haddad, A.; Waters, R.; Griffiths, H.; Chrzan, K.; Harid, N.; Sarkar, P.; Charalampidis, P. A new approach to anti-fog design for polymeric insulators. *IEEE Trans. Dielectr. Electr. Insul.* **2010**, *17*, 343–350. [CrossRef]
13. Waters, R.T.; Haddad, A.H.; Griffiths, H.; Harid, N.; Charalampidis, P.; Sarkar, P.K. Dry-band discharges on polluted silicone rubber insulation: Control and characterization. *IEEE Trans. Dielectr. Electr. Insul.* **2011**, *18*, 1995–2003. [CrossRef]
14. Charalampidis, P.; Albano, M.; Griffiths, H.; Haddad, A.; Waters, R. Silicone rubber insulators for polluted environments part 1: Enhanced artificial pollution tests. *IEEE Trans. Dielectr. Electr. Insul.* **2014**, *21*, 740–748. [CrossRef]
15. Slama, M.E.A.; Albano, M.; Haddad, A.; Waters, R.T. Dry-Band Discharges Dynamic at the Surface of SiR Textured Insulator During AC Inclined Plane Test. In *Proceedings of the 21st International Symposium on High Voltage Engineering*; Springer: Berlin/Heidelberg, Germany, 2019; Volume 2, pp. 504–517.
16. Working Group C4.303. *Artificial Pollution Test for Polymer Insulators. Results of Round Robin Test*; Cigré: Paris, France, 2013.
17. IEC 60507:1991. *Artificial Pollution Tests on High-Voltage Insulators to Be Used on A.C. Systems*; International Electrotechnical Commission: Geneva, Switzerland, 2014.
18. Albano, M.; Waters, R.T.; Haddad, A.; Griffiths, H. Electric field distortion at outdoor insulation in severe Environments. In Proceedings of the 19th International Symposium on High Voltage Engineering, Pilsen, Czech Republic, 23–28 August 2015.
19. Slama, M.E.-A.; Beroual, A.; Hadi, H. Analytical computation of discharge characteristic constants and critical parameters of flashover of polluted insulators. *IEEE Trans. Dielectr. Electr. Insul.* **2010**, *17*, 1764–1771. [CrossRef]

Article

Lessons to Learn from Post-Installation Pollution Levels Assessment of Some Distribution Insulators

Issouf Fofana [1,*], Janvier Sylvestre N'cho [2], Amidou Betie [2], Epiphane Hounton [1], Fethi Meghnefi [1] and Kouba Marie Lucia Yapi [1]

1 Research Chair on the Aging of Power Network Infrastructure (ViAHT), Université du Québec à Chicoutimi, 555 Boulevard de l'Université, Chicoutimi, QC G7H 2B1, Canada; epiphane.hounton1@uqac.ca (E.H.); fmeghnef@uqac.ca (F.M.); KoubaMarieLucia1_Yapi@uqac.ca (K.M.L.Y.)

2 Département Génie Électrique et Électronique, Institut National Polytechnique Houphouët Boigny (INP-HB), Yamoussoukro BP 1093, Cote D'Ivoire; janvier.ncho@inphb.ci (J.S.N.); betie.amidou1@uqac.ca (A.B.)

* Correspondence: ifofana@uqac.ca; Tel.: +1-418-545-5011; Fax: +1-418-545-5012

Received: 8 July 2020; Accepted: 4 August 2020; Published: 6 August 2020

Abstract: Among the main causes of outdoor insulation failures is their poor specifications in terms of leakage distances. This happens when the selected criteria are unable to cope with all the stresses imposed by the changes in environmental pollutions. Therefore, it is important for utilities to fully understand the actual pollution characteristics of the service environment in which the insulators are operating. In this paper, the pollution severity and performance of some 13.2 kV ceramic insulators, sampled in different areas of a Canadian aluminum factory, are assessed. The investigations were performed taking into account the influence of air humidity. Various characteristics were investigated to assess the pollution levels of the insulators, such as equivalent salt deposit density (ESDD) and non-soluble deposit density (NSDD), surface resistance, and leakage current characteristics (density, 3rd harmonic amplitude, and phase). It was witnessed that the insulators, collected around the factory, were much more polluted in comparison to the initial expectation. The pollution level should not be considered static due to the environmental parameters' dynamics. Lessons to learn: the reliability of an electrical grid is dependent on components whose own reliability is strongly affected by external factors, of which there is often a poor awareness. If care is not taken to re-evaluate the post-installation pollution levels of the insulators, the light may simply turn out!

Keywords: insulator; pollution; humidity; equivalent salt deposit density (ESDD); non-soluble deposit density (NSDD); leakage current; post-installation study

---

1. Introduction

Outdoor insulators are essential hardware of the power delivery system. They are a basic requirement of open-air outdoor switchgears, and a failure of them means a failure in the system. These essential hardware are found in the transmission and distribution of electricity from power stations to substations, where the voltage is stepped down and distributed to commercial and residential consumers. Due to the remoteness of power plants, the energy is generally transported over long distances using high voltage (HV) overhead transmission lines supported by pylons. Insulators provide the mechanical means by which high voltage transmission lines (composed of bare conductors) are suspended from transmission structures (pylons), while also providing the required electrical insulation [1,2]. These important hardware are exposed to various electrical and environmental stresses that affect their performance and increase their premature aging and degradation. When insulators fail, either in their mechanical or electrical role, the consequences are power outages and, in some cases, additional equipment and structure damages. Reducing the risk of failure through proper installation, inspection, and assessment practices is the focus of ongoing research and utilities engineer training.

The reduction in the performance of outdoor insulators occurs mainly by the pollution accumulation at the insulating surfaces. Surface discharges are precursors of flashover. Outdoor insulators are exposed to various pollution sources (sea salts, domestic pollution, dirt, and chemical residues in the industrial areas) [3,4]. Insulator pollution can lead to flashover. In polluted areas, overhead lines may see their reliability and performance decline due to pollution insulators. When the contaminated insulating surface is wet, it becomes a conductive electrolyte. The leakage currents then increase on the insulating surface with potential flashover [5]. Identification of the factors causing insulation surface pollution, pollution severity assessment, and tackling its unfavorable effects have an important role in increasing the grid reliability. This is highly remarkable, particularly in environments with high dynamics (wind, pollution, humidity, etc.). Significant differences within the seriousness of pollution between sites can be assessed through equivalent salt deposit density (ESDD), non-soluble deposit density (NSDD), and the leakage current measurements. One of the major problems faced by this factory is the pollution of ceramic insulators installed on the 13.2 kV distribution grid. The interaction between the air transporting dust and the insulators creates a pollution layer on the insulator's surface. Once this layer is moistened, the withstand voltage drops considerably, causing the insulator to flashover. The concomitant power outages are not acceptable because they lead to huge financial losses and loss of control of the production process cycle. Since the insulators were initially designed considering a light pollution level, it is suspected that the environmental parameters dynamics resulting from the factory's production have affected the type and amount of particle accumulation. In this paper, seven service-aged 13.2 kV ceramic insulators, surrounding a Canadian aluminum plant, are collected to examine the severity of the pollution. The assessment was conducted through the leakage current and the ESDD and NSDD (non-soluble deposit density) measurements. The results are useful for assessing the insulation performance in the different areas and to propose an updated pollution severity map.

2. Background on Insulators

Generally, insulators are made based on porcelain or tempered glass. They are composed of a wet mixture of four primary materials: feldspar, flint, ball clay, and talc [6]. The mixture is molded and heated at a temperature of 1200 °C–1400 °C, and then glazed. The excellent dielectric strength of porcelain, which is around 1.574–11.02 kV/mm, and its relative permittivity (between 5.1 and 5.9) allow improving the resistance to materials' aging due to electrical and environmental stresses [7]. The Achilles heel of porcelain insulators is, essentially, their hydrophobic surfaces [8]. Later on, polymer insulators were introduced because of their excellent performance against pollution. Polymer insulators are of two types: in resin or composite.

Composite insulators are used extensively for the levels of voltage distribution and high voltage transmission lines [1,9]. In contaminated environments, the leakage current at their level is much lower than that of ceramic insulators [10]. However, a composite insulator has some disadvantages: brittle fractures, erosion, tracking, and chemical changes on the surface due to weathering are the main reasons for failures [1].

A significant cause of both service interruptions and flashovers is due to polluted insulators. Their surface is mostly responsible. There are two types: hydrophilic (ceramic insulators) or hydrophobic (polymer insulators) [3,4]. The pollution flashover process for ceramic insulators can be seen in [5]. Pollution sources can be found in Table 1 [6].

In non-ceramic and porcelain insulators, the contamination process is the same, but non-ceramic insulators collect less pollution than ceramic insulators [6,11]. Recently, with nanotechnology and nanoscience, new materials with innovative properties have been proposed for several applications [12–14]. A wide range of such monitoring devices and techniques has been developed over the years. The most widely used ones are [3]: directional dust deposit gauge, NaCl, ESDD, environmental monitoring (air sampling, climate measurements), NSDD, surface conductance, insulator flashover stress, surge counting, and leakage current measurement. New diagnostic tools have emerged [1].

Table 1. Typical sources of pollution.

Pollution Type	Source of Pollutant	Deposit Characteristics	Area
Rural areas	Soil dust	Low conductivity layer, effective rain washing	Large areas
Desert	Sand	High conductivity	Large areas
Coastal area	Sea salt	Very high conductivity, easily washed by rain	10–20 km from the sea
Industrial	Steel mill, cocoa plants, chemical plants, generating stations, quarries	High conductivity, extremely difficult to remove, insoluble	Localized to the plant area
Mixed	Industry, Highway, desert	Very adhesive, medium conductivity	Localized to the plant area

3. Experimental Arrangement

The insulators shown in Figure 1 were sampled from different locations around the Rio Tinto's factory in Saguenay (Canada). Three methods were used to assess the pollution level of the insulators: ESDD, NSDD, and the leakage current.

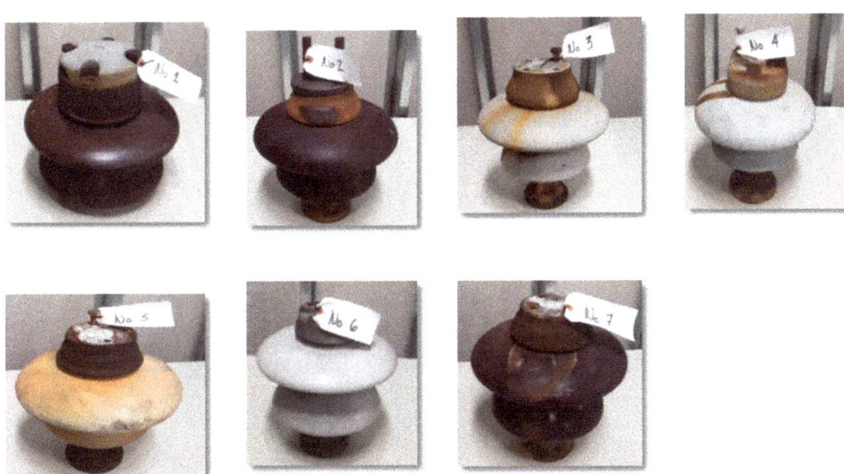

Figure 1. Samples of 13.2 kV ceramic insulators investigated in the study.

3.1. ESDD and NSDD Assessment

The assessment of the pollution of a site is possible by measuring both ESDD and NSDD on insulators generally in service. The measurements were made under standard IEC TS 60815-1 [4]. The probe of the YOKOGAWA Model SC 72 was used to measure the conductivities. These conductivities at different temperatures were corrected at 20 °C. The ESDD and NSDD were calculated according to IEC TS 60815-1 [4].

3.2. Leakage Current Measurement

The pollution of insulators can be assessed by measuring the leakage current to avoid any possible breakdowns [15]. Figures 2 and 3 show the measuring circuit and the leakage current test device, respectively. The high voltage (HV) AC source was connected to the insulator via a capacitive voltage divider. The circuit consisted of a 380 V/100 kV–10 kVA testing transformer, whose primary winding was connected to the autotransformer integrated in a semi-automatic or manual automatic control system. This later adjusted the voltage at the desired value. The system included a data acquisition that collected the main electrical characteristics. R' was a power resistor (500 Ω, 500 W), connected between the insulator and the ground that measured the leakage current. The voltage was measured via a capacitive divider.

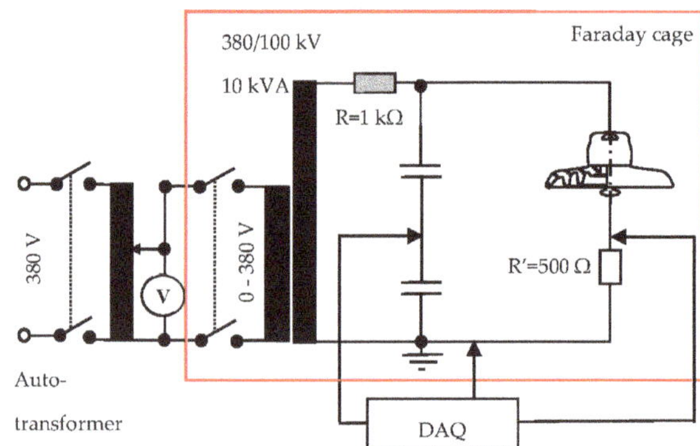

Figure 2. Schematic configuration of the leakage current-measuring circuit.

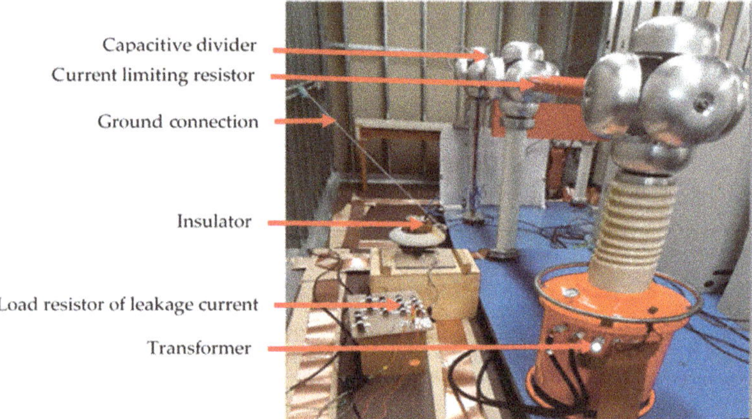

Figure 3. An overview of the experimental setup.

In order to evaluate the impact of humidification, the insulator samples were stored in a chamber where the relative air humidity was controlled by a water and glycerin mixture. The humidification concept can be found in the literature [16].

4. Results and Discussion

4.1. Leakage Current Assessment

Figure 4 shows the comparison of the measured leakage current on the humidified and non-humidified polluted insulators. The applied voltage was 7.62 ± 6% kV. From this figure, it can be observed that the leakage current increased with the relative humidity. This confirms the important role played by moisture on the leakage current flowing at the surface of polluted insulators.

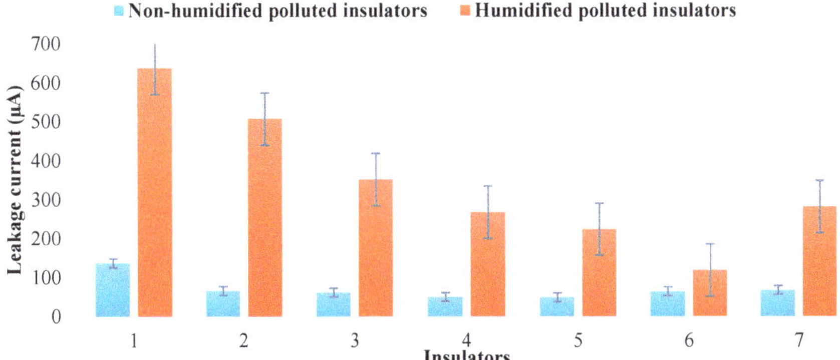

Figure 4. Leakage current on the humidified and the non-humidified polluted insulators.

In order to better explain the effect of humidification on the polluted insulators and the high values of the leakage current measured after humidification, the surface resistance of the non-humidified and humidified insulators was computed (Figure 5). This figure expresses the result differently for the sake of readers more familiar with viewing resistance. The results show that the non-humidified insulators have a very high surface resistance.

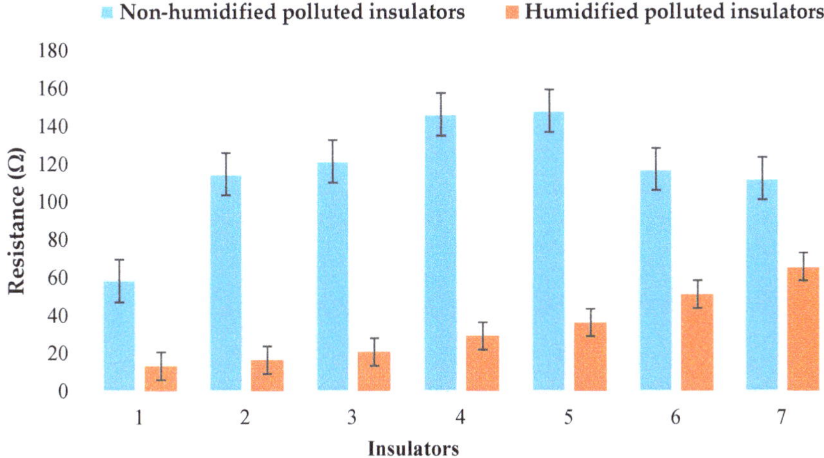

Figure 5. Resistance of non-humidified and humidified polluted insulators.

The leakage current density of the non-humidified and humidified polluted insulators is shown in Figure 6. It can be seen that the leakage current density of these insulators after humidification was much higher than that of non-humidified insulators. This increase can be explained by the fact that, due to contamination on the surface, the insulator can retain more moisture, thus increasing the leakage current. Moisture is a naturally occurring phenomenon and on a macro level will vary with the location and weather. High relative air humidity is very stressful to the insulator.

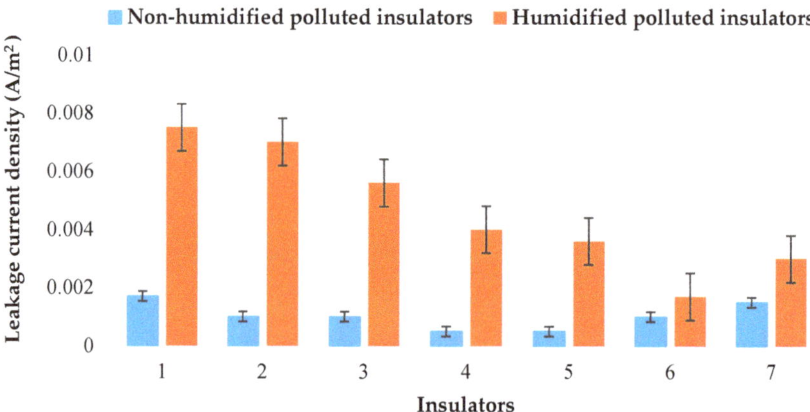

Figure 6. Leakage current density of non-humidified and humidified polluted insulators.

4.2. Phase Difference between Leakage Current and Applied Voltage

It is a well-known fact that the electric field distribution is greatly influenced by the pollution layer's characteristics (conductivity and thickness) [3]. With increasing pollution layer conductivity, the electric field intensity increases with a concomitant increase in the leakage current. In order to highlight the capacitive nature of these insulators, the calculation of the phase shifts between the measured leakage current and the applied voltage was performed for the humidified and the non-humidified polluted insulators (Figure 7). From the analysis of these results, it appears that the waveform of the leakage current had a phase shift of about 90° with respect to the applied voltage. This shows the dominance of the capacitive effect of these different non-humidified polluted insulators. The results clearly show that the phase shift between the applied voltage and the measured leakage current had decreased substantially for all humidified polluted insulators but did not tend toward zero (this is based on subtracting complex current in dry conditions from wet conditions and seeing the resulting phase, which lies between 35° and 64°, not near 0°).

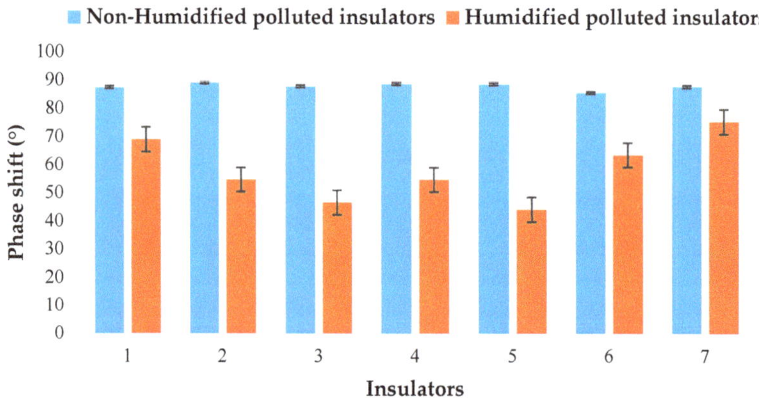

Figure 7. Comparison of phase shifts between non-humidified and humidified polluted insulators.

4.3. Harmonic Analysis

Polluted and humidified insulators circulate a leakage current on their surface leading, under certain conditions, to a flashover. Odd order harmonics in general, and in particular the

3rd and 5th, have adverse effects on the electrical network [17]. Indeed, by analyzing the leakage current harmonics, through the ratio of the 3rd/5th harmonic, it possible to predict the functional conditions of and occurrences of flashover in ceramic insulators. Furthermore, by analyzing the values of the leakage current and the harmonic components of clean insulators under normal operating conditions, it is observed that in all insulators, the 5th harmonic value of the leakage current is greater than the 3rd. Therefore, this criterion can be used to detect abnormal operating conditions in polluted insulators. For medium and severe pollution with fog and humidity, the value of the 3rd harmonic becomes larger than that of the 5th harmonic. The increase in the 3rd harmonic is faster and greater than that of the 5th harmonic with the imminence of the flashover. As a result, the condition of the insulator becomes critical and the probability of flashover occurrence increases. In this case, the ratio of the 3rd to the 5th harmonic is greater than one [17,18]. The harmonics measured during the various tests carried out on the non-humidified and humidified insulators are presented in Figure 8. In order to assess the effect of humidification on the polluted insulators, the 1st, 3rd, and 5th harmonics were measured, and the results are shown in Figure 8. After humidification, the harmonic of the leakage current was also computed in order to establish the ratio of the 3rd and 5th harmonics presented in Figure 9.

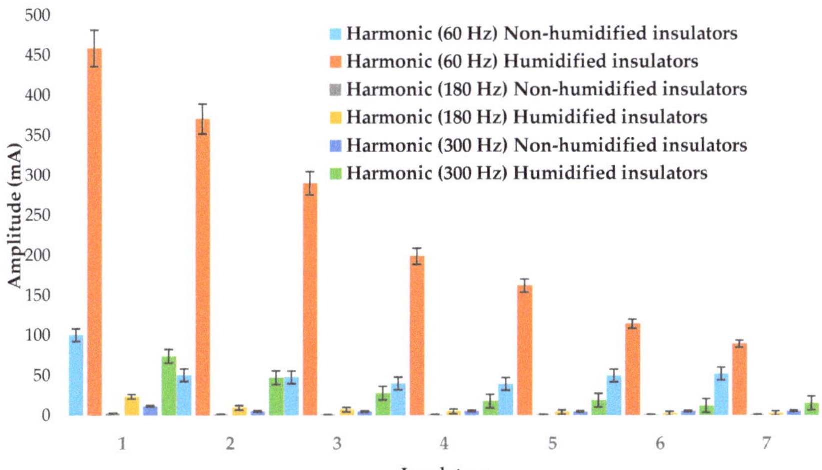

Figure 8. Leakage current harmonics of non-humidified and humidified polluted insulators.

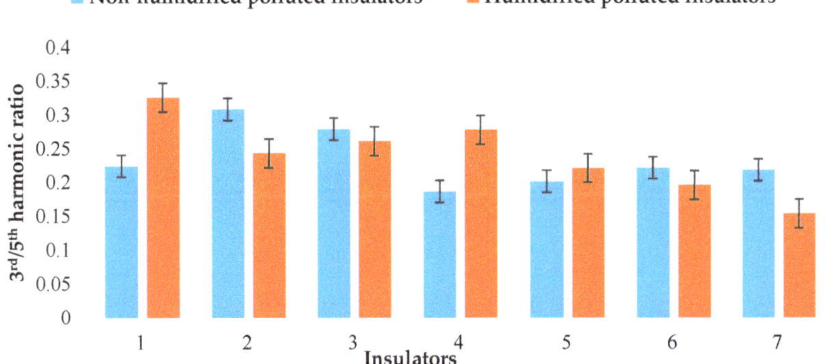

Figure 9. The 3rd/5th harmonic ratio of non-humidified and humidified polluted insulators.

From Figure 8, it can be seen that the 5th harmonic was greater than the 3rd one for all insulators. This allows deducing that these insulators, although in polluted conditions, are in normal operation. In real life conditions, adverse external conditions (such as rain, dew or fog) may wet the contaminated surface of insulators, hence increasing the levels of leakage current. The received samples must therefore be submitted to humidification to resume testing. This will make it possible to compare different results of the leakage current recorded and to draw accurate conclusions. It can be seen from Figure 9 that the amplitude of the 5th harmonic was always higher than the 3rd one. This indicates the non-imminence of flashover, because the ratio of the 3rd to the 5th harmonic was below one in this case (Figure 10). However, the 1st, 3rd, and 5th harmonics are higher than the harmonics measured on the non-humidified polluted insulators. The 3rd harmonic is mainly related to the electric discharge activities (corona, creepage, and so on), and it can be used to detect sensitively initial discharge voltage and discharge intensity of the samples. For better comparison, the 3rd harmonic of the non-humidified and humidified polluted insulators are accommodated on the same graph (Figure 10).

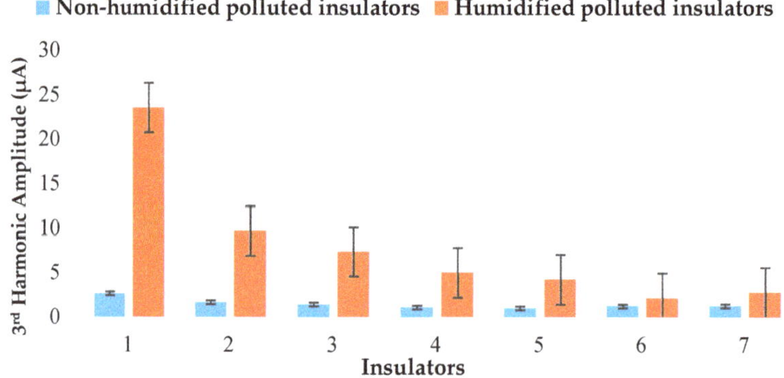

Figure 10. 3rd harmonic of non humidified and humidified polluted insulators.

From this figure, it can be seen that the 3rd harmonic increased when insulators were humidified. These results are in agreement with the investigations reported in the literature [19].

4.4. Pollution Level Assessment

The salinity, ESDD, and NSDD results for the insulators tested are shown in Table 2. The low values of salinity indicate that the contamination contained low levels of sodium chloride, but in large part, other contaminants [20,21]. To determine the pollution site severity (SPS), different limits are standardized by IEC [4], as shown in Table 3. Comparing the results in Table 2 with the IEC 60815 standard allows the determination of the pollution levels of the different insulators that are presented in Table 4.

Table 2. Values of calculated salinity, equivalent salt deposit density (ESDD) and non-soluble deposit density (NSDD).

Insulators/Areas	Salinity (mg/cm^3)	ESDD (mg/cm^2)	NSDD (mg/cm^2)
1	0.1944	0.1184	1.6800
2	0.2419	0.1408	5.0200
3	0.3050	0.2125	1.6300
4	0.0229	0.0307	0.0750
5	0.0525	0.0325	0.2800
6	0.0017	0.0010	0.0098
7	0.1662	0.1063	3.8100

Table 3. IEC Pollution Severity [4].

Site Pollution Severity	ESSD (mg/cm^2)	NSDD (mg/cm^2)
Light	0–0.03	0.03–0.06
Medium	0.03–0.06	0.10–0.20
High	0.06–0.10	0.30–0.60
Very High	>0.10	>0.80

Table 4. Pollution level of the tested insulators.

Insulators/Areas	Pollution Level
1	Very high
2	Very high
3	Very high
4	Medium
5	High
6	Light
7	Very high

It can be seen that the insulators suffered from light to very high pollution levels. The results indicate that the area where these insulators are located has a high level of pollution, with an equivalent salt deposit of 0.2125 mg/cm^2. This pollution should be attributed to the local environmental parameters dynamics, which provides different pollutant sources and amounts [22]. From the surrounding air analyses, it was found that the contaminants that may have been deposited on the surface of the insulators could have been: aluminum process bath mixes: artificial cryolite, trisodium hexafluoroaluminate, electrolytic bath with or without lithium, crushed electrolytic bath, electrolytic casting bath, enriched alumina (recycled) dry scrubbers, re-circulated alumina, fluorinated alumina, charged alumina, calcined alumina, dusts of anode butts (shot blasting machine), or solid tars-tailings of anode furnace ducts.

These results are in agreement with the recommendations described in the IEC standard, which mentions that [4]: "In the same environment, different types of insulators and even differently oriented accumulate pollution at different rates. Also, some insulators can be more efficient than others due to changes in the nature of the pollutant."

5. Conclusions

Seven service-aged 13.2 kV ceramic insulators, located in the surroundings of a Canadian aluminum plant, were collected for a post-pollution severity assessment. Various characteristics, such as ESDD, NSDD, and the leakage current, were investigated to assess the pollution level of the insulators. From these investigations, the main conclusions can be summarized as follows:

- The 1st, 3rd, and 5th harmonic components may allow monitoring the entire development trend of the leakage current.
- A number of observations by other published papers have been re-confirmed in this paper; the theoretical premises and expectations that the leakage current increases with the insulator's humidification is verified.
- The observed current is a combination of an increased conductive leakage plus a fairly steady capacitive part.
- The ESDD and NSDD assessments indicate that all the insulators, initially designed for light pollution level, are differently polluted. The pollution level should, therefore, not be considered static. These investigations have confirmed that the local environmental parameters dynamics should be considered for the grid reliability. Post-installation investigations are recommended whenever the surrounding insulator's area undergo changes (construction, habitation, changes in factory processes, etc.).

Author Contributions: This work was p under the supervision of I.F., who designed this research and gave the whole guidance. E.H., F.M. and K.M.L.Y. performed the experimental investigations. J.S.N. and A.B. collected all the data, carried out calculations, result display and analysis and wrote the manuscript. The final draft of paper was thoroughly reviewed by I.F. All authors have read and agreed to the published version of the manuscript.

Funding: This research was funded by Natural Science and Engineering Research Council (NSERC) grant number 513476-17.

Acknowledgments: The authors gratefully acknowledge the cooperation of Rio Tinto, Saguenay branch in Canada, for allowing the investigations and authorizing the publication of this manuscript.

Conflicts of Interest: The authors declare no conflict of interest.

References

1. Papailiou, K.O. Overhead Lines. In *CIGRE Green Books*; CIGRE: Paris, France, 2017.
2. EPRI. *EPRI AC Transmission Line Reference Book—200 kV and Above*, 3rd ed.; EPRI: Palo Alto, CA, USA, 2005.
3. CIGRE Working Group 33-04, TF 01. Polluted insulators: A review of current knowledge. *CIGRE Rep.* **2000**, *158*, 128–129.
4. IEC TS 60815-1. *Selection and Dimensioning of High-voltage Insulators Intended for Use in Polluted Conditions-Part 1: Definitions, Information and General Principles*, 1st ed.; International Electrotechnical Commission: Geneva, Switzerland, 2008.
5. Gençoglu, M.T.; Cebeci, M. The pollution flashover on high voltage insulators. *Electr. Power Syst. Res.* **2008**, *78*, 1914–1921. [CrossRef]
6. Grigsby, L.L. *Electric Power Generation, Transmission and Distribution*, 3rd ed.; CRC Press: Boca Raton, FL, USA, 2012.
7. Mishra, A.P.; Gorur, R.S.; Venkataraman, S. Evaluation of porcelain and toughened glass suspension insulators removed from service. *IEEE Trans. Dielectr. Electr. Insul.* **2008**, *15*, 467–475. [CrossRef]
8. Braini, S.M. Coatings for Outdoor High Voltage Insulators. Ph.D. Thesis, School of Engineering Cardiff, Cardiff University, Wales, UK, 2013.
9. Papailiou, K.O.; Schmuck, F. *Silicone Composite Insulators Materials, Design, Applications*; Springer: Berlin/Heidelberg, Germany, 2013.
10. EPRI. *Review for the State of the Art and Application of Polymer Materials Insulation used in Distribution Class. (12–46 kV class) Substations*; EPRI: Palo Alto, CA, USA, 2003.
11. Imakoma, T.; Suzuki, Y.; Fujii, O.; Kawamura, S. Electrical and mechanical characteristics of non-ceramic phase-to-phase spacers and akimbo insulator assemblies. In Proceedings of the SEE Conference, Paris, France, 7–8 June 1994.
12. Contreras, J.E.; Rodríguez, E.A. Nanostructured insulators–A review of nanotechnology concepts for outdoor ceramic insulators. *Ceram. Int.* **2017**, *43*, 8545–8550. [CrossRef]
13. Dave, V.; Dubey, P.; Gupta, H.O.; Chandra, R. Nanotechnology for outdoor high voltage insulator: An experimental Investigation. *Int. J. Chemtech Res.* **2013**, *5*, 666–670.
14. Portella, K.F.; Mengarda, P.; Bragança, P.M.; Ribeiro, S.; Santos de Melo, J.; Pedreira, C.D.; Pianaro, S.; Mazur, M.M. Nanostructured titanium film deposited by pulsed plasma magnetron sputtering (Pdms) on a high voltage ceramic insulator for outdoor use. *Mat. Res.* **2015**, *18*, 853–859. [CrossRef]
15. Li, J.Y.; Sun, C.X.; Sima, W.X.; Yang, Q.J. Stage pre-warning based on leakage current characteristics before contamination flashover of porcelain and glass insulators. *IET Gener. Transm. Distrib.* **2009**, *3*, 605–615. [CrossRef]
16. Fofana, I.; Borsi, H.; Gockenbach, E. Fundamental Investigations on Some Transformer Liquids under Various Outdoor Conditions. *Trans. Dielectr. Electr. Insul.* **2001**, *8*, 1–8. [CrossRef]
17. Palangar, M.F.; Mirzaie, M. Detection of Critical Conditions in Ceramic Insulators Based on Harmonic Analysis of Leakage Current. *Electr. Power Compon. Syst.* **2016**, *44*, 18541–18864. [CrossRef]
18. Ahmadi-Joneidi, I.; Shayegani-Akmal, A.A.; Mohseni, H. Leakage current analysis of polymeric insulators under uniform and non-uniform pollution conditions. *IET Gener. Transm. Distrib.* **2017**, *11*, 2947–2957. [CrossRef]

19. Zheng, Q.; Liu, S.; Yang, S.; Lei, J.; Peng, X.; Lin, C. Influence of insoluble substance hygroscopicity on insulator leakage current. In Proceedings of the International Conference on High Voltage Engineering and Application (ICHVE), Chengdu, China, 19–22 September 2016.
20. Hussain, M.M.; Chaudhary, M.A.; Razaq, A. Mechanism of Saline Deposition and Surface Flashover on High-Voltage Insulators near Shoreline: Mathematical Models and Experimental Validations. *Energies* **2019**, *12*, 3685. [CrossRef]
21. Vosloo, W.L. A Comparison of the Performance of High-Voltage Insulator Materials in a Severely Polluted Coastal Environment. Ph.D. Thesis, University of Stellenbosch, Stellenbosch, South Africa, 2002.
22. Chen, S.; Zhang, Z. Dynamic pollution prediction model of insulators based on atmospheric environmental parameters. *Energies* **2020**, *13*, 3066. [CrossRef]

 *energies* 

Article

Evolution of Countermeasures against Atmospheric Icing of Power Lines over the Past Four Decades and Their Applications into Field Operations

Stephan Brettschneider * and Issouf Fofana

International Research Center for Atmospheric Icing and Power Systems Engineering (CENGIVRE), Université du Québec à Chicoutimi, 555 Boulevard de l'Université, Chicoutimi, QC G7H 2B1, Canada; Issouf_fofana@uqac.ca
* Correspondence: stephan_brettschneider@uqac.ca; Tel.: +1-418-545-5011; Fax: +1-418-545-5012

Abstract: The reliability and efficiency of power grids directly contribute to the economic well-being and quality of life of citizens in any country. This reliability depends, among other things, on the power lines that are exposed to different kinds of factors such as lightning, pollution, ice storm, wind, etc. In particular, ice and snow are serious threats in various areas of the world. Under certain conditions, outdoor equipment and hardware may experience various problems: cracking, fatigue, wear, flashover, etc. In actual fact, a variety of countermeasures has been proposed over the past decades and a certain number have been applied by utilities in various countries. This contribution presents the status and current trends of different techniques against atmospheric icing of power lines. A snapshot look at some significant development on this topic over the last four decades is addressed. Engineering problems in utilizing these techniques, their applications, and perspectives are also foreseen. The latest up-to-date review papers on the applications and challenges in terms of PhD thesis, journal articles, conference proceedings, technical reports, and web materials are reported.

Keywords: overhead power lines; atmospheric icing; power outage; anti-icing; de-icing; line design; passive devices; coatings; mechanical methods; thermal methods

Citation: Brettschneider, S.; Fofana, I. Evolution of Countermeasures against Atmospheric Icing of Power Lines over the Past Four Decades and Their Applications into Field Operations. *Energies* **2021**, *14*, 6291. https://doi.org/10.3390/en14196291

Academic Editors: Abu-Siada Ahmed and Andrea Mariscotti

Received: 2 September 2021
Accepted: 27 September 2021
Published: 2 October 2021

1. Introduction

To reach the consumption areas, electricity often travels large distances through power transmission lines due to the remote location of most power plants, crossing different climatic environments. These transmission lines are often exposed to various stresses influencing their operation, causing in some cases different outages. Power grid infrastructures in many countries around the world are consequently impacted by ice and snow accretions. Some countries in the arctic region, such as Canada, the United States, Russia, Iceland, and Scandinavian countries, have been exposed to these problems since the deployment of electricity networks [1–3]. Additionally, power lines crossing mountainous areas are prone to ice and snow, for example, in China or Italy [4,5]. Furthermore, climate change leads to more and extreme weather events in various countries [3]. As the impact of ice and snow accretions is more intense for temperatures close to the freezing point, the occurrence of critical situations may not be limited to the known cold regions of our planet.

Associated to the ever-growing world's population and faster industry development, many power grid projects have been commissioned or are underway around the world. The impact of snow or ice events consequently gained importance. Extreme reliability is therefore demanded for electricity distribution. When failures occur, they inevitably lead to high repair costs, long downtime, and potential risks to human safety. Therefore, a variety of countermeasures against atmospheric icing of power lines were proposed in the past to avoid or at least to minimize power outages during such ice or snow events. Pohlman et al. [1] in 1982 firstly reviewed the various anti-icing and de-icing methods known at that moment. Since then, various other reviews and technical reports have been published [2,6–11]. In the framework of the present contribution, the evolution of the proposed countermeasures over the past four decades was analyzed by comparing these reviews. In the second part of the article, a few study cases focusing on the selection of methods for field applications or further investigations were examined based on the publications [4,5,12–16].

It may be noted that it is not the intention of this article to provide detailed information on each individual anti-icing and de-icing method. This information can readily be found in the various publications cited in the article.

For the general understanding of the importance and possible impact of snow and icing events, some information on two examples of important cases of ice storms are provided. Furthermore, some general information on the distinction of anti-icing and de-icing methods as well as on the general impacts of snow and ice on power lines are included.

1.1. Two Examples of Extreme Ice Storm Events

Similar to any meteorological event, snow and ice storms do not occur on a regular basis. However, some areas of the world may experience these events more or less recurrently. Rarely, extreme icing events occur with disastrous impacts on a larger region. Two of such catastrophic events may be recalled here:

- In the first week of January 1998, up to 110 mm of ice accumulated in some parts of eastern Canada (southern Quebec including the city of Montreal, some eastern parts of Ontario). About 1500 high voltage line towers and 17,000 distribution poles had to be replaced. Approximately, 3 million people had lost electricity at the peak of the ice storm, and it took one month to reconnect the last customers [17]. Figures 1 and 2 illustrate the damage on the transmission and distribution networks during this event.
- At the end of January 2008, heavy wet snow and freezing ice precipitations were experienced for several days in parts of southern and central China. Over 100 of the 500 kV power lines and about 340 of the 220 kV power lines were interrupted [4]. Several millions of people were affected, and it took two to three weeks in some areas to completely restore the electrical power supply.

1.2. Anti-Icing and De-Icing

The literature uses commonly the two terms: "anti-icing" and "de-icing". The first term, "anti-icing", is used for the approach where it is intended to prevent any accumulation of snow or ice on the power lines. The second term "de-icing" identifies the approach where existing ice or snow accumulations are to be removed [2,8]. Some methods are efficient for one or the other approach. For example, icephobic coatings are a measure against ice buildup [8], whereas mechanical methods such as manual scraping or roller wheels are used to remove the ice or snow that was already accumulated [2]. Other methods such as heating of the conductor may be applied for both approaches [5,9]. More details on the distinction between anti-icing and de-icing methods can be found in the appendix B of [6] and in reference [11].

Figure 1. Example of icing impact on a high voltage transmission line (1998 Eastern Canada ice storm, credit: Photo Hydro-Québec).

Figure 2. Example of icing impact on a medium voltage distribution line (1998 Eastern Canada ice storm, credit: private archive).

1.3. Consequences of Ice and Snow Accumulations on Power Lines

The impacts of ice or snow accumulations on power lines are already well documented by different authors, e.g., [1,8,11]. It should be mentioned here that the impacts for power lines are mainly of a mechanical nature. The increase in the static load is important due to the additional weight of ice or snow accumulations (for example, ref [15] mentions up to 100 kg/m) and it may push the supporting structures to their limits. Furthermore, dynamic forces may become problematic, either due to galloping of conductors or ground wires or during the shedding of the additional weight. If two adjacent spans of a standard tower shed their ice load at different moments, the tower may be subjected to lateral forces beyond its designed limit.

Electrical failure may also occur in some cases [11]. Conductors of different phases may get close or in contact due to unequal accumulation or dynamic wind forces. Induced corona discharges may increase power losses or electromagnetic interferences. Finally, ice-covered insulators may experience flashover, especially if the line is subject to pollution or salt deposit (natural from the sea or artificial near highways) [11]. A review of the present stage of knowledge of outdoor insulator flashover is presented in [3].

2. Comparison Approach

Various anti-icing and de-icing methods have been developed in different countries and important research efforts have been deployed for several decades [9]. This article presents a comparative analysis of the different countermeasures that have been proposed to protect the power line infrastructure based on the following two approaches:

First, a comparative study was carried by analyzing a certain number of publications and technical reports that reviewed the proposed countermeasures over the past four decades. This comparison was carried out with the following objectives:

- To provide a concise overview of various proposed countermeasures;
- To analyze the evolution of countermeasures in the past decades;
- To examine if a catastrophic event (such as the big ice storm in Eastern Canada in January 1998) had an impact on the development of countermeasures.

The following comprehensive references were considered for this part of the study. The selection includes publications before and after the 1998 Eastern Canada ice storm:

- Reference [1] published in 1982 is considered by several authors as the first known review focusing on anti-icing and de-icing methods for overhead power lines.
- Reference [2] summarizes a technical report that was prepared for Hydro-Québec before the 1998 ice storm in eastern Canada.
- References [6] is a detailed technical report that was published in 2002, about four years after the 1998 ice storm in eastern Canada.
- References [7,8] were published, respectively, in 2005 and 2008 by the same team of authors. They described and classified various anti-icing and de-icing methods as a result of the intense research activities during the years after the 1998 ice storm in Eastern Canada.
- Reference [9] was written by an international expert committee in the framework of a CIGRE working group in 2010. It can be seen as a milestone for the documentation of the anti-icing and de-icing methods reporting in detail the knowledge at the moment of publication.
- Reference [10] is a master thesis published in 2018 on the topic of this study, thus some years after the preceding analyzed documents.
- Reference [11] is a recent publication that provides technical content to the power industry. It offers a comprehensive overview of the present knowledge of various anti-icing and de-icing technologies. Besides the countermeasures for power lines, it also covers the possible impacts and countermeasures for insulators.

The results of this comparative analysis are presented in Section 3.

Second, the application and integration of countermeasures into field operations are reported by reviewing various references in order to identify those, which found real life applications. The actions of Hydro Québec after the 1998 Eastern Canada ice storm could be analyzed using various publicly available publications (e.g., annual reports). Other studies with recommendations for future applications were analyzed for China, Italy, and Norway.

The following references were considered for this part of the study:

- References [12–14] present the methods that were selected by Hydro-Québec for further analyses after the Eastern Canada ice storm in 1998. The references include a ranking of the selected methods according to various criteria, including cost, complexity, implementation delay and reliability. This evaluation was carried out in order to help Hydro-Québec deciding which methods should be implemented into the field operations.
- Reference [5] includes a concise overview of various methods and it reports details of the method that was selected for future applications in parts of the Italian power grid.
- Reference [15] includes a selection of methods that are considered applicable for the specific situation in Norway. It presents a ranking for some selected methods addressing recommendations for future applications in the Norwegian power grid.
- References [4,16] report overviews and research efforts in China after the 2008 ice storm. Each publication provides a recommendation on the method that should be investigated for further research and optimization.

The results for this second part are presented in Section 4.

3. Comparison of the Various Countermeasures Proposed over the Last Four Decades against Power Line Icing

3.1. Classification of Various Countermeasures

In order to compare various countermeasures that were proposed over the past four decades, different anti-icing and de-icing methods are compiled in Table 1. Each analyzed publication is represented by a column. The two catastrophic ice storms (1998 in Eastern Canada and 2008 in southern and central China) were also included as time reference.

Different approaches may be used to classify the anti-icing and de-icing technologies [7,9]. This study started by adopting four groups that were presented in reference [9]: passive methods, active coatings and devices, mechanical methods, and thermal methods. During compilation, a separate group was formed for line design considerations as previously done in references [6,11]. The group of passive methods was split into two separate groups for passive devices and passive coatings as their effects on the ice-and snow accumulations are different. A seventh group for miscellaneous methods was added in order to list some proposed countermeasures that did not seem to fit in the other groups (as previously done by reference [2]).

A brief description for each group of countermeasures is given next. The reader is encouraged to read references [1,2,6–11] for detailed information on each individual anti-icing and de-icing method.

3.1.1. Line Design Considerations

These considerations can be applied for new line constructions or the reconstruction of heavily damaged lines in order to prevent either ice or snow accumulations (by avoiding critical regions or by putting the lines underground) or to strengthen the withstand capability of the lines (stronger towers or increase in distance between phases). The choice of the conductor or bundle may have different effects; it can either influence the amount of accumulation, the speed of ice shedding, or the torsional strength of power lines.

Table 1. Comparison of various anti-icing and de-icing methods proposed over the last four decades (for time reference, the table also includes the two catastrophic icing event).

Countermeasure		[1]—1982	[2]—1996	1998 Ice Storm in Eastern Canada	[6]—2002	[7,8] 2005/—2008	2008 Ice Storm in Eastern/Southern China	[9]—2010	[10]—2018	[11]—2020
1. Line Design										
1.1 Improved line routing					X					X
1.2 Underground cables					X					X
1.3 Measures to avoid interphase-short-circuits	1.3.1 Using separate structures for each phase							X		
	1.3.2 Modifying crossarms (longer middle crossarms)							X		
	1.3.3 Modify circuit spacing							X		
	1.3.4 Rearranging phase configuration							X		
1.4 Mechanical reinforcement	1.4.1 Stronger towers				X			X		X
	1.4.2 Anti-cascading towers				X					X
1.5 No ground wire (but addition of special surge arrestors)					X					X
1.6 Choice of conductor or bundle	1.6.1 Compact design (trapezoidal or z-shaped strands)				X	X		X		X
	1.6.2 Optimised simple conductor									
	1.6.3 Increased number in bundle							X		X
2. Passive Devices										
2.1 Counterweights (anti-rotation device)			X		X	X		X	X	X
2.2 Snow rings, wires and Teflon tapes			X					X	X	X
2.3 Interphase spacers						X		X		X
2.4 Spacer dampers						X				X
2.5 Load reducing devices, mechanical fuses, fasteners			X		X					X
2.6 Aeolian shedding			X		X					
3. Passive Coatings										
3.1 Industrial viscous liquids		X	X		X	X			X	X
3.2 Heterogeneous polymer coatings					X	X		X	X	X
3.3 Freezing point depressant fluids		X	X		X	X				X
3.4 Icephobic coatings		X	X		X	X		X		X
3.5 Hydrophobic and superhydrophobic coatings			X			X		X		X
3.6 Black coatings / Heat absorbent coatings			X		X					X
3.7 Chemically reactive coating					X					
4. Active Coatings										
4.1 Addition of ferroelectric material (high frequency injection)					X	X		X	X	X
4.2 LC spiral rods (ferromagnetic spirals)			X		X	X		X	X	X
4.3 Heated tracers			X		X	X		X		
4.4 Ice electrolysis						X		X		X
4.5 Piezoelectric polymers (PVDF)					X	X				X
4.6 Shape memory alloys			X		X					
5. Mechanical Methods										
5.1 Line Maintenance	5.1.1 Roller wheels or scraping	X	X		X	X		X	X	X
	5.1.2 Manual shock wave	X	X			X		X		X
	5.1.3 Tensioning and releasing		X		X					
	5.1.4 De-icer Actuated by Cartridge (DAC)		X			X		X		X
	5.1.5 Helicopter (wooden pole or rope with knots)		X		X	X		X	X	
	5.1.6 Remotely Operated Vehicle (ROV) (Scraping)					X		X		X

Table 1. *Cont.*

Countermeasure		[1]—1982	[2]—1996	1998 Ice Storm in Eastern Canada	[6]—2002	[7,8] 2005/—2008	2008 Ice Storm in Eastern/Southern China	[9]—2010	[10]—2018	[11]—2020
5.2 Pneumatic Hammer			X		X	X		X		
5.3 Eletro-impulse method (EIDI)			X		X	X		X		
5.4 Ice-shedder device (induced vibrations)			X		X	X		X	X	X
5.5 Twisting device						X		X		
5.6 High-magnitude short circuit method (for bundled conductors)					X	X		X	X	X
6. Thermal Methods										
6.1 Increase of line current using the network	6.1.1 Load shifting (Load transfer) method	X	X		X	X		X	X	X
	6.1.2 On-load De-Icer method (ONDI) (phase shifting)					X		X	X	X
	6.1.3 Reactive Current De-icing								X	
6.2 External current source	6.2.1 Reduced (medium) voltage short-circuit method		X		X	X		X	X	X
	6.2.2 Ground wire de-icing method (med.volt.short circ.)					X		X		X
	6.2.3 DC current method for AC-lines				X	X		X	X	X
6.3 DC-Line De-icing					X					
6.4 High frequency electric field method					X	X		X	X	
6.5 Contactor load transfer method (bundle shifting / current re-distribution)					X	X		X	X	X
6.6 Pulse electrothermal de-icer method					X	X		X		X
6.7 Various external heat sources	6.7.1 Infrared waves		X							
	6.7.2 Radio waves radiation ext. heat source		X		X	X		X		
	6.7.3 Steam generating device	X				X		X	X	
	6.7.4 Ultrasonic De-icing		X						X	
	6.7.5 Hot Gas		X							
	6.7.6 Hot Water/Anti-icing fluids		X							
7. Miscellaneous Methods										
7.1 Drop freezing before impact			X		X					
7.2 Drop heating before impact			X		X					
7.3 Corona discharges			X		X					
7.4 Electro-congelation			X							
7.5 Development of ice-resistant conductors										
7.6 Laser De-Icing			X							
7.7 Pneumatic air boots		X	X		X					
7.8 Asymmetric operation of 3-phase power lines										

3.1.2. Passive Devices

These devices do not prevent or reduce the amount of ice or snow that may accumulate on power lines, but reduce the negative impacts of these accumulations. For example, counterweights will lead to non-cylindrical ice deposits that will shed faster. Interphase spacers will maintain the distance between phase conductors even if they are not loaded with the same amount of ice or snow. Due to these kinds of indirect effects, problems such as phase-to-phase short-circuits or tower collapse can be avoided.

3.1.3. Passive Coatings

Coatings are applied to the conductor to avoid or at least limit ice or snow accumulations. Passive coatings do not require any external energy, but modify the surface properties of the conductors with the aim of weakening the adhesion forces of water droplets, ice or snow on the conductor surface.

3.1.4. Active Coatings

Active coatings use external energy to prevent or limit the accumulation of ice or snow on the conductor surface. Various principles are proposed to prevent or reduce any accumulation, such as heating due to the increase in losses, addition of heating tracers or the reduction in ice adhesion forces through ice electrolysis.

3.1.5. Mechanical Methods

Mechanical methods use different principles to break off the accumulated ice or snow. The forces required to break the ice may be applied by linemen with isolated poles or tools pulled by ropes or also from helicopters. Other methods consider apparatus that are installed on the lines temporarily or on a permanent basis. For bundled conductors, the force to break the ice may be generated by a temporary short circuit.

3.1.6. Thermal Methods

Thermal methods use different principles to heat conductors and ground wires in order to avoid any accumulation of ice or snow or to melt off existing accumulations. Different energy sources may be used; for example, the energy of the power network itself through load shifting, external current sources such as DC injection or high frequency injection or other external heat sources.

3.1.7. Miscellaneous Methods

This last group covers methods that were reported in some reviews and that did not seem to fit in any of the six preceding groups [2]. Some methods are still at a conceptual stage and their feasibility was not yet validated. Other methods were mentioned, but their applicability to power lines does not seem to be efficient [2]. The method 7.8 for asymmetrical operation of a three-phase power line was not included in any of the analyzed review publications, but was mentioned during the public hearing process for the authorization of the construction of the Levis de-icer project by Hydro-Québec [18]. It was therefore included in the last group of this study.

3.2. Comparative Analysis over the Last Four Decades

Figure 3 shows the chronology of the analysed review papers in relation to the two catastrophic icing events. It can be recognized that the number of proposed anti-icing and de-icing methods increased significantly between the 1980s and 1990s, but there is no remarkable increase in the number of countermeasures during the last three decades. A large variety of methods for anti-icing and de-icing was already proposed in the 1990s, before the two catastrophic icing events in 1998 and 2008. However, the compilation in Table 1 does not show the degree of advancement of the individual methods at the moment of publication. For example, icephobic coatings were already mentioned as a potential countermeasure by Laforte et al. in 1996 [2]. At that time, the stage of advancement for this passive method was indicated as "in development". Intensive research efforts have led to important advancements in the field of icephobic coatings and surface treatments, as witnessed by recent CIGRE working group activities [19,20]. However, additional efforts in research and development are still required in order to come up with effective and durable solutions for practical applications [20].

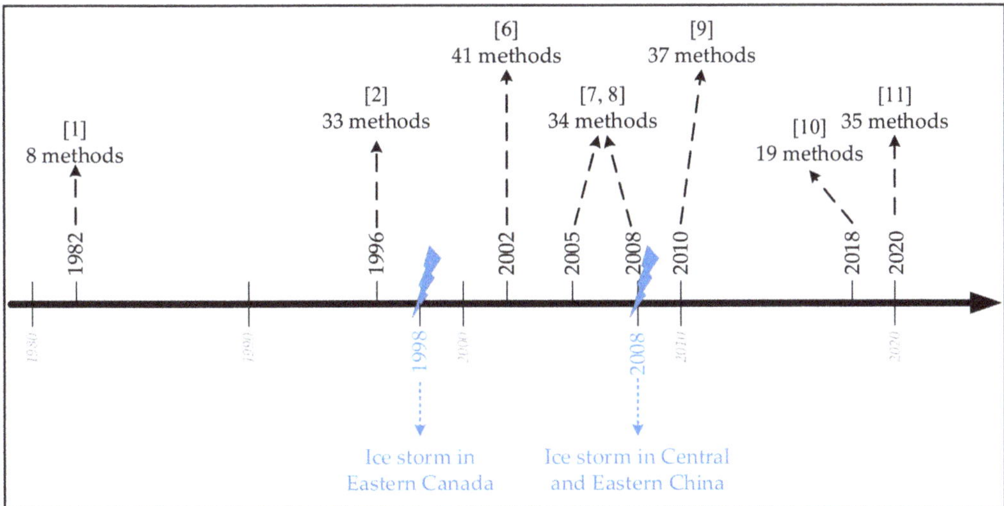

Figure 3. Chronology of the analysed review papers in relation to the two catastrophic icing events.

The compilation in Table 1 allows two more observations. The analyzed documents before the 1998 Eastern Canada ice storm [1,2] did not include any countermeasure in the field of line design (group 1 in Table 1). Perhaps these methods were initially not considered as countermeasures, because no additional equipment is required. Nevertheless, it can be noted that the knowledge on the positive effects of line design considerations has increased with the experiences gained from various icing events over the past few decades. As shown in the second part of the study (see Section 4.2), several line design measures have been implemented after the 1998 ice storm in Eastern Canada. Reference [11] indicates that these approaches present the best prevention technique for new constructions, but should be seen as complementary to the other countermeasures in order to obtain an optimized protection of power lines against icing events.

The second observation shows that in more recent articles, the miscellaneous methods (group 7 in Table 1) that were proposed in the two documents of 1996 and 2002 [2,6] were no longer included. One might understand that the applicability of these more theoretical approaches has not been demonstrated and that research institutions and electric utilities focused their efforts on the development of already known countermeasures that showed greater potential for practical applications in the field.

4. Selection of Countermeasures for the Integration into Field Operations or for Further Investigations

Four example cases were reviewed for the selection or recommendation of counter measures for field application. The results are compiled in Table 2. The first column is similar to the one of Table 1 and lists the countermeasures that were proposed over the past four decades. The other columns present the various methods that were considered in the four example cases. The bold and capital letters indicate the methods that were selected or recommended for field applications or further investigations.

4.1. Technology Integration within Hydro-Québec after the 1998 Eastern Canada Ice Storm

References [12–14] show the analysis performed by Hydro-Québec after the 1998 ice storm in Eastern Canada. Eleven countermeasures were considered and a detailed evaluation of each method using several criteria was carried out. This analysis allowed assigning a score to each method and establishing a ranking. The numbers in the second column of Table 2 represent the results of this ranking. The two methods with the highest scores were:

- The highest score was obtained for the thermal method by load shifting (method 6.1.1). However, this method can only be applied to a certain number of power lines with voltage levels varying from 49 kV to 315 kV, but it is not suitable for the 735 kV network of Hydro-Québec [14].
- The second score was obtained for the thermal method by DC current injection (method 6.2.3). The decision was made to install a large DC source at a centrally located substation that could inject DC current into the high voltage power lines and ground wires. A technology was chosen that could also be used as a static var compensator (SVC) throughout the year, which would help making the project economically viable. However, the implementation of this de-icer project was a complex task and it is still ongoing at the present time. The analysis of annual reports published by Hydro-Québec allowed tracing a timeline for the various steps:
  - Approval for the project was obtained in August 2004 [21].
  - The engineering for the de-icer project was completed in 2005 [22].
  - In 2008, the commissioning of the de-icer project was completed [23].
  - However, the de-icer could not be used until today. One of the main reasons for this blockage is the risk of ice pieces falling onto highways and damaging cars during the de-icing period. Therefore, Hydro-Québec started installing the LC-spirals (method 4.2) on the power lines above highways in 2017 [24]. These installations are relatively difficult to schedule as the highways have to be closed temporarily. The installation efforts are scheduled through 2023 in order to cover all critical highway crossings.

Hydro-Québec did not only rely on these two methods. The following other countermeasures were integrated into field operations on the Hydro-Québec power network:

- In 2010, Hydro-Québec stared to test the de-icer actuated by cartridge (DAC, method 5.1.4) for the de-icing of ground wires [25]. From 2012, this tool was adapted for general use [26].
- In 2012, Hydro-Québec reported that the robot LINEROV for mechanical de-icing of power lines (method 5.1.6) was used with success on several occasions [26].

Furthermore, several line design considerations are adapted to strengthen the network against impacts of future winter storms [27]:

- Reconstructed lines were built with improved mechanical strength in order to withstand higher ice loads (method 1.4.1).
- Anti-cascading towers were included in critical line sections in order to limit damage that could result from the collapse of towers (method 1.4.2).
- On the distribution network, poles and their anchoring were strengthened. In the case of a catastrophic event, only the crossarm and conductors could fall (method 2.5). This makes reconstruction easier and faster.
- Some additional new line sections were constructed in order to create additional meshed loops in the power network that provide increased redundancy for the power supply.
- The vegetation control program has been enhanced to prevent falling iced trees from causing power outages in the distribution network.

Another example for the improvement of network design is the construction of a new line for the interconnection of a windfarm in 2011. This line had to be constructed in an area with particular weather conditions that favour the accumulation of important ice loads. The following line design considerations were applied in this particular case [28]:

- The type of towers selected can withstand higher ice loads (method 1.4.1).
- The line does not have a ground wire, but surge arresters with external spark gap were installed (method 1.5).
- The type of conductor selected had superior mechanical strength (method 1.6).

Table 2. Summary of the countermeasures that were selected for application or future investigation in four different countries.

Countermeasure		[12–14]—Canada—2004	[5]—Italy—2019	[15]—Norway—2019	[4,16,29,30]—China—2010–2015
1. Line Design					
1.1 Improved line routing					
1.2 Underground cables					
1.3 Measures to avoid interphase-short-circuits	1.3.1 Using separate structures for each phase				
	1.3.2 Modifying crossarms (longer middle crossarms)				
	1.3.3 Modify circuit spacing				
	1.3.4 Rearranging phase configuration				
1.4 Mechanical reinforcement	**1.4.1 Stronger towers**	7			
	1.4.2 Anti-cascading towers				
1.5 No ground wire (but addition of special surge arrestors)				X	
1.6 Choice of conductor or bundle	1.6.1 Compact design (trapezoidal or z-shaped strands)	5		X	
	1.6.2 Optimised simple conductor	4		X	
	1.6.3 Increased number in bundle			X	
2. Passive Devices					
2.1 Counterweights (anti-rotation device)			**X**	X	
2.2 Snow rings, wires and Teflon tapes			X	X	
2.3 Interphase spacers			**X**		
2.4 Spacer dampers					
2.5 Load reducing devices, mechanical fuses, fasteners		10			
2.6 Aeolian shedding					
3. Passive Coatings					
3.1 Industrial viscous liquids					
3.2 Heterogeneous polymer coatings					
3.3 Freezing point depressant fluids					
3.4 Icephobic coatings		8	X	X	
3.5 Hydrophobic and superhydrophobic coatings					
3.6 Black coatings / Heat absorbent solid coatings					
3.7 Chemically reactive coating					
4. Active Coatings					
4.1 Addition of ferroelectric material (high frequency injection)			X		
4.2 LC spiral rods (ferromagnetic spirals)			X	**X**	
4.3 Heated tracers					
4.4 Ice electrolysis					
4.5 Piezoelectric polymers (PVDF)					
4.6 Shape memory alloys					
5. Mechanical Methods					
5.1 Line Maintenance	5.1.1 Roller wheels or scraping		X		
	5.1.2 Manual shock wave		X		
	5.1.3 Tensioning and releasing				
	5.1.4 De-icer Actuated by Cartridge (DAC)				
	5.1.5 Helicopter (wooden pole or rope with knots)			X	
	5.1.6 Remotely Operated Vehicle (ROV)			X	**X**
5.2 Pneumatic Hammer					
5.3 Electro-impulse method (EIDI)					X
5.4 Ice-shedder device (induced vibrations)			X		
5.5 Twisting device			X		
5.6 High-magnitude short circuit method (for bundled conductors)		6			

Table 2. *Cont.*

Countermeasure		[12–14]—Canada—2004	[5]—Italy—2019	[15]—Norway—2019	[4,16,29,30]—China—2010–2015
6. Thermal Methods					
6.1 Increase of line current using the network	**6.1.1 Load shifting (Load transfer) method**	1	X	X	X
	6.1.2 On-load De-Icer method (ONDI) (phase shifting)				X
	6.1.3 Reactive Current De-icing		X		X
6.2 External current source	6.2.1 Reduced (medium) voltage short-circuit method	3	X	X	X
	6.2.2 Ground wire de-icing method (MV short circuit)			X	
	6.2.3 DC current method for AC-lines	2		X	X
6.3 DC-Line De-icing					X
6.4 High frequency electric field method				X	X
6.5 Contactor load transfer (bundle shifting / current re-distribution)		11		X	X
6.6 Pulse electrothermal de-icer method					X
6.7 Various external heat sources	6.7.1 Infrared waves				X
	6.7.2 Radio waves radiation ext. heat source				
	6.7.3 Steam generating device				
	6.7.4 Ultrasonic De-icing				
	6.7.5 Hot Gas				
	6.7.6 Hot Water/Anti-icing fluids				
7. Miscellaneous Methods					
7.1 Drop freezing before impact					
7.2 Drop heating before impact					
7.3 Corona discharges					
7.4 Electro-congelation					
7.6 Development of ice-resistant conductors		9			
7.7 Laser De-Icing					X
7.8 Pneumatic air boots					
7.9 Asymmetric operation of 3-phase power lines					

4.2. Recent Applicability Studies in Italy

Even if Italy is not an arctic country, it experiences major power outages from time-to-time, mainly due to wet snow accretions. Reference [5] presented an analysis of various anti-icing measures for the Italian power system including recommendations for the implementation of a new method. Until now, two passive methods have already been in use: in 2019, about 20,000 counterweights (method 2.1) were installed on 2500 km of 132 kV lines, which led to a significant reduction in power outages due to snow overload. As the second countermeasure, interface spacers (method 2.3) were first installed on some critical spans of 132 kV lines in 2007. Due to the success in eliminating phase-to-phase faults, about 28 km of double circuits were equipped in 2019 with these devices.

As there are some local areas that have recurrent service disruptions due to wet snow accretion, the applicability of thermal anti-icing was studied. Detailed load flow analysis showed that load shifting (method 6.1.1) would not allow obtaining the current magnitudes that would be needed to create sufficient heat in the conductors. Moreover, the installation of shunt reactors (method 6.1.3) at critical locations allowed the creation of sufficient current flow. Thus, the Italian network operator started a program for the installation of shunt reactors at different critical sites. It is noted that the addition of these shunt reactors might also be beneficial for voltage regulation in the local networks.

4.3. Recent Applicability Studies in Norway

A recent applicability study of anti-icing and de-icing technologies was presented by [15] for the Norwegian power grid. Presently, the main method for de-icing of power lines consists of the use of helicopters that strike the lines with a pole attached to an insulated rope (method 5.1.5). Several technologies were investigated in order to verify if they could be applied in the context with the specific conditions that prevail in Norway: mostly in-cloud icing, extremely non-uniform ice loads, sometimes only on a few spans, with difficult or no terrestrial access to the lines due to complex topography and remote areas.

A workshop was organized in October 2018 where about 80 international specialists gathered and presented the current development state of various methods. Nine methods were retained for further analyses as these were considered potentially applicable to the Norwegian conditions. The following preliminary recommendations were formulated in [15] in order to identify the methods that should be further investigated:

- For ground wires, the use of remotely controlled external current sources that are permanently installed on critical line sections (method 6.2.2).
- For phase conductors, two methods are identified—current redistribution by contactors load transfer in bundled lines (method 6.5) and use of high frequency injection with generators installed on critical line sections (method 6.4).
- For fjord crossings, LC spirals seem to be an interesting option (method 4.2).

4.4. Recommendations after the Winter Storm of January 2008 in China

The important icing disaster in large parts of southern and central China in January and February 2008 led to intense research efforts and an important number of publications similar to the development that could be observed in Québec after the 1998 ice storm. As examples, two reviews [4,16] were included in the present analysis. It should be mentioned that China experienced an increase in the high voltage DC line projects during the last years. Therefore, Li et al. [16] elaborated various technologies designed for thermal de-icing of DC power lines (method 6.3). Lv and He [4] concluded that the development of de-icing methods for power lines are expected to follow a trend towards mechanical de-icing based on robots (method 5.1.6).

In another publication by a joint team of experts from university and power companies [29], Jiang et al. reported on the strategies and status of anti-icing methods that were studied in China after the 2008 ice storm. Various countermeasures applying coatings, laser technologies, and robots as well as AC and DC de-icing techniques were tested. It was found that the thermal de-icing method using DC voltage (methods 6.2.3 and 6.3) is a convenient solution with relatively low power demand and high efficiency. According to Jiang et al. [29], more than 200 substations have been equipped with fixed DC de-icing systems and mobile ice-melting devices are also employed by various utilities. Two years later, Jiang et al. reported ongoing research efforts with emphasis on the load current transfer method of bundled conductors (method 6.5) [30].

5. Conclusions

The comparative analysis of several review papers and technical reports show that a large number of technologies were already proposed decades ago, before the catastrophic ice storms of 1998 in Eastern Canada and 2008 in central and southern China. The intense work within research institutions as well as utilities in the wake of these events led to the increase in the understanding of the meteorological phenomena and their impacts on the power grids. Rather than finding new technologies, efforts have been focused on developing already known countermeasures with the potential for practical application in field operations.

There is no single trend for the deployment of anti-icing or de-icing technologies around the world. Passive, thermal, and mechanical methods are used by utilities in various regions of the world, depending on their local needs and their local experiences in the past. Additionally, the recommendations for future applications are diverse. Regarding the construction of new lines or the reconstruction of damaged lines, several lessons have been learned since the catastrophic events in 1998 and 2008 and actions are taken today to improve the line design.

Once one or several anti-icing or de-icing methods have been selected for the integration into field operations, several years may pass before the deployment is accomplished. This should be considered in the deployment strategies.

6. Future Outlook

There are three reasons why we should be seeking for affordable and long-term solutions against icing and snowing of power grids, a major problem that has attracted a global effort.

- With climate changes, power grids are poised to experience unpredicted icing and snowing events that may not be limited only to the so-called "cold regions".
- With increasing age, most installed transmission lines are more fragile.
- The ever-growing demand for electricity is another major concern that adds to the problem.

While active (heating, chemicals, and mechanical methods) solutions are widely used to remove ice or snow, passive solutions by nanotechnology are expected to affect the anti-icing industry. Although coating solutions are already available, research in this area still needs to be pursued. Ongoing research will increase our knowledge of the fundamental mechanisms through which nanoparticles interact with freezing rains.

Key challenges include icing prevention for extended periods thereby decreasing the risk of potential damage to infrastructure and outdoor structures and concomitant lowering of related maintenance and replacement costs. Encapsulated phase change materials and superhydrophobicity [31] achieved by texturing a surface to develop novel coating platforms are promising candidates for developing robust anti-icing materials. Smart materials with a self-sustainable lubricating layer, achieved via modifying solid substrates or self-healing ones are also interesting alternatives [32]. The impact can be significant as various areas from the power grids to aerospace and sea vessels can benefit from the research results.

Author Contributions: Data collection and analysis, original draft preparation, Sections 2–5, S.B.; abstract, Sections 1 and 6, general review and editing, I.F. All authors have read and agreed to the published version of the manuscript.

Funding: This research received no external funding.

Institutional Review Board Statement: Not applicable.

Informed Consent Statement: Not applicable.

Data Availability Statement: Not applicable.

Conflicts of Interest: The authors declare no conflict of interest.

References

1. Pohlman, J.C.; Landers, P. Present State-of-the-Art of Transmission Line Icing. *IEEE Trans. Power Appar. Syst.* **1982**, *PAS-101*, 2443–2450. [CrossRef]
2. Laforte, J.L.; Allaire, M.A.; Laflamme, J. State-of-the-Art on Power Line De-Icing. In Proceedings of the 7th International Workshop on Atmospheric Icing of Structures (IWAIS), Chicoutimi, QC, Canada, 3–7 June 1996.
3. Farzaneh, M. (Ed.) *Atmospheric Icing of Power Networks*, 1st ed.; Springer: Berlin/Heidelberg, Germany, 2008.
4. Lv, X.; He, Q. Analysis of Dicing Techniques and Methods of Overhead Transmission Line. *Procedia Eng.* **2011**, *15*, 1135–1139. [CrossRef]
5. Quaia, S.; Marchesin, A.; Rampazzo, D.; Mancuso, C. Anti-Icing Measures in the Italian Power System. *IET Gener. Transm. Distrib. J.* **2019**, *13*, 5577–5585. [CrossRef]
6. CEATI. *De-Icing Techniques before, during, and Following Ice Storms*; Technical Report T003700-3303; CEA Technologies Inc.: Montreal, QC, Canada, 2002.
7. Volat, C.; Farzaneh, M.; Leblond, A. De-icing/Anti-icing Techniques for Power Lines: Current Methods and Future Direction. In Proceedings of the 11th International Workshop on Atmospheric Icing of Structures (IWAIS), Montréal, QC, Canada, 13–16 June 2005.
8. Farzaneh, M.; Volat, C.; Leblond, A. Anti-icing and de-icing techniques for overhead lines. In *Atmospheric Icing of Power Networks*, 1st ed.; Farzaneh, M., Ed.; Springer: Berlin/Heidelberg, Germany, 2008; pp. 229–268.
9. CIGRÉ WG B2.29. *Systems for Prediction and Monitoring of Ice Shedding, Anti-Icing and De-Icing for Power Line Conductors and Ground Wires*; Technical Brochure No. 438; International Council on Large Electric Systems: Paris, France, 2010.

10. Solangi, A.R. Icing Effects on Power Lines and Anti-Icing and De-Icing Methods. Master's Thesis, UiT The Arctic Univeristy of Norway, Tromsø, Norway, 2018.
11. Farzaneh, M. Impact and Mitigation of Icing on Power Network Equipment. Available online: https://www.inmr.com/impact-mitigation-icing-power-network-equipment/ (accessed on 9 June 2021).
12. Régie de l'Énergie du Québec, Public Hearing R-3522-2003, HQT-12 Document 1, Responses to Request for Information no. 1 from the Régie de l'Énergie, 15 April 2004. Available online: http://www.regie-energie.qc.ca/audiences/3522-03/RepDemRens/HQT-12_doc1_3522-03_16avr04.pdf/ (accessed on 28 May 2021).
13. Régie de l'Énergie du Québec, Public Hearing R-3522-2003, HQT-12 Document 3, Responses to Request for Information from the Union of Consumers, 15 April 2004. Available online: http://www.regie-energie.qc.ca/audiences/3522-03/RepDemRens/HQT-12_doc3_3522-03_15avr04.pdf/ (accessed on 28 May 2021).
14. Prud'Homme, P.; Dutil, A.; Laurin, S.; Lebeau, S.; Benny, J.; Cloutier, R. Hydro-Québec TransÉnergie Line Conductor De-Icing Techniques. In Proceedings of the 11th International Workshop on Atmospheric Icing of Structures (IWAIS), Montréal, QC, Canada, 13–16 June 2005.
15. Gutman, I.; Lundengard, J.; Naidoo, V.; Adum, B. Technologies to Reduce and Remove Ice from Phase Conductors and Shield Wires: Applicability for Norwegian Conditions. In Proceedings of the 18th International Workshop on Atmospheric Icing of Structures (IWAIS), Montréal, QC, Canada, 19–23 June 2019.
16. Li, S.; Wang, Y.; Li, X.; Wei, W.; Zhao, G.; He, P. Review of De-icing Methods for Transmission Lines. In Proceedings of the 2010 International conference on Intelligent System Design and Engineering Applications, Changsha, China, 13–14 October 2010. [CrossRef]
17. Twenty Years Ago, Québec Was Battered by an Ice Storm. Available online: http://news.hydroquebec.com/en/press-releases/1313/twenty-years-ago-quebec-was-battered-by-an-ice-storm/ (accessed on 22 August 2021).
18. Régie de l'Énergie du Québec, Public Hearing R-3522-2003, Expert Report by G. Olivier for the Union of Consumers, 4 May 2004. Available online: http://www.regie-energie.qc.ca/audiences/3522-03/PreuveInterv/UC_Preuve_RapportExpertise_3522_05mai04.pdf/ (accessed on 27 May 2021).
19. CIGRÉ WG B2.44. *Coatings for Protecting Overhead Power Network Equipment in Winter Conditions*; Technical Brochure No. 631; International Council on Large Electric Systems: Paris, France, 2015.
20. CIGRÉ WG B2.69. *Coatings for Protecting Overhead Power Networks against Icing, Corona Noise, Corrosion and Reducing Their Visual Impact*; Technical Brochure No. 838; International Council on Large Electric Systems: Paris, France, 2021.
21. Régie de l'Énergie du Québec, Public Hearing R-3522-2003, D-2004-175, Decision of the Régie de l'Énergie du Québec, 20 August 2004. Available online: http://www.regie-energie.qc.ca/audiences/decisions/D-2004-200.pdf/ (accessed on 2 August 2021).
22. Hydro-Québec. *Rapport Annuel 2005*; © Hydro-Québec: Montreal, QC, Canada, 2006; ISBN 2-550-46386-2. (In French)
23. Hydro-Québec. *Rapport Annuel 2008*; © Hydro-Québec: Montreal, QC, Canada, 2009; ISBN 978-2-550-55045-7. (In French)
24. Hydro-Québec. *Rapport Annuel 2017*; © Hydro-Québec: Montreal, QC, Canada, 2018; ISBN 978-2-550-80380-5. (In French)
25. Hydro-Québec. *Rapport Annuel 2010*; © Hydro-Québec: Montreal, QC, Canada, 2011; ISBN 978-2-550-60867-7. (In French)
26. Hydro-Québec. *Rapport Annuel 2012*; © Hydro-Québec: Montreal, QC, Canada, 2013; ISBN 978-2-550-66871-8. (In French)
27. After 1998: A More Robust Grid. Available online: https://www.hydroquebec.com/ice-storm-1998/after-the-storm.html (accessed on 22 August 2021).
28. Hydro-Québec. *Rapport sur le Développement Durable 2011*; © Hydro-Québec: Montreal, QC, Canada, 2012; ISBN 978-2-550-63874-2. (In French)
29. Jiang, X.; Dong, B.; Dong, L.; Lu, J.; Zhang, J. Strategies and Status of Anti-Icing and Disaster Reduction for Power Grid after the 2008: Ice Storm in China. In Proceedings of the 15th International Workshop on Atmospheric Icing of Structures (IWAIS), St. Johns, NL, Canada, 8–11 September 2013.
30. Jiang, X.; Wang, Y.; Shu, L.; Zhang, Z.; Hu, Q.; Wang, Q. Control scheme of the de-icing method by the transferred current of bundled conductors and its key parameters. *IET Gener. Transm. Distrib. J.* **2015**, *9*, 2198–2205. [CrossRef]
31. Amir, A.Y.; Anahita, A.; Khosrow, M.; Reza, J.; Gelareh, M. Potential anti-icing applications of encapsulated phase change material–embedded coatings; a review. *J. Energy Storage* **2020**, *31*, e101638.
32. Gong, H.; Yu, C.; Zhang, L.; Xie, G.; Guo, D.; Luo, J. Intelligent lubricating materials: A review. *Compos. Part B Eng.* **2020**, *202*, 108450. [CrossRef]

Article

Image Characteristic Extraction of Ice-Covered Outdoor Insulator for Monitoring Icing Degree

Yong Liu [1,*], Qiran Li [1], Masoud Farzaneh [2] and B. X. Du [1]

[1] School of Electrical and Information Engineering, Tianjin University, 92 Weijin Road, Tianjin 300072, China; qrli@tju.edu.cn (Q.L.); duboxue@gmail.com (B.X.D.)

[2] Department of Applied Sciences, Université du Québec à Chicoutimi, Chicoutimi, QC G7H2B1, Canada; masoudfarzaneh@uqac.ca

* Correspondence: tjuliuyong@tju.edu.cn; Tel.: +86-136-8201-8949

Received: 30 August 2020; Accepted: 8 October 2020; Published: 12 October 2020

Abstract: Serious ice accretion will cause structural problems and ice flashover accidents, which result in outdoor insulator string operating problems in winter conditions. Previous investigations have revealed that the thicker and longer insulators are covered with ice, the icing degree becomes worse and icing accident probability increases. Therefore, an image processing method was proposed to extract the characteristics of the icicle length and Rg (ratio of the air gap length to the insulator length) of ice-covered insulators for monitoring the operation of iced outdoor insulator strings. The tests were conducted at the artificial climate room of CIGELE Laboratories recommended by IEEE Standard 1783/2009. The surface phenomena of the insulator during the ice accretion process were recorded by using a high-speed video camera. In the view of the ice in the background of the picture of fuzzy features and high image noise, a direct equalization algorithm is used to enhance the grayscale iced image contrast. The median filtering method is conducted for reducing image noise and sharpening the image edge. The maximum entropy threshold segmentation algorithm is put forward to extract the insulators and its surface ice from the background. Then, the modified Canny operator edge detection algorithm is selected to trace the boundaries of objects through the extraction of information about attributes of the endpoints of edges. After we obtained the improved Canny edge detection image for both of the ice-covered insulators and non-iced insulators, the icing thickness can be obtained by calculating the difference between the edge of the non-iced insulators image and the edge of the iced insulator image. Besides, in order to identify the icing degree of the insulators more accurately, this paper determines the location of icicles by using the region growth method. After that, the icicle length and Rg can be obtained to monitor the icing degree of the insulator. It will be helpful to improve the ability to judge the accident risk of insulators in power systems.

Keywords: ice-covered insulator; characteristics extraction; image processing method; median filtering method; entropy threshold segmentation; modified Canny operator; region growth method; icing degree

---

1. Introduction

Ice accumulation on insulator strings has been recognized as a serious threat for power systems operating in many atmospheric icing regions [1–5]. These hazards can be mainly divided into two categories: one is serious ice accretion, which will cause structural problems, for instance, wire breakage, tower collapse, etc., and the other is insulation problems, for instance, the icicle will change the distribution of insulators electric field significantly, which reduces its insulation performance and can lead to an ice flashover accident easily. Therefore, it is necessary to improve the ability for monitoring the operation of iced outdoor insulator strings for preventing structural accidents and icing flashover.

Over the past decades, many investigations have researched monitoring methods to reduce icing accidents [6–16]. In these investigations, the characteristics of iced insulators surface phenomena were extracted by image processing method for monitoring.

Liu et al. studied the performance of insulators under icing conditions, recorded the test process with a high-speed camera and analyzed the flashover characteristics of iced insulators and the growth characteristics of ice pillars based on image processing technology [17,18]. According to the characteristic value of a flashover image, the flashover process is divided into different stages, and the quantitative analysis method of flashover risk value of iced insulators is proposed. At the same time, the growth characteristics of ice pillars at different edges and the variation characteristics of surface discharge are calculated and analyzed. The research results can be used to evaluate the main hazards of iced outdoor insulators and improve the safety of iced suspension insulators. Hao et al. used the image processing method to study the natural icing of glass insulator strings in service. Based on the grab segmentation method, by identifying the convex defects of Icelandic contour, the algorithm of graphical shed spacing and graphical shed overhanging is proposed [19,20]. This method can identify the most serious icing situation when the insulator cover is completely bridged. The bridge position can also be detected, including the left, right, or both sides of the insulator string in the image. Yang et al. proposed a method for identifying the ice coating type of an in-service glass insulator based on the texture feature description operator [21]. A uniform local binary model (ULBP) and an improved uniform local binary model (IULBP) are used to extract the texture features of ice cover types. The experimental results show that, due to the different texture features of each kind of ice, IULBP has a good recognition effect on six kinds of ice. Zhu et al. proposed an image recognition algorithm for monitoring the icicle length and insulator ice bridging condition. The saliency analysis algorithm is applied to the extract region of the insulator ice layer and the length of the insulator icicle was calculated by the Fourier transform of the pixel distribution curve [22]. Pernebayeva et al. studied a Gabor filtering algorithm for extracting a set of Gabor phase congruency features from insulator images for the presence or absence of snow, ice, and water droplets by utilizing the minimum distance nearest neighbor classifier [23]. Vita, V. et al. constructed different neural network models for insulator contamination identification using different structures, learning algorithms, and transfer functions. All the models are compared and analyzed, the best model is found, and the calculation results match the experimental results [24]. Chen et al. applied digital image processing technologies such as gray level transformation, image sharpening, image segmentation, and edge detection to the research of structural images, and effectively extracted the effective information in the image [25]. Gilboa, G. et al. use the free Schrodinger equation and extended the linear and nonlinear scale space generated by the intrinsic real value diffusion equation to the complex diffusion process, thus obtaining two examples of nonlinear complex processes which play an important role in image processing: One is the regularized impact filter for image enhancement, the other is the denoising process keeping slope [26]. From comparative analysis of the research on other aspects, few investigations have been conducted on monitoring and diagnostic of insulator strings in extreme weather environments, which need further research to decrease icing accidents.

Therefore, in this paper, for improving the ability to judge the icing degree risk of outdoor insulators and reduce icing accidents caused by ice-covered insulators, the features of insulator surface performance was extracted by an image processing method in order to monitor icicle length and the ice bridging state of iced outdoor insulator strings. The tests were conducted at CIGELE Laboratories. The test specimen was the five units' suspension ceramic insulators, which were artificially accreted with wet-grown ice in the cold-climate room of CIGELE. The procedure of ice accumulation was recommended by IEEE Standard 1783/2009. The surface phenomena of the insulators during the icing accretion were recorded by a high-speed video camera with a rate of six thousand frames per second.

2. Test Setup and Procedures

The test specimen is the five units' suspension ceramic insulators. The picture and parameters are shown in Table 1.

Table 1. Configuration, dimensions, and parameters of each unit of the test specimen.

Main Dimension and Parameters	Configuration
Diameter = 254 mm	
Height = 146 mm	
Leakage distance = 305 mm	
Number of units = 5 units	
Arcing distance for 5 units = 809 mm	

The tests were conducted in an artificial climate room with a length of 4.8 m, a width of 2.8 m, and a height of 3.5 m of CIGELE Laboratories, as shown in Figure 1 [27]. By using the proportional integral and differential system, the freezing devices can make the ambient temperature drop to −12 °C after the test setup was fixed. The spray device mainly consists of a water supplying system and wind blowing equipment. Ice was formed from super-cooled droplets produced by the former system through 4 oscillating nozzles. The latter system produced a relatively uniform airflow by using a series of fans with a diffusing honeycomb panel. The test power source was supplied by an AC test transformer with a rated capacity of 240 kV·A.

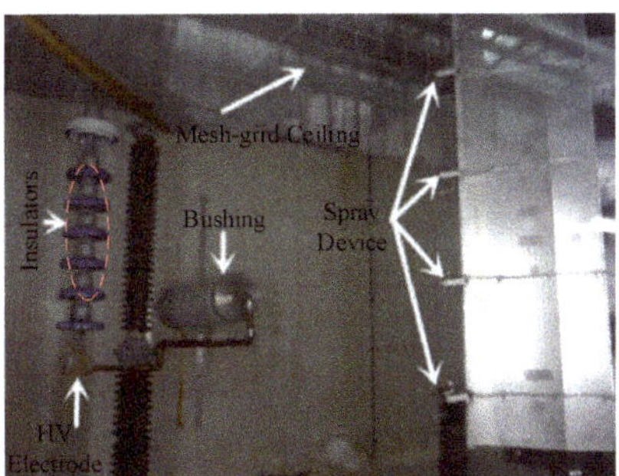

Figure 1. Artificial cold-climate room.

The surfaces of the insulator sheds were cleaned by deionized water before the ice accretion. The insulators needed under the setting ambient temperature last about sixteen hours to give all the experimental setup enough time to reach the same temperature as that of the test environment. The AC voltage of 75 kVrms (15 kVrms per unit) was energized on insulators during ice accretion for simulating the operating environment. Meanwhile, the water supply system started to spray freezing droplets on the insulators' surface. The water conductivity was set at 30 μS/cm by mixing deionized water and sodium chloride. The wind speed was fixed at 3.3 m/s to blow on the windward side of insulators in the ice accumulation period. When ice accumulation duration reached 90 min, the applied voltage and spray device were turned off immediately and the icing process is stopped. The ice accretion process on insulators was photographed during the whole experiment [28,29].

3. Image Processing of Ice-Covered Insulator

The iced insulator image of recording was influenced by various factors, such as the glazed icing, which is a smooth and transparent structure with unobvious transverse volume change. It is difficult to identify the overall state of iced insulators; and therefore, low quality images were attained. Therefore, in order to improve the image quality, an image processing method is proposed for recognizing the insulators' bridged state and extracting the characteristic values for the warning of icing accidents.

3.1. Enhancement of Image

(1) Image grayscale

Each pixel point of an iced insulator image is essentially composed of components in three directions of RGB (red, green, and blue). The grayscale is to convert the three-component value of RGB into a single gray value, so that the calculation of the subsequent image processing would be simplified. The calculation is shown in Equation (1). The result is shown in Figure 2.

$$M = 0.3R + 0.59G + 0.11B \tag{1}$$

here, M, R, G, and B are the values of the pixel, respectively representing gray value, red value, green value, and blue value. The coefficients in the formula are derived from the sensitivity of human eyes to color.

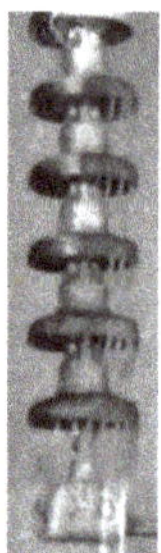

Figure 2. Color image and grayscale image.

(2) Gray Stretch

The image of the iced insulator has the characteristics of fuzziness and noise in the image background. The direct equalization method was selected to make a gray value of the image uniform distribution for enhancing contrast and highlighting the iced insulator image details. The result is shown in Figure 3.

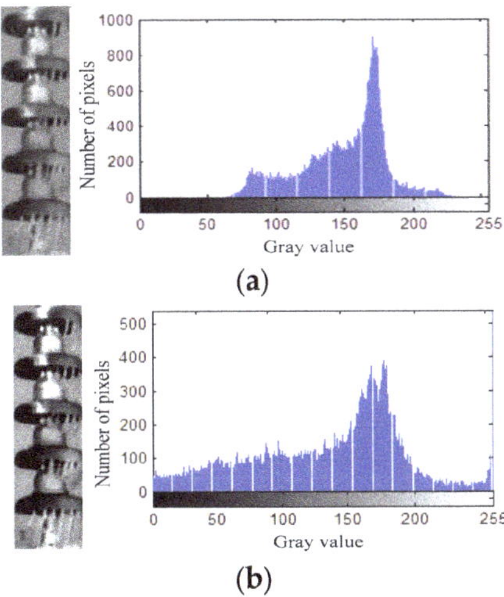

Figure 3. The grayscale image and the image after direct equalization and its histogram of grayscale distribution. (**a**) Original grayscale image and its histogram. (**b**) Grayscale image and its histogram after direct equalization.

(3) Image denoise

In fact, after the direct equalization processing, the noise interference is still present in the iced insulator image. To remove the small bright spot and improve the definition of image, the median filtering algorithm method was chosen to diminish the gap of the image. The iced insulator image edge can be sharpened, and the obvious background noise can be decreased through enhancing the filtering effect. The result is shown in Figure 4.

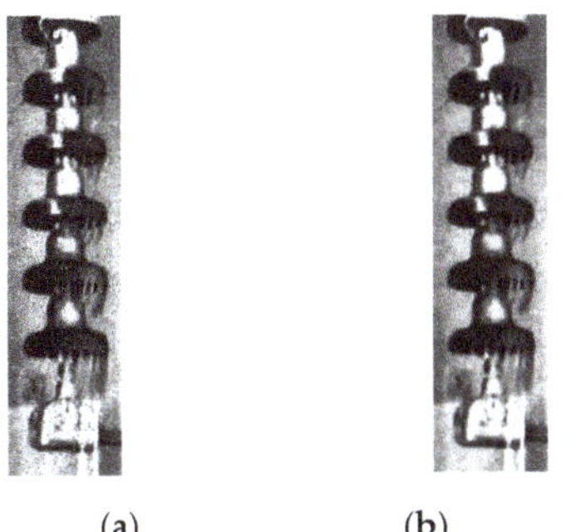

Figure 4. The image with noise interference and the image after median filtering processing. (**a**) The image with noise interference. (**b**) The image after median filtering processing.

3.2. Image Segmentation

(1) Maximum entropy threshold segmentation

The key of image processing is image segmentation, which is to segment an image into meaningful regions by extracting some target area of image characteristics, and then obtain the binarization image [16,17]. The maximum entropy threshold segmentation algorithm was proposed for acquiring excellent efficacy of segmentation and the characteristics of recorded images during the ice regime.

This method is essentially using the images' regional features to segment images based on the similarity between the pixels. The higher entropy value of the segmentation image can indicate the more information it contains, and it is beneficial to the effect of division [18]. The calculation formula of entropy is

$$H(S) = -P_1 \ln P_1 - P_0 \ln P_0 \tag{2}$$

where $H(s)$ is the statistical value of the amount of information that the binarization image contained after segmentation. P_1 and P_0 represent the probability that the output value of the segmentation image is one and zero, respectively.

According to the effective segment, the gray value of the image is compressed and transformed into 0 or 255 pixel values. Then, the approximate edge contours are obtained and the process of binarization is completed. The image processing result is shown in Figure 5. It can be observed that the maximum entropy threshold segmentation method can extract the object points of the image and remove the redundant information. Therefore, the proposed algorithm ensures the segmentation more efficient and segments insulator images with intensity inhomogeneity correctly.

Figure 5. The binarization image after maximum entropy threshold segmentation processing.

(2) Edge detection

Figure 5 shows that there is still background noise in the image and the objectives and background of the segmentation image have low contrast. Thus, the modified Canny operator edge detection algorithm is selected to trace boundaries of objects through extraction of information about attributes of endpoints of edges, in particular orientation and neighborhood relationships [19,20]. In the algorithm, the image is smoothed by Gaussian filter which is used to determine the adjustable parameters based on the characteristics of the image. Most of the background is divided according to the information of the image edge for reducing imprecise background and objective. Then, the image edge can be detected by using Canny operator. The image processing result is shown in Figure 6.

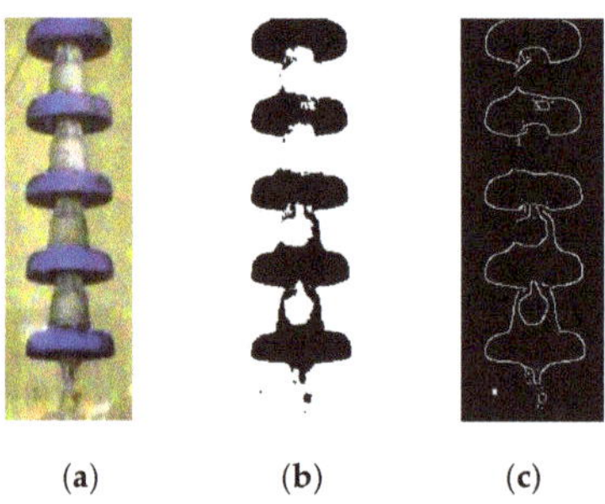

(a) (b) (c)

Figure 6. The comparison diagram of the original image and the binarization image and the image after edge detection processing. (**a**) Original image. (**b**) Binarization image. (**c**) Image after edge detection processing.

The modified Canny algorithm is used to calculate the amplitude of the gradient through the directional derivatives for pixels of the image G(i,j) in the selected neighborhood. Equations (3)–(9) are as follows:

The calculation of the X directional derivative:

$$G_x(i,j) = F(i+1,j) - F(i-1,j) \tag{3}$$

The calculation of the Y directional derivative:

$$G_y(i,j) = F(i,j+1) - F(i,j-1) \tag{4}$$

The calculation of the 45° directional derivative:

$$G_{45}(i,j) = F(i+1,j+1) - F(i-1,j-1) \tag{5}$$

The calculation of the 135° directional derivative:

$$G_{135}(i,j) = F(i-1,j+1) - F(i+1,j-1) \tag{6}$$

The calculation of the first partial derivatives:

$$E_x = G_x(i,j) + \frac{G_{45}(i,j) + G_{135}(i,j)}{2} \tag{7}$$

$$E_y = G_y(i,j) + \frac{G_{45}(i,j) - G_{135}(i,j)}{2} \tag{8}$$

The calculation of the gradient magnitude:

$$A(i,j) = \sqrt{{E_x}^2 + {E_x}^2} \tag{9}$$

(3) Region growth method

After obtaining the improved Canny edge detection images, the icing thickness can be obtained by calculating the difference between the edge of the non-iced insulator image and the edge of the iced insulator image. Besides, in order to obtain the icicles and air gap parts of the ice-covered insulators' image for identifying the icing degree of the insulators more accurately, this paper determines the location of icicles by using the regional growth method, the schematic diagram as shown in Figure 7.

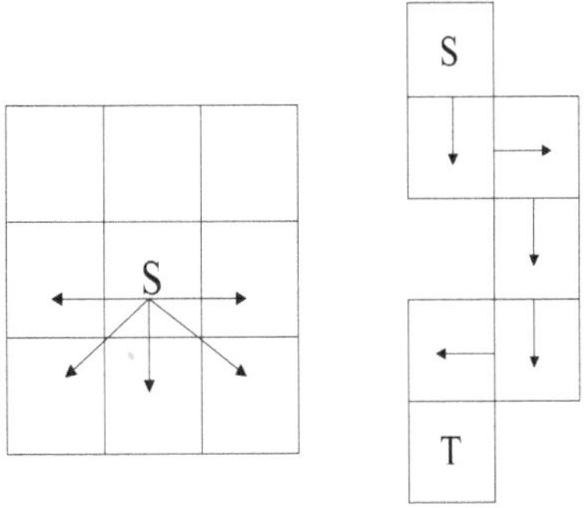

Figure 7. Regional growth method diagram.

Regional growth corresponds to the process of development of a set of pixels and regions extend to a larger area. Each pixel of the edge of the non-ice shed parts of ice-covered insulators is set as the seed pixel, which is used as the starting point of growth. By inspecting the pixel value of all direction point of neighborhoods, when the pixel value of neighborhood point is the same as the pixel value of seed point, this point is defined as a new seed point. The neighborhood point will be searched continuously until it cannot satisfy the above condition. Due to the fact that the direction of the edge region of the iced insulators image is downward, the point of the top directional can be eliminated. After the process of regional growth method, it can be considered that the selected points are the tip of icicle, and then it separates the tip of icicle part and the air gaps according to the location point of the tip of icicle.

The distance between the tip of icicle and the edge of insulator is considered the air gap length. However, the pixel value of the air gap length obtained by the Canny edge detection image method needs to be transformed in millimeters for further accurate calculation. However, the air gap length cannot be used as the only parameter to indicate the icing degree of insulators because of it is impacted by the different insulator models and angles of camera recording. Thus, the icing degree is indicated by the Rg (the ratio of the air gap length to the insulator length) for avoiding the influence of these factors. Then, we can establish the relationship between the icing degree and the Rg.

4. Analyze the Results of Characteristic Extraction

The results of ice thickness, icicle length, and Rg are as shown in Table 2.

Table 2. The results of ice thickness, icicle length, and Rg.

Image of Ice-Covered Insulators		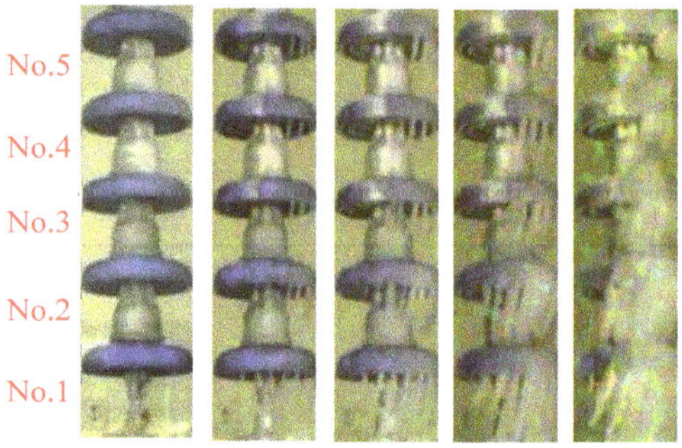 No.5 No.4 No.3 No.2 No.1				
Ice accretion time (min)		0	10	20	30	50
Ice thickness (mm)		0	3	7	24	88
NO.5	Icicle length (mm)	0	13	21	38	79
	R_g(%)	100	91.4	85.7	74.3	45.7
NO.4	Icicle length (mm)	0	0	8	40	91
	R_g(%)	100	100	94.6	72.9	37.8
NO.3	Icicle length (mm)	0	31	52	84	146
	R_g(%)	100	78.6	64.3	42.8	0
NO.2	Icicle length (mm)	0	24	77	73	146
	R_g(%)	100	83.3	47.6	50	0
NO.2	Icicle length (mm)	0	42	57	146	146
	R_g(%)	100	71.4	61.2	0	0
NO.1	Icicle length (mm)	0	142	211	402	624
	R_g(%)	100	80.5	71.2	44.9	14.6

As shown in Table 2, it can be found that non-uniformity distribution of ice accretion on the surface of insulators and the ice is mainly accumulated on the windward side of insulators. When the cold room maintains the ambient temperature at −12 °C, the low-temperature droplets frozen on the surface of insulator sheds under the experimental condition. The ice thickness shows the nonlinear increasing tendency on the sheds of insulators during the process of ice accumulation. From 0 min to 20 min, the ice thickness increased slowly, only 7 mm. While from 30 min to 50 min, ice thickness increased 64 mm, the increment is about 9 times higher than the former, which indicates that the ice layer grew rapidly and the insulators' icing degree became more serious during this period.

The variation of icicle length and Rg during ice accretion regime is shown in Figure 8. The super-cooled droplet formed ice, and simultaneously, the test voltage produced joule heat which can cause the melting and dripping of accreted ice. Hence, the ice accreted on the insulators is a dynamically varying phenomenon. However, each shed of the ice-covered insulators shows an obvious decreasing tendency of the variation of icicle length and Rg because of the serious degree of icing effect. Note that the icicle length of the second unit of insulators slightly decreases and Rg slightly increases between 20 min and 30 min. In this period, leakage current generated by heat energy can cause the melting of icicles and can even make some of them fall down to the ground. Although the electric field distortion of the insulator surface will increase further and the thermal effect of the arc discharge will have a negative influence on the ice accumulation, the freezing influence of precipitated droplets dominates the shed surface so the ice layer can maintain growth and the icicle length can keep increasing and Rg still can decrease at this time. Therefore, the variations of icicle length and Rg correspond well to the ice accretion process, which can be used to judge the risk of the icing degree.

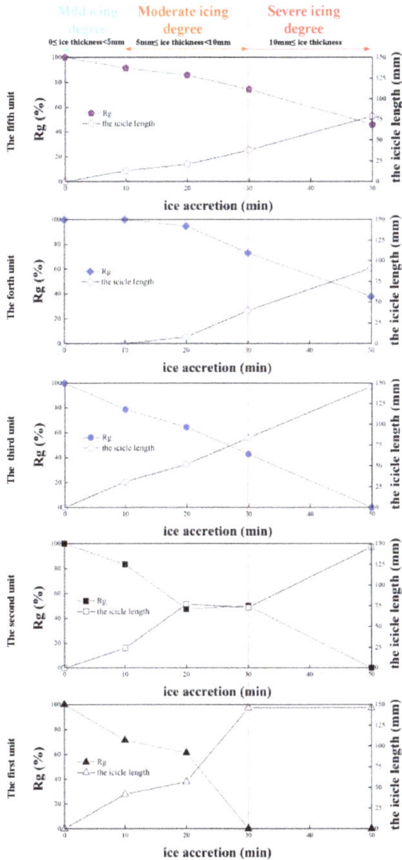

Figure 8. The variation of icicle length and Rg during the ice accretion regime.

In addition, from the HV (high voltage) electrode to the ground side, the shed separation between cap-and-pin insulators is significantly reduced by the increasing of icicles. Because the super-cooled droplet was subjected to the effect of gravity, the ice bridging condition of the bottom side of the insulator string units are heavy. At 30 min, the average ice thickness just reaches 24 mm, the value of the Rg between the second unit and the fifth unit of the ice-covered insulators is more than 40%, the first unit of ice-covered insulators Rg is 0%, which indicates the icing degree from the first unit to HV electrode is extremely serious. At 50 min, the first unit and the third unit are fully bridged. While the fourth and fifth units are without complete bridging, the value of Rg is 37.8 and 45.7% respectively. Meanwhile, the partial arcs constantly burn at the air gaps, which inhibits the growth of the icicles. Finally, the ice growth rate and the melting rate will reach a balance state that causes these units to be unable to be bridged completely. By analyzing the icing degree of every unit of the insulator string, we can evaluate the hazards of ice-covered insulators accurately.

In this way, through the independent analysis of the icing degree of different insulator sheds, the icing degree of insulators can be evaluated more accurately, which avoids the simple generalization of icing conditions of different types of insulators under the same environmental conditions, enhances the detailed judgment of the icing degree of insulators, and reduces the estimation error of the icing degree.

Because most of the transmission lines are exposed in the field, the monitoring equipment also needs to be exposed in the field for a long time, which will inevitably be affected by severe weather such as strong wind, high temperature, and rainstorms. There are still a series of problems in security protection, energy consumption, wireless communication, data encryption, and video image compression. In short, there will be some obstacles to the implementation of the method in this paper, but this is mainly a technical problem. With the development of science and technology, these obstacles will be solved one by one. At the same time, when dealing with different background noise, we can improve the denoising method. We can judge the type of noise by intelligent algorithms and select the corresponding denoising method automatically. The practical application of this method needs further research.

5. Conclusions

The image processing technology is used to process the icing image of the insulator, extract the characteristics of the icing degree of the iced insulator, and analyze the icing characteristics of the insulator. The proposed method will be helpful to the monitoring of the icing degree of the iced insulator. The main conclusions are as follows

(1) Aiming at the problems of fuzzy background and noise interference of the iced insulator image, the direct equalization method, and median filter method are proposed to preprocess the image. The method can effectively reduce the noise and enhance the contrast of the image, which is of great significance for further processing the image of the iced insulator.

(2) The maximum entropy threshold segmentation algorithm is used to extract the insulator and its surface ice from the image, and the key information is accurately screened out from the image. An improved Canny operator edge detection algorithm is used to track the edge of the non-icing insulator image and calculate the icing thickness of icing insulator image edge, which can accurately detect the insulator edge smoothly.

(3) A regional growing method is proposed to determine the location of icicles, so as to obtain the ice column and air gap in the image of the iced insulator, and then extract the icicle length and RG as the indication value to evaluate the ice bridge state of the insulator string. The analysis results show that once the total length of ice pole exceeds 402 mm and the RG value is lower than 44.9%, the higher the accident probability of iced insulator.

Author Contributions: Y.L., Q.L. conceived and designed the study. Y.L., Q.L. analyzed the results. Q.L. drafted this article. M.F., and B.X.D. made the critical and final reviewing. All authors have read and agreed to the published version of the manuscript.

Funding: This research was funded by Chinese National Natural Science Foundation grant number 51277131, National Basic Research Program of China grant number2014CB239501 and 2014CB239506, and Natural Science Foundation of Tianjin grant number 16JCQNJC06800.

Conflicts of Interest: The authors declare no conflict of interest.

References

1. Farzaneh, M. Insulator Flashover Under Icing Conditions. *IEEE Trans. Dielectr. Electr. Insul.* **2014**, *21*, 1997–2001. [CrossRef]
2. Farzaneh, M. 50 Years in Icing Performance of Outdoor Insulators. *IEEE Electr. Insul.* **2014**, *30*, 14–24. [CrossRef]
3. Jiang, X.; Meng, Z.; Zhang, Z.; Hu, J. DC Ice-Melting and Temperature Variation of Optical Fibre for Ice-Covered Overhead Ground Wire. *IET Gener. Transm. Distrib.* **2016**, *10*, 352–358. [CrossRef]
4. Liu, Y.; Du, B. Recurrent Plot Analysis of Leakage Current on Flashover Performance of Rime-Iced Composite Insulator. *IEEE Trans. Dielectr. Electr. Insul.* **2010**, *7*, 465–472. [CrossRef]
5. Yang, H.; Pang, L.; Li, Z.; Zhang, Q.; Yang, X.; Tang, Q.; Zhou, J. Characterization of Pre-Flashover Behavior Based on Leakage Current Along Suspension Insulator Strings Covered with Ice. *IEEE Trans. Dielectr. Electr. Insul.* **2015**, *22*, 941–950. [CrossRef]
6. Ale-Emran, S.; Farzaneh, M. Experimental Design of Booster Shed Parameters for Post Station Insulators Under Heavy Icing Conditions. *IEEE Trans. Power Deliv.*. **2015**, *30*, 488–496. [CrossRef]
7. Shu, L.; Shang, Y.; Jiang, L.; Hu, Q.; Yuan, Q.; Hu, J.; Zhang, Z.; Zhang, S.; Li, T. Comparison Between AC and DC Flashover Performance and Discharge Process of Ice-Covered Insulators Under the Conditions of Low Air Pressure and Pollution. *IET Gener. Transm. Distrib.* **2012**, *6*, 884–892. [CrossRef]
8. Taheri, S.; Farzaneh, M.; Fofana, I. Empirical Flashover Model of EHV Post Insulators Based on ISP Parameter in Cold Environments. *IEEE Trans. Dielectr. Electr. Insul.* **2016**, *23*, 403–409. [CrossRef]
9. Hu, Q.; Wang, S.; Shu, L.; Jiang, X.; Liang, J.; Qiu, G. Comparison of AC Icing Flashover Performances of 220 Kv Composite Insulators With Different Shed Configurations. *IEEE Trans. Dielectr. Electr. Insul.* **2016**, *23*, 995–1004. [CrossRef]
10. Farzaneh, M.; Drapeau, J. AC Flashover Performance of Insulators Covered With Artificial Ice. *IEEE Trans. Power Deliv.* **1995**, *10*, 1038–1051. [CrossRef]
11. Yan, B.; Wang, B.; Zhu, L.; Liu, H.; Liu, Y.; Ji, X.; Liu, D. A Novel, Stable, And Economic Power Sharing Scheme for an Autonomous Microgrid in the Energy Internet. *Energies* **2015**, *8*, 12741–12764. [CrossRef]
12. Phan, C.; Hara, M. Leakage Current and Flashover Performance of Iced Insulators. *IEEE Trans. Power Appl. Syst.* **1979**, *98*, 849–859. [CrossRef]
13. Hong, Y.Y.; Wei, Y.H.; Chang, Y.R.; Lee, Y.D.; Liu, P.W. Fault Detection and Location by Static Switches in Microgrids Using Wavelet Transform and Adaptive Network-Based Fuzzy Inference System. *Energies* **2014**, *7*, 2658–2675. [CrossRef]
14. Meghnefi, F.; Volat, C.; Farzaneh, M. Temporal and Frequency Analysis of the Leakage Current of a Station Post Insulator during Ice Accretion. *IEEE Trans. Dielectr. Electrl. Insul.* **2007**, *14*, 1381–1389. [CrossRef]
15. Volat, C.; Meghnefi, F.; Farzaneh, M.; Ezzaidi, H. Monitoring Leakage Current of Ice-Covered Station Post Insulator using Artificial Neural Networks. *IEEE Trans. Dielectr. Electr. Insul.* **2010**, *17*, 43–450. [CrossRef]
16. Liu, Y.; Gao, P. Icing Flashover Characteristics and Discharge Process of 500 kV AC Transmission Line Suspension Insulator Strings. *IEEE Trans. Dielectr. Electr. Insul.* **2010**, *17*, 434–442. [CrossRef]
17. Liu, Y.; Farzaneh, M.; Du, B. Investigation on Shed Icicle Characteristics and Induced Surface Discharges Along a Suspension Insulator String During Ice Accretion. *IET Gener. Transm. Distrib.* **2017**, *11*, 1265–1269. [CrossRef]
18. Li, Q.; Liu, Y.; Farzaneh, M.; Du, B. Image Characteristic Extraction of Surface Phenomena for Flashover Monitoring of Ice-Covered Outdoor Insulator. In Proceedings of the International Conference on Electrical Materials and Power Equipment, Guangzhou, China, 7–10 April 2019; pp. 431–434.

19. Hao, Y.; Wei, J.; Jiang, X. Icing Condition Assessment of in-Service Glass Insulators Based on Graphical Shed Spacing and Graphical Shed Overhang. *Energies* **2018**, *11*, 318. [CrossRef]
20. Wei, J.; Hao, Y.; Fu, Y. Detection of Glaze Icing Load and Temperature of Composite Insulators Using Fiber Bragg Grating. *Sensors* **2019**, *19*, 1321. [CrossRef] [PubMed]
21. Yang, L.; Jiang, X.; Hao, Y. Recognition of Natural Ice Types on in-Service Glass Insulators Based on Texture Feature Descriptor. *IEEE Trans. Dielect. Elect. Insul.* **2017**, *24*, 535–542. [CrossRef]
22. Zhu, Y.; Liu, C.; Huang, X.; Zhang, X. Research on Image Recognition Method of Icicle Length and Bridge State on Power Insulators. *Access* **2019**, *7*, 183524–183531. [CrossRef]
23. Pernebayeva, D.; James, A.; Bagheri, M. Live line snow and ice coverage detection of ceramic insulator using Gabor image features. In Proceedings of the IET International Conference on Resilience of Transmission and Distribution Networks, Birmingham, UK, 26–28 September 2017; pp. 1884–2021.
24. Vita, V.; Ekonomou, L.; Chatzarakis, G.E. Design of artificial neural network models for the estimation of distribution system voltage insulators' contamination. In Proceedings of the Wseas International Conference on Mathematical Methods. World Scientific and Engineering Academy and Society (WSEAS), Sousse, Tunisia, 3–6 May 2010; pp. 3–6.
25. Chen, Y.; Liu, H. Study on the meso-structure image of shale based on the digital image processing technique. In Proceedings of the 2009 International Conference on Image Analysis and Signal Processing, Taizhou, China, 11–12 April 2009; pp. 150–153. [CrossRef]
26. Gilboa, G.; Sochen, N.; Zeevi, Y.Y. Image Enhancement and Denoising by Complex Diffusion Processes. *IEEE Trans. Patt. Anal. Mach. Intel.* **2004**, *26*, 1020–1036. [CrossRef] [PubMed]
27. IEEE. *Standard 1783/2009TM, Guide for Test Methods and Procedures to Evaluate the Electrical Performance of Insulators In Freezing Conditions*; IEEE: Piscataway, NJ, USA, 2009.
28. Farzaneh, M.; Zhang, J.; Chen, X. Modeling of the AC Arc Discharge on Ice Surfaces. *IEEE Trans. Power Deliv.* **1997**, *12*, 325–338. [CrossRef]
29. Farzaneh, M.; Baker, T.; Bernstorf, A.; Brown, K. Insulator Icing Test Methods and Procedures: A Position Paper prepared by the IEEE Task Force on Insulator Icing Test Methods. *IEEE Trans. Power Deliv.* **2003**, *18*, 1503–1515. [CrossRef]

 energies MDPI

Article

Common Perceptions about the Use of Fillers in Silicone Rubber Insulation Housing Composites

Refat Atef Ghunem *, Yazid Hadjadj and Harold Parks

Metrology Research Center, National Research Council Canada, 1200 Montreal Rd., Ottawa, ON K1A0R6, Canada; Yazid.Hadjadj@nrc-cnrc.gc.ca (Y.H.); Harold.Parks@nrc-cnrc.gc.ca (H.P.)
* Correspondence: Refat.Ghunem@nrc-cnrc.gc.ca

Abstract: This paper discusses the lessons learned about fillers that are added to silicone rubber insulation housing composites, in order to improve the erosion resistance against the dry-band arcing are presented and common perceptions. Common practices that employ alumina tri-hydrate are reviewed, including the additional influential effect of the water of hydration in the suppression of the dry-band arcing. The effect of the water of hydration is shown to be dependent on various factors, such as the hydrated filler level and the type of the hydrated filler. More recent paradigms in which hydrated fillers have not been employed are also reviewed. Volume and shield actions of fillers are essential aspects that need to be understood in the design of silicone rubber insulation housing composites for new applications such as HVDC. In addition, the thermal degradation mechanisms of silicone rubber and the corresponding suppression effects of the added fillers are summarized.

Keywords: silicone rubber; outdoor insulators; transmission and distribution; pollution performance; tracking; erosion resistance

Citation: Ghunem, R.A.; Hadjadj, Y.; Parks, H. Common Perceptions about the Use of Fillers in Silicone Rubber Insulation Housing Composites. *Energies* **2021**, *14*, 3655. https://doi.org/10.3390/en14123655

Academic Editors: Pawel Rozga and Zhijin Zhang

Received: 28 April 2021
Accepted: 16 June 2021
Published: 19 June 2021

1. Introduction

Silicone rubber insulators are replacing the conventional ceramic insulators due to their proven performance when exposed to environmental pollutants. Silicone rubber has shown a unique ability to wash off contaminants and retain the hydrophobicity of the insulator, due to a low-molecular-weight (LMW) siloxane diffusing from the insulator housing bulk to the surface. Therefore, silicone rubber insulators have become backbone components of the overhead power delivery infrastructure. However, after aging in service, leakage currents may still develop under the effect of voltage, moisture and contamination, causing the depletion of the LMW siloxane, and thus, deteriorating the hydrophobicity recovery of the insulator [1]. Eventually dry-band arcing arises, causing heat ablation, which exposes the insulator fiberglass core to weathering and the consequent failure of the entire insulator due to tracking of the core (Figure 1) [2].

Fillers are added to silicone rubber outdoor insulation housing composites in order to enhance the erosion resistance, but the primary purpose of using fillers is cost reduction achieved by replacing the expensive silicone with the filler particles. Cost reductions have been a great incentive for loading silicone rubber outdoor insulation composites with as much filler as possible. However, this approach also reduces the amount of the LMW siloxane responsible for recovering the hydrophobicity of silicone rubber [1]. In addition, the cost of the fillers may vary depending on the type of filler. Therefore, achieving an optimized composite is a complex problem.

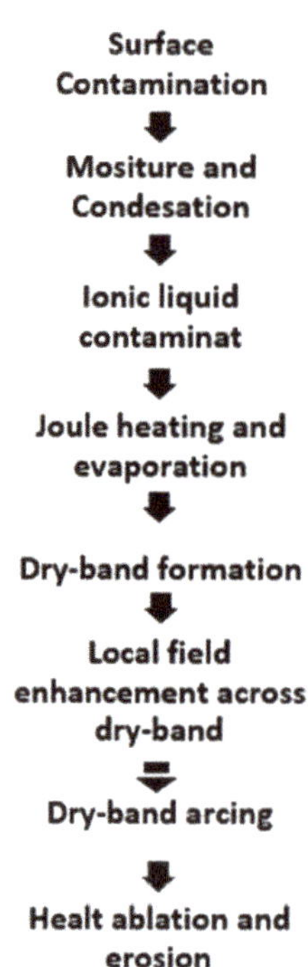

Figure 1. Erosion mechanism on the surface of silicone rubber outdoor insulation housing.

The application of fillers in silicone rubber outdoor insulator composites is usually based on experience, which has been so far successful as demonstrated by decades of proven performance in AC insulation applications. This successful experience has been used to justify the application of existing insulator composite designs for newer applications such as high voltage direct current (HVDC) outdoor insulation. However, recent finding raise concerns about this approach for new and different application under HVDC. Experimental findings and field observations have reported a more severe erosion under DC voltages as compared to AC voltages [3,4]. The HVDC power system has become an essential building block of the power transmission and distribution infrastructure in the green electricity grid. Therefore, it is essential to address concerns related to DC insulator aging by investigating the suitable use of fillers at the material development stage. In general, little attention has been given to the housing material design aspects as compared to the geometry aspect of silicone rubber insulators. While a geometric design has been mainly correlated with the insulator flashover performance, the use of fillers mainly controls the erosion performance of the insulator [1]. This paper summarizes the evolution of the design paradigms in the development of erosion-resistant silicone rubber insulation housing composites by shedding light on the common perceptions about the application of inorganic fillers. In particular, critical discussion of re-applying perceptions, developed under AC conditions for new applications, such as HVDC outdoor insulation is presented and lessons learned are summarized.

2. Tracking and Erosion of Silicone Rubber

The terms tracking and erosion refer to two different surface failure mechanisms of insulators and therefore should not be interchangeably used. Tracking usually refers to the formation of a carbonaceous path that could be fully or partially conductive on the surface of the insulation housing. Erosion of silicone rubber insulators may be defined as heat ablation of the insulation housing as a result of surface discharges in general or dry-band arcing in particular impinging the surface. It must be emphasized that silicone rubber has shown a tracking-resistant behavior during material screening tests [3,5]. Unlike many other outdoor insulating materials, silicone rubber possesses a backbone that does not intrinsically promote a significant amount of carbon residue when exposed to surface discharges in wet conditions [5]. Although it is reported to be susceptible to tracking in dry-conditions, silicone rubber without hydrated fillers have shown acceptable tracking performance in wet conditions [5]. In addition, silicone rubber does not leave a significant amount of carbonaceous residue during dry-band arcing [6]. Both tracking and erosion have been shown to be stimulated as a result of the dry-band arcing heat, but they are a result of two different thermal degradation mechanisms of silicone rubber. Erosion has been shown initiated with the depolyemrization of the silicone rubber, through secession of Si-O (siloxane) bond, yielding cyclic oligomers with small molecular weights and low flashpoints in the gas phase [7]. Tracking is promoted in silicone rubber as a result of the radical-based crosslinking [8], initiated with the CH_3-Si bond scission in silicone rubber [9]. Therefore, tracking as an aging mechanism must not be conflated with erosion for silicone rubber insulators. Yet, this understanding has not been emphasized in the literature despite decades of experience.

Understanding the aging mechanisms leading to erosion is an essential task to support the proper application of inorganic fillers in silicone rubber insulators. The two main thermal degradation mechanisms of silicone rubber can be identified using thermogravimetric analysis (TGA) and the derivative thermogravimetric analysis (DTGA). Figure 2 is an illustrative example shown for a TGA-DTGA conducted on RTV silicone rubber composite in Nitrogen atmosphere (test details in the caption). Two main weight loss stages are observed in Figure 2 for silicone rubber, indicating two distinct mechanisms governing erosion. The first weight loss stage begins at 350 °C and is due to depolymerization, i.e., due to scission, of the siloxane bonds. The second weight loss stage, beginning at 600 °C, is associated to another distinct peak in the DTGA, indicating occurrence of another degradation mechanism, i.e., radical-based crosslinking [8,9]. Depolymerization is mainly responsible for weight loss and thus erosion during the dry-band arcing [7]. Figure 3 illustrates a simplified diagram for the tracking and erosion mechanisms of silicone rubber.

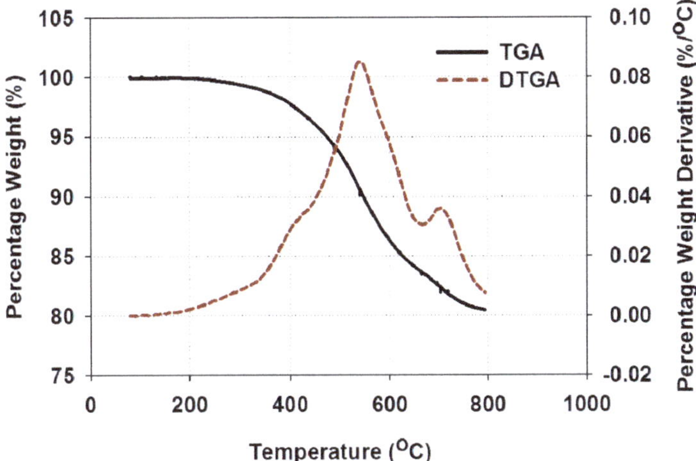

Figure 2. *TGA* of *RTV* silicone rubber filled with natural silica to 50 wt.%, performed in Nitrogen atmosphere under a temperature ramping protocol between 80 and 800 °C at a rate of 10 °C/min.

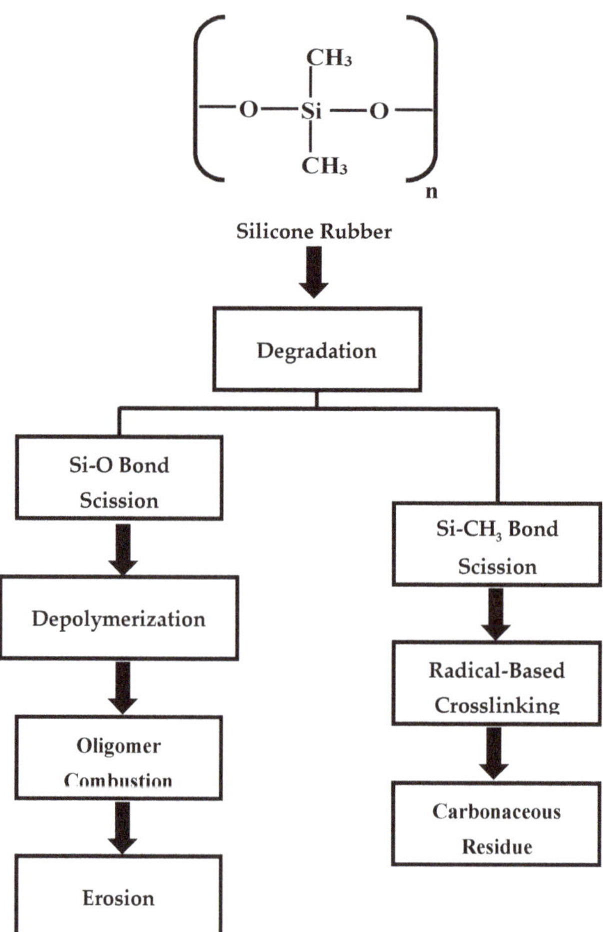

Figure 3. Thermal degradation mechanisms promoting carbonaceous residue and erosion of silicone rubber outdoor insulation.

The relative degree of crosslinking can be estimated for the silicone rubber composite, given the filler level added should be known, using the final residue level in the TGA. This estimation is useful in understanding the filler protective mechanisms against the degradation mechanisms of silicone rubber [8]. For example, the total amount of residue obtained after the TGA for the silicone rubber composite shown in Figure 2 is about 80 wt.%. Given the silicone rubber base is unfilled [10], this residue constitutes of the filler amount, i.e., 50 wt.% and the remaining 30 wt.% is additional residue formed due to radical-based crosslinking. Note that this estimation method assumes a negligible amount of other constituents, such as crosslinking agents or pigments. It should be noted that the water of hydration amount should be taken into consideration in the estimation if the filler is hydrated. Figure 4 shows the TGA-DTGA conducted in Nitrogen atmosphere for RTV silicone rubber filled with ATH to 50 wt.% (test details in caption). Similar to the silica-filled RTV silicone rubber, the ATH-filled RTV silicone rubber also showed a depolymerization stage beginning at 350 °C and a crosslinking beginning at 550 °C. An additional weight loss stage is seen with ATH before depolymerization, starting at 200 °C, which has been attributed to the dehydration of ATH [11]. Typically, ATH in silicone rubber releases about

35 wt.% of the total filler amount during TGA [11], so the amount of the water of hydration released from the composite is estimated to be about 17.5 wt.% and the remaining amount of alumina particles are 32.5 wt.%. The final TGA residue level is about 66 wt.%. With alumina particles estimated to be about 32.5 wt.%, the amount of the crosslinked silicone rubber is estimated to be 33.5 wt.%.

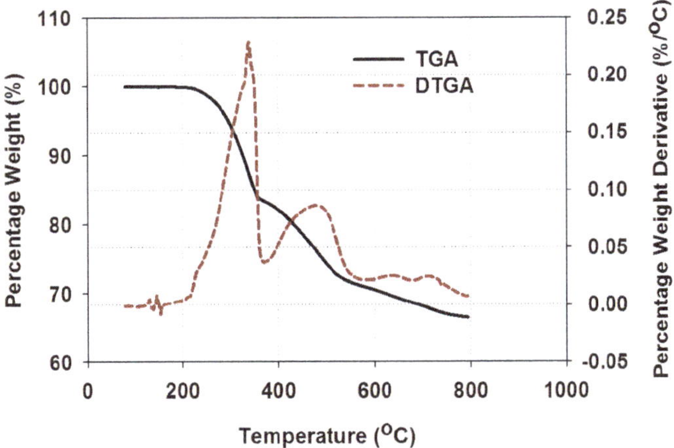

Figure 4. *TGA-DTGA* of *RTV* silicone rubber filled with ATH to 50 wt.%, performed in Nitrogen atmosphere under a temperature ramping protocol between 80 and 800 °C at a rate of 10 °C/min.

Often, fumed silica is added to the silicone base, in order to enhance the mechanical properties of the silicone rubber insulation housing. This fumed silica has a different purpose and is added in smaller amount than either ground silica or ATH, which are added to reduce the cost and enhance the erosion resistance. If the silicone base contains fumed silica, the corresponding base residue would include fumed silica particles and some crosslinked residue is formed as a result of polymer-filler interactions between the base and the fumed silica. As such, this base residue needs to be taken into account, by running first TGA for silicone base without the ground silica or ATH filler and the amount of the base residue should be noted. Accordingly, the percentage of the crosslinked residue due to ground silica or ATH in the composite can be estimated by subtracting from the percentage of the final TGA residue of the composite: (1) Percentage of the ground silica (or alumina level if ATH is added), (2) percentage of the water of hydration released if ATH is added and the percentage amount of the base residue obtained from the first TGA multiplied by the fraction of the actual base silicone rubber in composite. Details about this estimation method can be found in [12].

3. Fillers in Silicone Rubber Outdoor Insulation Composites

Mineral fillers were initially employed in polymeric insulating materials in order to reduce the organic content that promotes tracking. This reduction in organic content inhibits the tendency of the insulator to form a carbonaceous residue during thermal decomposition when the insulator was impinged by dry-band arcs [5]. However, different outcomes were reported under wet as compared to dry conditions [5]. In particular, mineral fillers were not shown effective in inhibiting tracking in wet conditions. It should be noted that wet laboratory conditions better simulate the outdoor insulation environment than dry conditions. The use of hydrated fillers such as alumina tri-hydrate (ATH) was later proposed. It was shown that the released water of hydration during surface discharges stimulates an internal oxidation between the organic backbone of the insulation and the hydroxyl groups released from ATH, yielding carbon oxides and hydrocarbons in the gas phase, instead of leaving a carbonaceous track on the surface [13,14]. In addition, the released water of hydration as vapor was reported to physically promote sputtering of any traces of carbonaceous tracks formed on the surface, thereby preventing the formation of a continuous conductive path that may lead to insulator failure [15].

Some concerns were initially raised about the erosion taking place in addition to tracking of insulating materials, but the suppression of tracking was clearly the main concern at the early times. The focus on tracking was probably because most of the outdoor insulation materials had organic backbones. Even material screening methods were primarily developed as tests for tracking with less emphasis on erosion [16]. Later on, the need to evaluate erosion was highlighted [17].

Today, ATH is the main filler added to silicone rubber and makes up about 60 wt.% of the formulation, in order to achieve acceptable performance in the outdoors. This performance is verified at the material development stage in the inclined plane tracking and erosion test (IPT) at 4.5 kV as per ASTM D2303 or IEC 60587, i.e., the critical AC_{rms} voltage level [7,18]. It is unclear why, specifically, 60 wt.% of ATH was decided upon, but this approach seems to be come from the use of ATH as a flame retardant in silicone rubber formulations for different applications [19].

The use in ATH at 60 wt.% has been successfully used for decades in various HVAC outdoor insulation applications. Therefore, there has been little motivation for the development of improved or cost-reduced designs of silicone rubber composites for HVAC insulation and only a few alternatives to ATH in silicone rubber have been proposed. However, hotspots causing erosion has been highlighted as an issue that needs to be suppressed. As such, the increase in the thermal conductivity of the overall composite has been used as an indicator for the efficacy of the fillers to suppress erosion. Silica been shown to offer a similar increase in thermal conductivity of silicone rubber to ATH, but at fraction of the overall cost of the composite [20]. The use of ATH was shown to promote a layered-residue on the surface during surface discharges, which led to the failure in the IPT at the critical voltage [21]. The layered residue consists of mullite formed when hotspots temperatures exceeding 1200 °C on the surface [6,21]. The UL94 burning test was also employed to test silicone rubber composites filled with alternative fillers to ATH. Both silica and melamine cyanurate were shown to enhance the ability of silicone rubber to extinguish arcs on the surface. In addition, three main design aspects were highlighted as areas that need to be taken into account: Filler level, good dispersion and bonding between host matrix and filler particles [21].

These existing paradigms need to be critically examined when designing formulas for applications where there is limited experience from the field such as silicone rubber insulators for HVDC lines. The nature of the dry-band arcing under DC has been shown in different tests to be more continuous and intense than under AC conditions, leading to deep erosion [22]. Therefore, the need to develop specialized silicone rubber composites for HVDC has become an important issue. The use of fillers as flame-retardants has been discussed for HVDC silicone rubber insulation [23], highlighting two main aspects; (1) the suppression of the heat source, i.e., surface discharge causing erosion and (2) the suppression of the thermal degradation leading to erosion of silicone rubber. The first aspect has been investigated extensively, and it is known that ATH impedes the induction hotspots in both the solid and the gas phase of silicone rubber [23,24]. However, the influential of ATH on the second more important aspect, the suppression of thermal degradation, has yet to be investigated.

4. Design Aspects

4.1. Volume Effect of Fillers

The term "volume effect" refers to the filler volume replacing the amount of the polymeric fuel prone to degradation during the dry-band arcing. This paradigm may be understood as analogous to the application of fillers as flame retardants in polymers. It has been understood that adding as much filler as possible would reduce the amount of the carbonaceous content that can lead to tracking of the polymer. However, tracking of polymeric insulating materials has not been shown to necessarily depend on the amount of the organic content [25]. In addition, applying fillers to inhibit tracking may not necessarily suppress erosion failure. For example, silica-filled silicone rubber impinged with the dry-band arcing did not track in the IPT, but failed due to erosion paths [10]. In particular, the combustion of silcione rubber oligomers produced from depolymerization may have promoted electrical breakdown in the gas phase [6]. The electrical breakdown in the gas phase has been detected using wavelet-based analysis of the leakage current measured in-situ during the IPT [22].

On the other hand, this failure pattern has not been observed for the RTV silicone rubber containing same amount of ATH [10]. It has been shown that a critical ATH level of 45 wt.% is needed in silicone rubber in order to promote an internal oxidation that could help suppress failure in the IPT [7]. At the same loading glevel, both silica and ATH have been shown to increase the thermal conductivty of silicone rubber to a similar extent. On the ther hand, an additional influence could be expected for the water of hydration in the gas phase. The use of ATH may inturupt the combustion cycle of silicone rubber by dilluting the gas phase of silicone rubber through internal oxidation [7,23]. The internal oxdiation effect has been detected in the DTA with a clear suppression in the exothermic hump obtained in TGA performed in air after the depolymerization of silicone rubber [7].

4.2. Residue Shield

Ground silica is not a hydrated filler and therefore cannot dilute the gas phase of silicone rubber during the dry-band arcing. Therefore, it cannot be accurately concluded that ATH would be have been more effective, and thus, better in preventing failure of RTV silicone rubber in the IPT than silica. However, when a larger particle size of silica was used at the same level of 50 wt.%, the RTV silicone rubber composite withstood the 4.5 kV AC voltage in the IPT [8]. Increasing the size of silica has been shown to lead to a more coherent residue shield against the dry-band arcing [8]. It has also been shown that the coherence of the residue shield against the dry-band arcing is an essential factor for the suppression of the DC dry-band arcing [8]. Also, the water arising from the hydration released from ATH filler in silicone rubber may leave porous residue during thermal degradation [11,19]. It has been suggested that using an ATH filler with a small particle size about 0.3 μm promotes hydrolysis that could disadvantageously cause erosion during DC dry-band arcing [11].

Another important example is the comparison of the water of hydration effect of ATH and magnesium hydroxide (MH). It has been widely understood that the enthalpy of dehydration of hydrated fillers such as ATH and Magnesium hydroxide fillers would be an important property, governing the flame retardancy of the hydrated filler. The MH has shown comparable enthalpy of dehydration to ATH [11]. The MH in silicone rubber dehydrates during the depolymerization of silicone rubber. Whereas, ATH dehydrates before the depolymerization [11]. Figure 5 illustrate an example of TGA-DTGA conducted for RTV silicone rubber filled with MH to 50 wt.% in Nitrogen atmosphere (test details in the caption), showing only one weight loss stage before radical-based crosslinking (at <600 °C), which indicates both dehydration of MH and depolymerization of silicone rubber take place during at the same weight loss stage. Based on such an understanding, similar or even better erosion suppression would be expected for MH, as compared to ATH in silicone rubber. This outcome has been reported under AC voltages [10]. However, silicone rubber filled with ATH has led to a significantly better erosion resistance than silicone rubber filled with same level of MH under DC voltages [11]. This finding could be a result of dehydration of ATH taking place prior to depolymerization of silicone rubber at 400 °C has allowed a more coherent protective shield of alumina to form against the dry-band arcing as compared to Magnesium hydroxide that leaves a more porous shield against the DC dry-band arcing after dehydration [11]

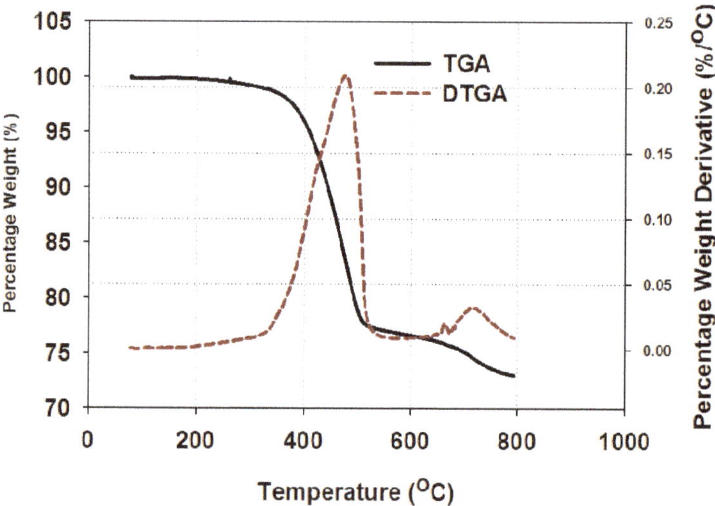

Figure 5. *TGA-DTGA* of *RTV* silicone rubber filled with MH to 50 wt.%, performed in Nitrogen atmosphere under a temperature ramping protocol between 80 and 800 °C at a rate of 10 °C/min.

Therefore, it must be emphasized that different filler actions may be differently utilized in order to achieve acceptable erosion performance. In addition, the claim that a filler type is superior to another type must not be made while only looking at a single filler action or design aspect. It is important to understand the different types of filler actions that suppress the erosion mechanisms of silicone rubber.

5. Screening Methods at the Material Development Stage

The IPT is the most widely accepted test for the evaluation of erosion resistance of silicone rubber composites for outdoor AC insulation applications when conducted as per ASTM D2303 or IEC 60587. The test has been modified to suit testing under DC voltages. However, the proper implementation of IPT test methods is still an essential task in the evlauation of silicone rubber composites. Otherwise, the outcomes reported may be misleading [15]. In particular, the constant voltage method is a more suitable method when testing for tracking-resistant materials such as silicone rubber. Whereas, the stepwise voltage method would be a more suitable method to test for tracking, progressive erosion or to determine the suitable test voltage for the costant voltage method [3,15].

Moreover, it is useful to develop other test methods for the erosion resistance that can provide more insight into the filler actions, particualy at the material development stage. Such considerations are very important for testing under DC voltages, given the relatively limited experience with testing under DC voltages. The application the IPT has been extended by applying a leakage current-based technique to compare the initation of eroding hotspots due to the DC dry-band arcing [22]. With the effect of the filler on the depolymerization and radical-based crosslinking of silicone rubber, the use of the TGA and DTA, thermal conducivity measurements and the scanning electron microspcopy (SEM) has been reported [8,11,12].

6. Conclusions

Perceptions and design paradigms about the development of erosion resistant silicone rubber insulators for outdoor insulation using inorganic fillers have been discussed and lessons learned are summarized. The concept of tracking resistance must be recognized as distinct from erosion resistance for silicone rubber. Two aspects must be taken into account when designing erosion resistant silicone rubber insulation housing composites, namely the volume effect of the fillers and the shield effect. In addition the methods of testing the design must be well considered. Supplementary material characterization tests and measurements such as TGA, SEM, leakage current measurements and thermal conductivity measurements are useful in understanding the protective filler actions in enhancing the erosion resistance of silicone rubber.

Author Contributions: Conceptualization, R.A.G., Y.H. and H.P; writing—original draft preparation, R.A.G.; writing—review and editing, R.A.G., H.P., and Y.H.; visualization, R.A.G.; supervision, H.P.; project administration, R.A.G.; All authors have read and agreed to the published version of the manuscript.

Funding: This research received no external funding.

Conflicts of Interest: The authors declare no conflict of interest.

References

1. Gorur, R.; Burnham, J.T.; Cherney, E.A. *Outdoor Insulators;* Ravi S. Gorur Inc.: Phoenix, AZ, USA, 1999.
2. Ghunem, R. A Study of the Erosion Mechanisms of Silicone Rubber Housing Composites. Ph.D. Thesis, ECE University Waterloo, Waterloo, ON, Canada, 2014.
3. Ghunem, R.; Jayaram, S.H.; Cherney, E.A. Erosion of silicone rubber composites in the AC and DC inclined plane tests. *IEEE Trans. Dielectr. Electr. Insul.* **2013**, *20*, 229–236. [CrossRef]
4. Liang, X.; Li, S.; Gao, Y.; Su, Z.; Zhou, J. Improving the outdoor insulation performance of Chinese EHV and UHV AC and DC overhead transmission lines. *IEEE Electr. Insul. Mag.* **2020**, *36*, 7–25. [CrossRef]
5. Sommerman, G.M.L. Electrical Tracking Resistance of Polymers. *Trans. AIEE Part III Power Appar. Syst.* **1960**, *79*, 969–974. [CrossRef]
6. Kumagai, S.; Wang, X.; Yoshimura, N. Solid residue formation of RTV silicone rubber due to dry-band arcing and thermal decomposition. *IEEE Trans. Dielectr. Electr. Insul.* **1998**, *5*, 281–289. [CrossRef]
7. Kumagai, S.; Yoshimura, N. Tracking and erosion of HTV silicone rubber and suppression mechanism of ATH. *IEEE Trans. Dielectr. Electr. Insul.* **2001**, *8*, 203–211. [CrossRef]
8. Koné, D.; Ghunem, R.A.; Cissé, L.; Hadjadj, Y.; El-Hag, A.H. Effect of residue formed during the AC and DC dry-band arcing on silicone rubber filled with natural silica. *IEEE Trans. Dielectr. Electr. Insul.* **2019**, *26*, 1620–1626. [CrossRef]
9. Delebecq, E.; Hamdani-Devarennes, S.; Raeke, J.; Lopez Cuesta, J.M.; Ganachaud, F. High Residue Contents Indebted by Platinum and Silica Synergistic Action During the Pyrolysis of Silicone Formulations. *ACS Appl. Mater. Interfaces* **2011**, *3*, 869–880. [CrossRef] [PubMed]
10. Ghunem, R.A.; El-Hag, A.H.; Hadjadj, Y.; Cherney, E.A. A study on the effect of inorganic fillers on the dry-band arcing erosion of silicone rubber composites. In *2017 IEEE Conference on Electrical Insulation and Dielectric Phenomenon (CEIDP)*; IEEE: Piscataway, NJ, USA, 2017; pp. 664–667.
11. Ghunem, R.; Koné, D.; Cissé, L.; Hadjadj, Y.; Parks, H.; Ambroise, D. Effect of hydrated fillers in silicone rubber composites during AC and DC dry-band arcing. *IEEE Trans. Dielectr. Electr. Insul.* **2020**, *27*, 249–256. [CrossRef]
12. El-Hag, A.; Simon, L.C.; Jayaram, S.H.; Cherney, E.A. Erosion resistance of nano-filled silicone rubber. *IEEE Trans. Dielectr. Electr. Insul.* **2006**, *13*, 122–128. [CrossRef]
13. Norman, R.; Kessel, A. Mechanism for nontracking organic insulations. *Electr. Eng.* **1958**, *77*, 699. [CrossRef]
14. Norman, R.; Kessel, A. Internal Oxidation Mechanism for Nontracking Organic Insulations. *Trans. AIEE Part III Power Appar. Syst.* **1958**, *77*, 632–636. [CrossRef]
15. Ghunem, R. Using the inclined-plane test to evaluate the resistance of outdoor polymer insulating materials to electrical tracking and erosion. *IEEE Electr. Insul. Mag.* **2015**, *31*, 16–22. [CrossRef]
16. Mathes, K.; Mcgowan, E. Surface electrical failure in the presence of contaminants: The inclined-plane liquid-contaminant test. *Trans. AIEE Part I Commun. Electron.* **1961**, *80*, 281–289. [CrossRef]
17. Mathes, K.; Mcgowan, E. Erosion—Inclined plane, liquid contaminant tracking test. In Proceedings of the Annual Report 1961 Conference on Electrical Insulation, Pocono Manor, PA, USA, 23–25 October 1961; pp. 121–124.
18. Krivda, A.; Krivda, A.; Schmidt, L.E.; Kornmann, X.; Ghorbani, H.; Ghorbandaeipour, A.; Eriksson, M.; Hillborg, H. Inclined-plane tracking and erosion test according to the IEC 60587 standard. *IEEE Electr. Insul. Mag.* **2009**, *25*, 14–22. [CrossRef]
19. Hamdani, S.; Longuet, C.; Perrin, D.; Lopez-Cuesta, J.M.; Ganachaud, F. Flame retardancy of silicone-based materials. *Polym. Degrad. Stab.* **2009**, *94*, 465–495. [CrossRef]
20. Meyer, L.; Cherney, E.A.; Jayaram, S.H. The role of inorganic fillers in silicone rubber for outdoor insulation alumina tri-hydrate or silica. *IEEE Electr. Insul. Mag.* **2004**, *20*, 13–21. [CrossRef]
21. Schmidt, L.; Kornmann, X.; Krivda, A.; Hillborg, H. Tracking and erosion resistance of high temperature vulcanizing ATH-free silicone rubber. *IEEE Trans. Dielectr. Electr. Insul.* **2010**, *17*, 533–540. [CrossRef]
22. Ghunem, R.; Jayaram, S.H.; Cherney, E.A. Investigation into the eroding dry-band arcing of filled silicone rubber under DC using wavelet-based multiresolution analysis. *IEEE Trans. Dielectr. Electr. Insul.* **2014**, *21*, 713–720. [CrossRef]

23. Ghunem, R.; Jayaram, S.H.; Cherney, E.A. The DC inclined-plane tracking and erosion test and the role of inorganic fillers in silicone rubber for DC insulation. *IEEE Electr. Insul. Mag.* **2015**, *31*, 12–21. [CrossRef]
24. Ghunem, R.; Jayaram, S.H.; Cherney, E.A. Suppression of silicone rubber erosion by alumina trihydrate and silica fillers from dry-band arcing under DC. *IEEE Trans. Dielectr. Electr. Insul.* **2015**, *22*, 14–20. [CrossRef]
25. Billings, M.; Smith, A.; Wilkins, R. Tracking in Polymeric Insulation. *IEEE Trans. Electr. Insul.* **1967**, *EI-2*, 131–137. [CrossRef]

Article

A Novel Framework to Study the Role of Ground and Fumed Silica Fillers in Suppressing DC Erosion of Silicone Rubber Outdoor Insulation

Alhaytham Y. Alqudsi [1,*], Refat A. Ghunem [1,2] and Eric David [1]

[1] École de Technologie Supérieure, Montréal, QC H3C 1K3, Canada; refat.ghunem@nrc-cnrc.gc.ca (R.A.G.); eric.david@etsmtl.ca (E.D.)
[2] Metrology Research Center, National Research Council Canada, Ottawa, ON K1A 0R6, Canada
* Correspondence: Alhaytham-yousef-j.alqudsi.1@ens.etsmtl.ca

Abstract: This paper investigates the effect of ground and fumed silica fillers on suppressing DC erosion in silicone rubber. Fumed silica and ground silica fillers are incorporated in silicone rubber at different loading levels and comparatively analyzed in this study. Outcomes of the +DC inclined plane tracking erosion test indicate a better erosion performance for the fumed silica filled composite despite having a lower thermal conductivity compared to the ground silica composite. Results of the simultaneous thermogravimetric and thermal differential analyses are correlated with inclined plane tracking erosion test outcomes suggesting that fumed silica suppresses depolymerization and promotes radical based crosslinking in silicone rubber. This finding is evident as higher residue is obtained with the fumed silica filler despite being filled at a significantly lower loading level compared to ground silica. The surface residue morphology obtained, and the roughness determined for the tested samples of the composites in the dry-arc resistance test indicate the formation of a coherent residue with the fumed silica filled composite. Such coherent residue could act as a barrier to shield the unaffected material underneath the damaged surface during dry-band arcing, thereby preventing progressive erosion. The outcomes of this study suggest a significant role for fumed silica promoting more interactions with silicone rubber to suppress DC erosion compared to ground silica fillers.

Keywords: HVDC outdoor insulators; silicone rubber; fumed silica; ground silica; dry-band arcing; erosion performance

Citation: Alqudsi, A.Y.; Ghunem, R.A.; David, E. A Novel Framework to Study the Role of Ground and Fumed Silica Fillers in Suppressing DC Erosion of Silicone Rubber Outdoor Insulation. *Energies* **2021**, *14*, 3449. https://doi.org/10.3390/en14123449

Academic Editor: Pawel Rozga

Received: 28 April 2021
Accepted: 7 June 2021
Published: 10 June 2021

1. Introduction

With the rising awareness of the impacts of fossil fuel-based electricity generation on climate change, solutions for integrating renewable energy sources into the existing electric grid infrastructure have been investigated. Utilizing a high voltage direct current (HVDC) transmission system would facilitate such integration by enabling an efficient transmission of electric power over long distances from remote renewable energy sources such as hydro, wind and solar farms to load centers [1,2]. Accordingly, HVDC outdoor insulators should be designed to ensure the reliability of the power transmission system. Silicone rubber's (SiR) characteristic hydrophobicity makes it highly desirable for use as a housing material in polymeric outdoor insulators. SiR, however, is susceptible to erosion caused by dry-band arcs sustained under heavily polluted conditions. Incorporating silica fillers in composite formulations of SiR was considered for enhancing the thermal conductivity and, in turn, the erosion performance of SiR.

Meyer et al., in [3], highlighted the correlation between the thermal conductivity and the erosion resistance of their silica filled SiR composites. It was concluded that the increase in silica filler loading from 10 to 50 wt% (percent by weight) caused a significant increase in the thermal conductivity, which resulted in lower eroded masses in the inclined plane-tracking and erosion test (IPT). El-Hag et al., in [4], illustrated that adding fumed silica by 10 wt% to SiR would result in a comparable erosion performance with 50 wt% micro silica filled SiR. This observation was attributed to the role of fumed silica in favorably bonding with the silicone rubber matrix. Nazir et al., in [5], reported an improvement in the IPT erosion performance of their hybrid SiR composites containing nano silica and aluminum nitride fillers with an increase in the nano silica loading level. An increase in the composite thermal conductivity was observed with an increased loading of nano silica fillers. Ramirez et al., in [6], explained that the high specific surface area of the fumed silica filler facilitates a better interaction with SiR as a result of the increased concentration of the silanol groups interacting with the siloxane chains of the polymer. Ansorge et al., in [7], highlighted the effect of an additional factor influencing the erosion performance of silica filled SiR that is related to the material curing temperature. It was reported in [7] that the erosion performance of micro silica filled room temperature vulcanized (RTV) SiR showed higher erosion depths under the IPT compared to high consistency silicone rubber. This outcome was attributed to the improved filler-polymer bonding at high temperature curing. Similar findings were reported in [3], highlighting the effect of high temperature curing on the erosion performance of silica filled SiR composites.

Several research studies conclude that the erosion performance of SiR worsens under DC voltage, particularly +DC, compared to AC [8,9]. The dry-band arcing exhibited by the insulators under DC voltage is of higher relative severity compared to that of AC in terms of arc discharge duration and leakage current magnitude [9]. Ghunem et al., in [10], illustrated the effect of increasing silica filler loading and, subsequently, the composite thermal conductivity on delaying the inception of stable eroding DC dry-band arcs in SiR. Comparable magnitudes of the third detail wavelet component of leakage currents were obtained between the SiR composites filled with silica and alumina trihydrate (ATH) at 30 wt%, despite the additional effect of the water of hydration of ATH in suppressing erosion [10]. These findings suggest the presence of erosion suppression mechanisms associated with silica's interaction with the silicone matrix. Kone et al., in [11], demonstrated the effect of the silica filler size and loading level on the integrity of the silica residue produced under the IPT. The coherency and porosity of such residue could shield the SiR material against progressive erosion under DC dry-band arcing [11]. In an earlier study [12], fumed silica was found to have a significant effect in suppressing DC erosion as compared to nano ATH and sub-micron boron nitride (BN) in SiR composites, despite the comparable thermal conductivities reported for all of the composites. Accordingly, the literature suggests that the role of silica fillers in suppressing the DC erosion in SiR is more than simply improving the composite thermal conductivity.

The literature indicates that the use of silica fillers in SiR composites, with their different particle sizes and loading levels, would improve the erosion performance of the composites as a result of an increase in the thermal conductivity of the composite. Merely using the +DC IPT as a means to rank the erosion performance of such composites without considering any additional analytical tools would overshadow the true role of the silica size in suppressing the DC erosion of SiR. Moreover, this would limit the role of the silica filler in merely improving the composite thermal conductivity. This paper introduces a framework to thoroughly investigate the role of fumed silica and ground silica fillers on suppressing the DC erosion of SiR. The study would enable using a number of analytical tools with outcomes that could be correlated with the IPT outcomes. This study, in turn, could ultimately support the developments in SiR outdoor insulators for their reliable use in the HVDC electric grid.

2. Materials and Methods

Fumed silica and ground silica, whose properties are shown in Table 1, are used as the fillers for this study. Based on the literature, fumed silica was selected due to its high specific surface area facilitating a favorable interaction with silicone at small weight fractions in the composite. Ground silica, on the other hand, can be filled at much higher loading levels to replace a significant portion of the SiR material and, subsequently, reduce the cost of the composite. A two-part RTV SiR is used in the study, where Part A is the main potting compound and Part B is the crosslinking agent. Part A and Part B are maintained at a weight ratio of 10:1, respectively. Weighed portions of the filler are added to Part A and mixed using a ROSS high shear mixer until all of the filler is added to the mixture. Part B is then added and mixed for one minute to be later poured into IPT specimen molds and degassed under a vacuum. The mixture is cured at room temperature for a 24 h time period, followed by thermal treatment at 85 °C for 3 h. An unfilled SiR was also prepared for selected tests in this study.

Table 1. Filler properties and prepared composites.

Filler Type	Supplier	Filler Code	Particle Size (μm)	Specific Surface Area (m^2/g)	Specific Gravity	Composite Formulation
Fumed silica	Sigma Aldrich	FS07	7×10^{-3}	390	2.3	SiR + 5 wt% FS07
Ground silica	US Silica	GS10	10.5 [1]	NA [2]	2.65	SiR + 30 wt% GS10

[1] Median particle size. [2] Not applicable for micro-sized fillers.

The +DC IPT is used in this study as part of the electrical analysis to assess the erosion performance of the prepared composites. The test setup is set as per the IEC 60587 standard [13] and modified for +DC testing as per the recommendations in [9]. The test voltage was set at +3.5 kV for a 6 h run time with a contaminant flow rate of 0.3 mL/min and a contaminant conductivity value of 2.5 mS/cm. A digital Mitutoyo 571-200 micrometer with an accuracy of 0.1 mm was used to measure the erosion depth of 10 specimen samples for each composite.

Leakage currents for the tested specimens during the IPT are acquired using a National Instruments NI USB-6356 data acquisition device at a sampling rate of 7 kHz. A sample window is then applied to capture the first 468 samples from each second. The root-mean-square value (RMS) of the leakage current is computed and stored as a single value for the 468 acquired samples at every second. Computing the RMS value for every second of the test run would suffice for representing the change in the leakage current values during the test and would provide a practical approach for acquiring the data with smaller storage requirements instead of saving the entire current waveform of 468 samples per second. Following the analytical approach presented in an earlier study [12], a statistical boxplot method is used to observe the distribution of the RMS leakage current values acquired during the IPT in 20 min time intervals. This statistical analysis allows one to observe the evolution of the dry-band arc from the intermittent state to the stable severe state, which is reflected in the changing distribution of the RMS leakage current values between consecutive time intervals in the boxplot. An increase in the dry-band arcing stability and severity is reflected in a reduction in the non-conducting periods of the leakage current. This corresponds to a transition from the initial intermittent state of the dry-band arc, which is characterized by a large number of nonconducting periods and frequent RMS leakage current values below 1 mA. Figure 1 shows the +DC IPT test setup used in the study [10].

The dry-arc resistance test is used in this study as a method to produce a controlled and quick heat ablation through the sample thickness rather than progressive erosion on the surface, as is the case with the IPT. The test will enable the fast production of tested composites whose surface residue could be analyzed for drawing preliminary conclusions regarding the role of the filler on defining the surface residue characteristics that could be observed using microscopy. The test setup is set as per the ASTM D 495 [14] standard, which utilizes tungsten electrodes to generate low current arcs under high voltages on the tested composites. The arc is generated in current steps that define the current magnitude and the duty cycles of the rms AC voltage applied, as per the schedule set in [14] and with each current step lasting for one minute. For this study, only the first 4 current steps of the 7 steps in [14] are used in the dry-arc resistance test; i.e. total test run of 4 min. Detailed description of the on and off duration of the current in these 4 current steps are described in details in [14]. The first 3 current steps of the test (denoted here as cycles 1 to 3) are analogous to an intermittent state of the dry-band arc. The degree of intermittency decreases as test goes from the first current step (cycle 1) to the third (cycle 3). The fourth current step (cycle 4), on the other hand, is analogous to the stable eroding state of the dry-band arc without intermittency. All the current steps used have a constant current magnitude of 10mA. Modifying the existing test setup to generate a DC dry-band arc could impair the proper functionality of the power electronics of the setup controlling the duty cycles. Figure 2 shows the dry-arc resistance test setup.

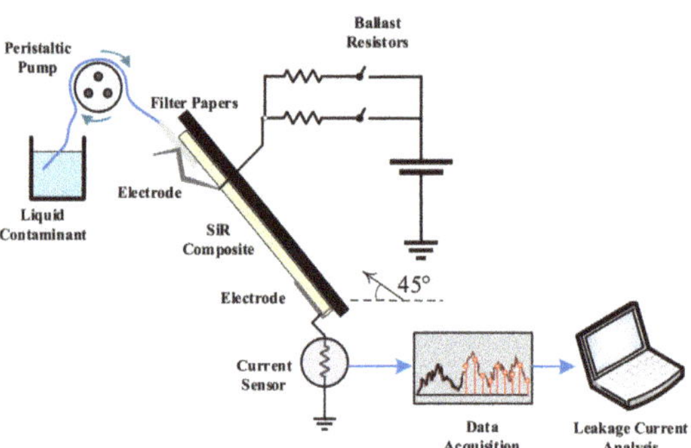

Figure 1. +DC inclined plane-tracking and erosion test (IPT) setup with the leakage current acquisition system.

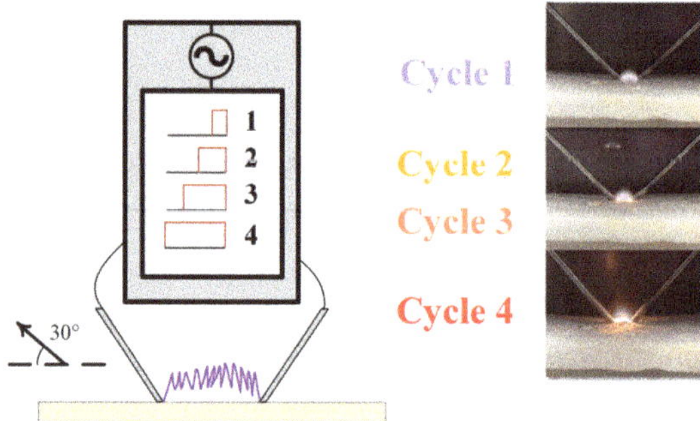

Figure 2. Dry-arc resistance test with images of the test during the 4 cycles of operation.

Simultaneous thermogravimetric–differential thermal analysis (TGA–DTA) was performed on the prepared composites under nitrogen (N_2) and air atmospheres to understand the thermal decomposition characteristics of the composites. The heating rate was set at 25 °C/min for a temperature span from 80 to 800 °C. Thermal conductivity measurements for the prepared composites were acquired using a thermal conductivity analyzer instrument as per the ASTM D7984 standards [15], which enables the acquisition of measurements in short test times without the use of a vacuum chamber. These measurements are necessary to understand the relationship between the composite thermal conductivity and the erosion performance of the composites.

Surface residue on post tested specimens of the +DC IPT, dry-arc resistance test and TGA are analyzed using two various methods. The surface morphology of the samples is observed using scanning electron microscopy (SEM) on 20 nm gold sputter coated surfaces and a laser confocal microscope. Surface roughness analysis on the samples is performed using a Keyence VR-5000 optical microscope. Quantitative representation of the surface roughness is performed by means of the average roughness parameter, R_a, whose computation details can be found in [16]. All of the aforementioned tools serve to observe the effect of the filler on the residue characteristics of the composite in terms of coherency and roughness.

3. Results and Discussion

3.1. Erosion Performance

Figure 3 shows the +DC IPT outcomes of the study. The results preliminarily indicate a better erosion performance of the FS07 filled composite compared to that of GS10. Figure 3a shows a higher average erosion depth obtained for the GS10 filled samples compared to the FS07 filled samples. Figure 3b further illustrates the inferior erosion performance of the GS10 filled composite through images of the post tested specimen, indicating larger eroded areas with the composite compared to those filled with FS07. It is important to note that the better erosion performance of the FS07 filled SiR does not necessarily conclude a superior erosion performance for the composite as compared to the GS10 filled SiR. The outcomes simply show that the FS07 filled SiR had a better or comparable erosion performance to the GS10 filled SiR, despite being filled at one sixth of the filler loading level of the GS10 filled SiR. This highlights a significant role for fumed silica in suppressing erosion and potentially facilitates its use as a co-filler with ground silica in practical formulations of SiR composites of high filler loadings that could be used in industry.

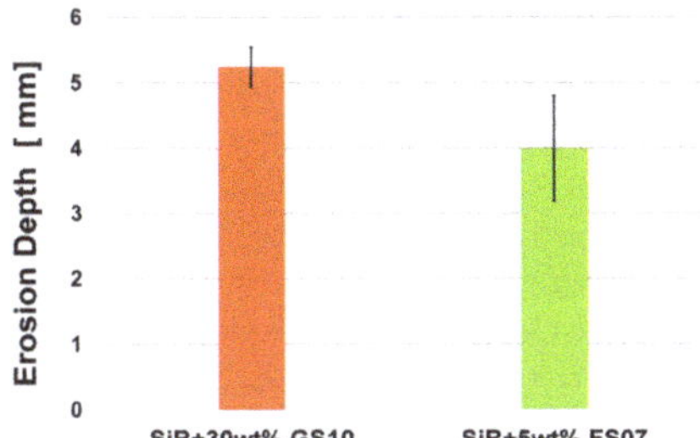

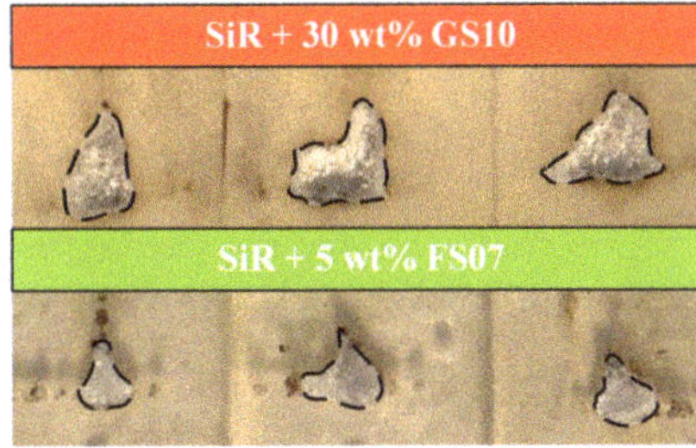

Figure 3. (**a**) +DC IPT erosion depth outcomes for the tested composites. (**b**) Images of the post tested +DC IPT composite specimens.

Table 2 shows the measured thermal conductivity of the prepared composites. The thermal conductivity values acquired were consistent with those found in [3,4]. The increase in the weight fraction of the GS10 and FS07 fillers in the composite leads to a significant increase in the composite thermal conductivity [3]. It is important to note that despite having twice the thermal conductivity of the FS07 filled SiR, the GS10 filled SiR showed inferior erosion performance, which suggests that that the thermal conductivity is not the main governing factor in suppressing the erosion of SiR under DC voltage. In an earlier study [12], it was found that the favorable interaction of fumed silica with the SiR matrix was more decisive in determining the erosion performance of SiR under the +DC IPT than the improvement of the composite thermal conductivity using BN fillers. The difference in DC and AC erosion in silica filled SiR was thoroughly investigated and discussed in studies such as [11] and is not the subject of this work. Rather, this study presents a practical framework for highlighting the prominent role of the fumed silica-silicone interactions on suppressing the DC erosion of SiR. According to Hshieh in [17], the silica-ash layer formed during the combustion of silicones produces a barrier effect that shields the silicone material against the influx of heat, preventing further combustion of the material. Accordingly, the presented framework aims to highlight the role of the silica filler size and its interaction with SiR in promoting the formation of a coherent residue with a barrier shielding effect that enhances the erosion performance of SiR, as shown in the outcomes of Figure 3.

Table 2. Thermal conductivity measurements (k) of the composites based on the 15 acquired measurements of each composite with a precision of ±1%.

Composite	Minimum k (W/m·K)	Maximum k (W/m·K)	Average k (W/m·K)
SiR + 5 wt% FS07	0.169	0.205	0.188
SiR + 30 wt% GS10	0.400	0.430	0.409

A statistical boxplot representation of the RMS leakage current of the composites during the +DC IPT is illustrated. Figure 4 shows the RMS leakage current waveform obtained for one of the GS10 filled SiR composites and its corresponding boxplot analysis. The boxplot shows the leakage current distribution for the first 3 h of the test, which was comprised of 12 20-min time intervals. Each bar shown in the boxplot represents the value distribution of the RMS leakage current values acquired during that time interval of the test. For example, the bar in the third time interval represents the RMS leakage current values acquired from minute 60 to minute 80 of the IPT. The bar width in the boxplot represents the distribution of the leakage current values during any given time interval. The top and bottom of the bar and the circled marker represent the 75th and 25th percentile and the median value of the RMS leakage current during that time interval, respectively. In Figure 4, the DC dry-band arc is shown to develop through two distinct stages in terms of arc stability and severity. The reduction in the bar width is an indication of the changing nature of the dry-band arc, from intermittent to stable with less nonconducting periods. As illustrated in [10,12], the initial stage of the dry-band arc is intermittent with inconsiderable erosion noted on the composite surface. The subsequent stage, however, is stable with reduced nonarcing periods leading to severe erosion of the composite. The inception of the stable dry-band arc stage was suggested to be dependent on the rate of formation of surface residue promoted with thermo-oxidation at temperatures just below 200 °C by Si-C bond scission, as illustrated in the Andrinov mechanism [18]. This residue would reduce the rate of evaporation of the liquid contaminant in the IPT, leading to the development of a stable dry-band arc [10].

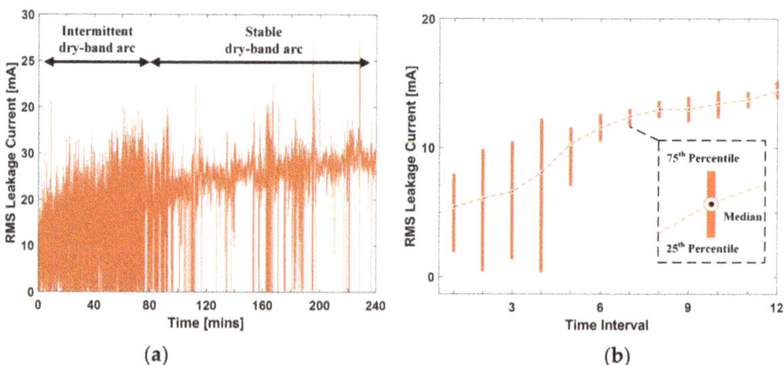

Figure 4. (**a**) Root-mean-square (RMS) leakage current for a GS10 filled silicone rubber (SiR) sample during the +DC IPT and (**b**) corresponding statistical boxplot representation for the first 12 20-min time intervals, first 240 min, of the test.

Figure 5 shows the statistical boxplots for a number of tested composites. The results indicate a faster inception of a stable eroding dry-band arc for the GS10 filled SiR compared to that of FS07 by 40–60 min (about 20% of the total testing duration). Similar outcomes were found in [12] for fumed silica filled composites against BN filled composites, which were attributed to the favorable interaction of fumed silica with SiR and a possible delay in the formation of the early surface residue by thermo-oxidation of the silicone volatiles.

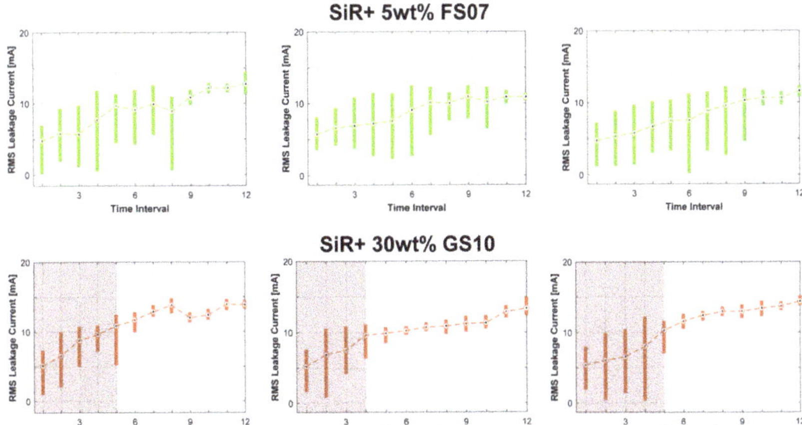

Figure 5. Statistical boxplot outcomes for selected samples of the +DC IPT tested composites during the first 12 20-min time intervals of the test.

3.2. Thermogravimetric–Differential Thermal Analysis

Figure 6 shows the TGA–DTA outcomes of the study. The TGA plot shown in Figure 6a is conducted for the prepared silica composites and the unfilled SiR. All of the composites and the unfilled SiR begin depolymerization at 400 °C, which, according to Camino et al. in [19], represents the scission of the Si-O bonds in SiR to produce cyclic oligomer volatiles. The rapid depolymerization of the unfilled SiR eventually leaves a low remnant residue of about 14.5 wt%. The TGA plot illustrated in [4] showed similar low remnant residues, while other studies [20,21] have shown the complete depolymerization of an unfilled SiR at the end of a TGA test. This variation could be attributed to a number of issues, such as the difference in suppliers and material preparation methods. For this study, 14.5 wt% was considered as the additional residue, possibly fused or crosslinked residue, produced for an unfilled SiR under TGA in an N_2 atmosphere. Beyond 400 °C, the FS07 composite decomposes at a slightly higher rate than that of GS10, which is still considered comparable despite having a much lower filler loading.

Figure 6b shows the DTGA plot for both of the composites under an N_2 atmosphere. The DTGA plot suggests the presence of multiple decomposition peaks, with the second one starting at temperatures higher than 500 °C. Camino et al., in [19], reported a radical-based crosslinking mechanism involving the homolytic scission of Si-CH_3 bonds in SiR, which competes with depolymerization during the second decomposition stage at elevated temperatures. At the onset of the 500 °C temperature, the decomposition rate of the FS07 filled composite becomes lower than that of the GS10 filled composite, as observed in the shaded region of the DTGA plot in Figure 6b. This observation may suggest the influence of FS07 on suppressing depolymerization and promoting radical based crosslinking, despite being filled in SiR at one sixth of the filler loading level of the GS10 composite. In other words, the DTGA peaks appearing more distinctively with FS07 as compared to the GS10 filled SiR suggest interactions between fumed silica and the SiR matrix to promote radical-based crosslinking to a greater extent as compared to the interactions between ground silica and SiR. The rate of SiR depolymerization was found to be subject to the mobility and flexibility of the SiR siloxane chains, as indicated by Delebecq et al. in [22] and Hamadani et al. in [23]. It was reported in [6] that the high silanol group concentration on the fumed silica's surface favorably interacts with the siloxane chains of SiR. Accordingly, this interaction could suppress the depolymerization and volatilization of SiR, as explained in [23]. To further support this conclusion, the DTA of both composites was conducted in an air atmosphere. The DTA plot shown in Figure 6c indicates the exothermic peaks obtained for both composites, which represent the combustion of the volatile SiR oligomers produced in depolymerization. Clearly, the suppressed depolymerization of the FS07 filled SiR leads to a lower exothermic peak compared to that of GS10.

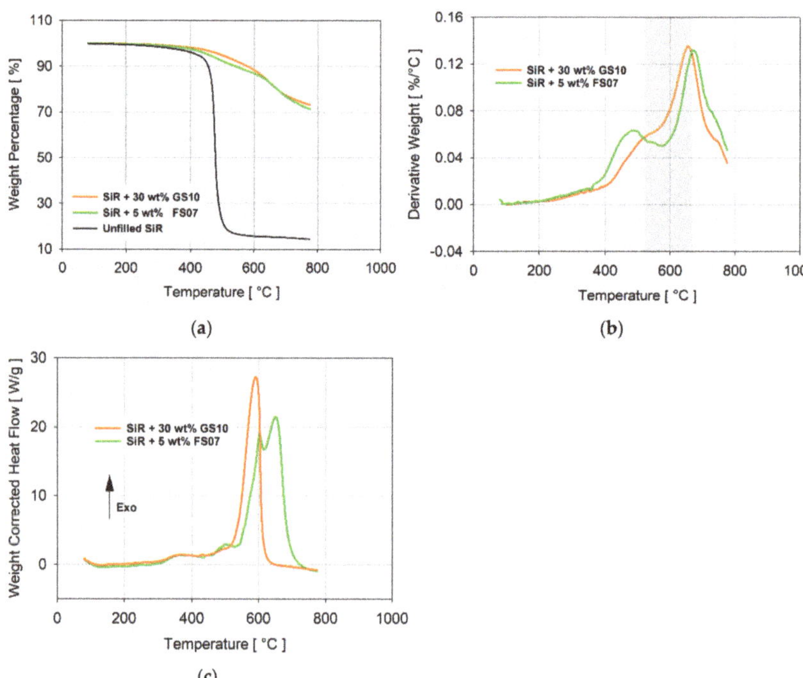

Figure 6. (**a**) Thermogravimetric analysis (TGA) for the prepared composites and the unfilled SiR in an N_2 atmosphere. (**b**) Corresponding differential thermogravimetric analysis (DTGA) plot for the silica filled composites. (**c**) DTA for the prepared composites in an air atmosphere.

The final wt% of TGA remnant residues (R_{TGA}) obtained for both composites was found to be comparable, with the GS10 filled SiR having a slightly higher remnant residue by a difference of only 1.8%, despite being loaded at six times the filler loading of the FS07 filled SiR. This further indicates the role of FS07 in suppressing SiR depolymerization and promoting radical-based crosslinking. To clarify this quantitively, based on the computation illustrated in [4,11], the final assumed residue R_{asm} if the composite polymer and filler components independently decompose without interaction is calculated as follows:

$$R_{asm} = (14.5\% \times W_{SiR}) + W_{filler} \quad (1)$$

where W_{SiR} and W_{filler} are the weight fractions of the SiR and the filler in the composite, respectively. As mentioned earlier, the 14.5% represents the undecomposed portion of SiR that was found in the TGA plot of Figure 6a. The additional residue R_{add} obtained, which accounts for the role of the filler interaction with the SiR polymer, is calculated as follows:

$$R_{add} = R_{TGA} - R_{asm} \quad (2)$$

Table 3 shows the calculated additional residue for each composite. Clearly, the additional residue obtained for the FS07 filled SiR, 52.6%, is much higher than that of GS10, 33%, by a factor of 1.6. This difference in the additional residue differentiates between the effect of each filler and its interaction with SiR on suppressing depolymerization and promoting crosslinking. The higher additional residue for the FS07 filled SiR could indicate a better suppression of depolymerization and a higher degree of crosslinking exhibited by the composite during TGA.

Table 3. Calculation of the additional residue R_{add} of the composites.

Composite	W_{SiR} (%)	W_{filler} (%)	R_{TGA} (%)	R_{asm} (%)	R_{add} (%)
SiR + 5 wt% FS07	95	5	71.4	18.8	52.6
SiR + 30 wt% GS10	70	30	73.2	40.2	33

Figure 7 shows an SEM image of the obtained TGA residues for both of the composites. The GS10 filled SiR TGA residue surface was observed to be of a coarser nature in comparison to the FS07 filled SiR TGA residue. This could in part be a result of the higher particle size of GS10, as shown in the image, or a result of the lower crosslinking and higher volatilization leading to a more porous residue compared to that of the FS07 filled SiR. The FS07 filled SiR, on the other hand, appears to promote a coherent residue with radical-based crosslinking. The weakness of the GS10 filled SiR residue can be certainly observed in terms of the propagating surface fractures shown in Figure 7, which are not present in the FS07 filled SiR residue, indicating coherency in the residue characteristics of the latter. The coherency of the residue was proposed to have a barrier shielding effect on the SiR material against the progressive erosion of silica filled SiR composites under DC voltage [11]. These observations shown in the TGA plots and residues could be correlated with the erosion performance of the composites illustrated earlier, indicating the role of the FS07 filler and its interaction with SiR in promoting a more coherent residue, which, in turn, suppresses the progressive erosion of SiR and enhances the erosion performance of SiR.

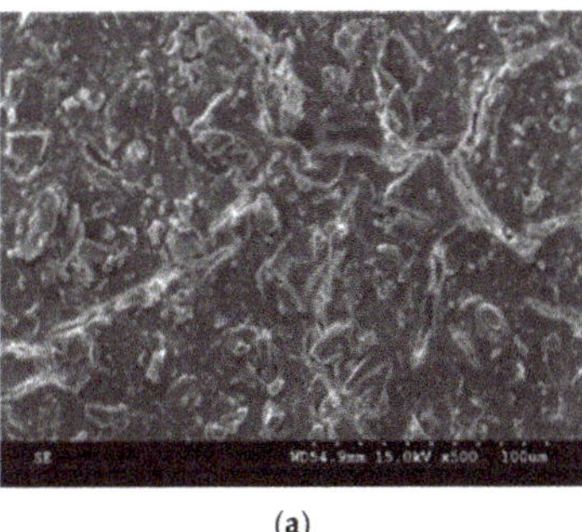

(a)

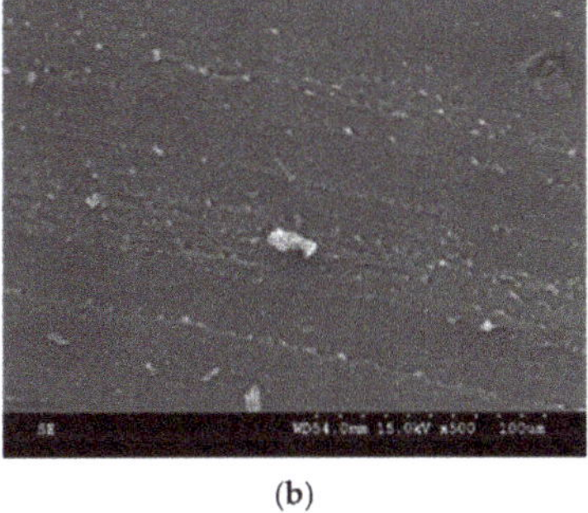

(b)

Figure 7. (**a**) Scanning electron microscopy (SEM) images for the TGA residue for the SiR + 30 wt% GS10 and (**b**) TGA residue for the SiR + 5 wt% FS07 under an N_2 atmosphere.

3.3. Residue Morphology Using the Dry-Arc Resistance Test

The formation of high additional residue with SiR composites could improve the erosion performance of SiR composites as a result of the formation of a coherent residue that shields the composite against an influx of heat from dry-band arcing. To better investigate this possible correlation, the dry-arc resistance test is utilized as a fast and controllable test for preparing eroded samples of the composites whose residues can be observed. Figure 8 shows the microscopic images obtained for the eroded pits of the post-tested silica composites of the dry-arc resistance test. Figure 8a,b clearly shows that the residue obtained from the FS07 filled SiR is more coherent, with less cracks and surface splitting compared to the GS10 filled SiR residue, which is seemingly rougher with porous surfaces. This observation is also confirmed by the SEM images shown in Figure 8c,d. Through SEM, Nazir et al., in [24], reported similar observations with corona-aged SiR composites showing less cracks with nano silica filled SiR compared to micro silica. The integrity of the residue could be attributed to the role of the radical-based crosslinking promoted by FS07 interacting with SiR, leading to a more stable residue with coherency characteristics similar to that shown in the TGA residue of Figure 7. These observations could explain the better erosion performance obtained for the FS07 filled SiR, as shown in Figure 3.

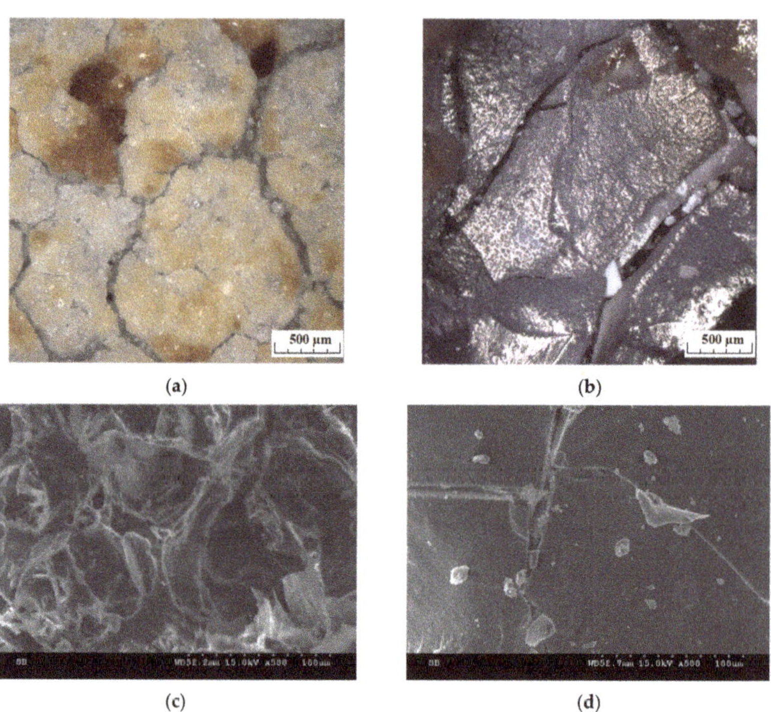

Figure 8. Microscopic images of dry-arc resistance post-tested samples at a magnification of 50 for (**a**) SiR + 10 wt% GS10 and (**b**) SiR + 5 wt% FS07 composites, and SEM imaging at a magnification of 15 k for (**c**) the SiR + 30 wt% GS10 and (**d**) SiR + 5 wt% FS07 composites.

To further validate the applicability of using the dry-arc test for observing the residue morphology of eroded composites, SEM was used to observe the eroded residue of the post IPT tested composites for comparison against those obtained under the dry-arc resistance test. Figure 9 shows the images obtained using SEM for the IPT tested composites. As can be seen in Figure 9, the surface morphology of both of the composites under the IPT are similar to their counterparts in the dry-arc resistance test in terms of roughness and coherency. This similarity further justifies the use of the dry-arc test as part of this mechanistic study. Though the experimental conditions involved in both of the tests are completely different, such as the absence of a wet contaminant in the dry-arc resistance test, the interest of this study is to observe the heat ablation effect of the arc on the composites and analyze the eroded residue characteristics accordingly. Creating this joule heating effect using either test does not necessarily dictate that similar testing methods or experimental conditions are to be followed. With this understanding, the dry-arc resistance test is advantageous in terms of the higher degree of controllability obtained with stimulating fast SiR erosion as a result of sustaining the arc at one fixed location above the sample during the test. The difference between the testing conditions of the tests has no effect on changing the residue characteristics of the composites, as can be seen in Figure 9, which further justifies the use of the dry-arc resistance test.

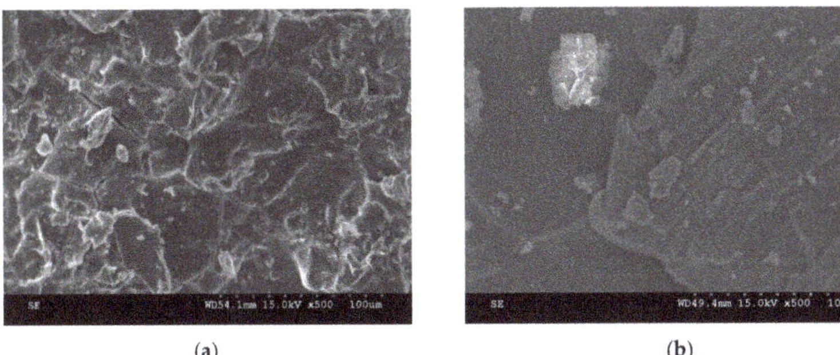

Figure 9. SEM imaging at a magnification of 15 k for (**a**) the SiR + 30 wt% GS10 and (**b**) SiR + 5 wt% FS07 composites tested using the +DC IPT.

Though the formation of coherent residue could enhance the erosion performance of the silica filled SiR composites, Delebecq et al., in [21], did explain that the carbon content of the residue could increase with increased SiR crosslinking. This increase in carbon content, however, would not significantly impact the composite during the IPT to cause a tracking failure, as explained in [11]. According to Kumagai et al., in [25], the analysis of the residue formed as a result of dry-band arcing in RTV SiR was found to contain 1 wt% of elemental carbon, which was considered insignificant for tracking. A simple demonstration of this would be achieved by testing the composites using the dry-arc resistance test for 10 s during the 1st current step of the test in cycle 1. Figure 10 shows the surface residue obtained for both composites after 10 s of the test in cycle 1 with equal electrode spacing. The FS07 filled composite shows a tendency to form a slightly higher burnt residue, possibly containing carbon, during the test compared to the GS10 filled composite. Still, however, the difference is insignificant, which is in line with [11,25].

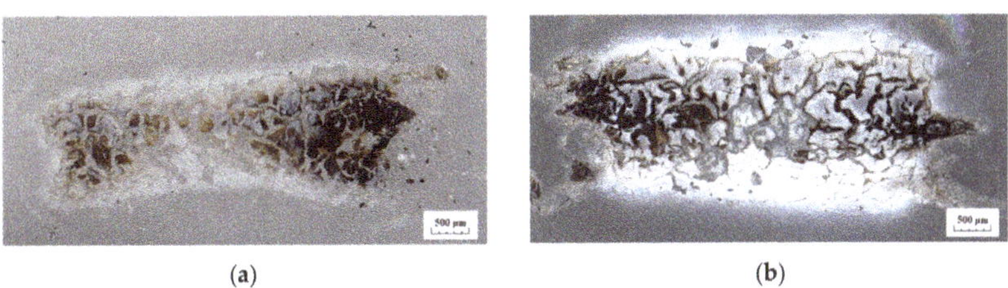

Figure 10. Surface residue of (**a**) the SiR + 30 wt% GS10 and (**b**) SiR + 5 wt%FS07 composites tested using the dry-arc resistance test for the first 10 s of cycle 1.

3.4. Surface Roughness of Eroded Composites

Analyzing the surface roughness of eroded silica filled SiR composites could further elaborate on the role of silica fillers in the DC erosion performance of SiR composites. The two elements that are associated with the roughness analysis are waviness and average roughness. Waviness describes the texture of the overall surface profile along a defined displacement axis, while average roughness describes the short-wavelength (high frequency) variations superimposed on the waviness along the same displacement axis [16]. Figure 11 shows the 3D topography and corresponding waviness profiles for the silica filled samples eroded using a dry-arc test. Clearly, the waviness of the FS07 filled SiR composite indicates a smoother surface with lower variations in the peak heights and valley depths within different segments of the profile. It is important to highlight that the erosion depth of the fumed silica filled SiR composite was found to be of higher value compared to that of the ground silica filled composite in the dry-arc resistance test, as shown in Figure 11c,d. This contrasts the +DC IPT outcomes of this study, which have been highlighted in Figure 3a. This implies that the use of the dry-arc resistance test in its standard testing conditions to assess the erosion performance of the composites in terms of erosion depth would not suffice. Rather, further modification of the testing conditions of the dry-arc resistance test are required to produce erosion depth ranking outcomes similar to those of the IPT for these specific composites. As mentioned earlier, the dry-arc resistance test is only used for the purpose of producing controlled arcing for post-testing residue analysis and not as means to rank the erosion performance of the composites in terms of the erosion depths. Using the dry-arc test in its standard form as a means to compare the erosion performance of the composites could work if both of the composites had equivalent filler loadings. An example of this would be in using the dry-arc resistance setup to test the erosion performance of a 30 wt% GS10 filled SiR against a 5 wt% FS07 + a 25 wt% GS10 filled SiR or a 10 wt% FS07 + a 20 wt% GS10 filled SiR composites.

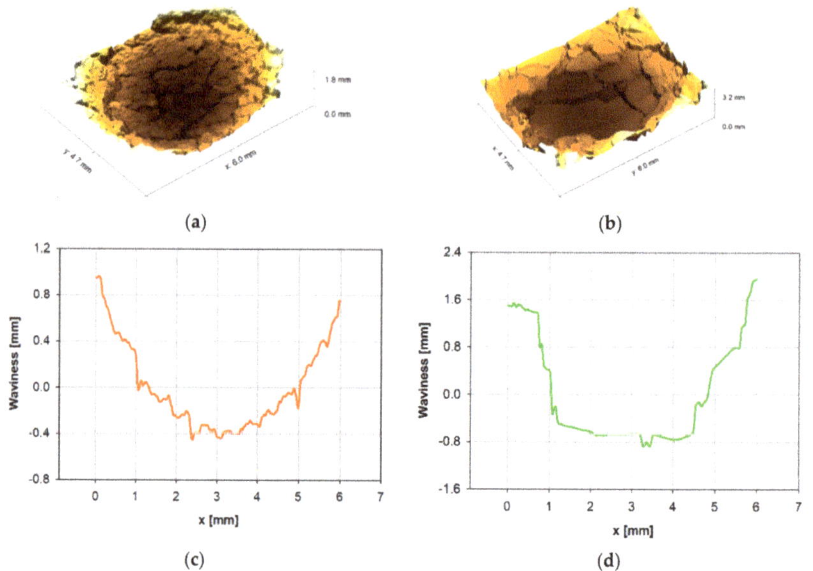

Figure 11. Three-dimensional topography of the (**a**) the SiR + 30 wt% GS10 and (**b**) SiR + 5 wt% FS07 composites tested using the dry-arc resistance test. Corresponding waviness profiles for (**c**) SiR + 30 wt% GS10 and (**d**) SiR + 5 wt% FS07 composites.

Figure 12 shows the statistical boxplot for the average roughness value distribution, R_a, of the sampled areas shown in the plot. A total of 10 profile lines were taken for each sampled area. Accordingly, each boxplot shows the median, 25th and 75th percentile and the minimum and maximum values for R_a within the 10 profiles. The wider variation and higher median value of R_a for the GS10 filled composite indicates a rough profile within small segments of the composites, which could indicate a higher degree of surface porosity compared to the FS07 filled SiR composite.

Based on Figures 11 and 12, the GS10 filled composite surface residue is of higher roughness with the overall surface waviness and within the smaller segments of the surface. Surface roughness could significantly impact the erosion performance of the composite due to a number of reasons. Under the salt-fog test, Deng et al., in [26], explained that rough SiR surfaces with large filler particle sizes tend to cause a higher impairment of the hydrophobicity retention properties of SiR. In their study, leakage currents developed at higher magnitudes with SiR composites having small filler particle sizes and less surface roughness. On the other hand, Kozako et al., in [27], illustrated that the addition of nano silica to their SiR composites did not have much of an influence on changing the hydrophobic properties of SiR, despite a slight increase in the surface roughness with respect to the unfilled SiR. Moreover, it is possible that the rough surface texture of the residue at early stages of the IPT could interrupt the smooth flow of liquid contaminant during the test, leading to more localized dry-band arcs being formed on the insulator surface as a result of liquid contaminant being trapped within small, eroded pits.

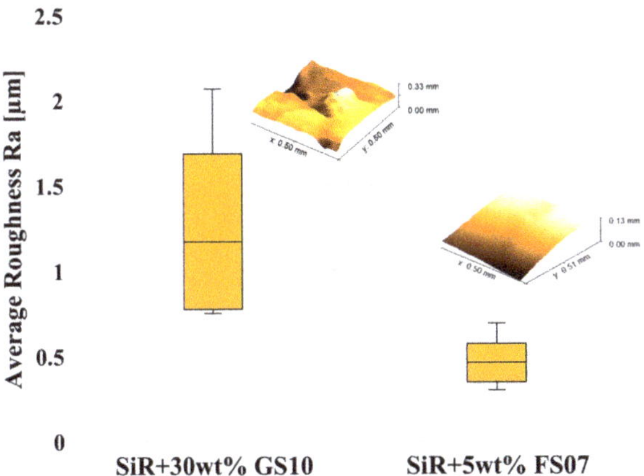

Figure 12. Statistical boxplot representation of 10 values of R_a for the sampled areas shown for each composite. The center bar represents the median value, the top and bottom box edges represent the 25th and 75th percentile values, respectively, while the top and bottom markers represent the minimum and maximum values of R_a, respectively.

4. Conclusions

The presented paper illustrated the role of fumed silica and ground silica fillers in suppressing the DC erosion of SiR through a novel framework. The erosion performance outcomes suggest that fumed silica and its interaction with SiR were effective in promoting the formation of a coherent shielding residue, which resulted in suppressing the DC erosion of SiR. This was found despite the higher composite thermal conductivity of the ground silica filled composite, which further supports the influence of the filler's interface interactions over enhancements in the thermal conductivity on suppressing DC erosion. Simultaneous TGA–DTA analysis shows the significant influence of fumed silica in suppressing depolymerization and promoting radical-based crosslinking at high temperatures in SiR as a result of its favorable interaction with the siloxane chains of the polymer tethering their flexibility and mobility during depolymerization. This results in the formation of a much higher additional residue with the fumed silica filled composite, despite being filled at one sixth of the loading level of the ground silica filled composite. The formation of a higher additional residue could result in a higher carbon content, which still would not be enough to promote a tracking failure during the IPT. The microscopy conducted on the eroded composites from the dry-arc resistance test shows coherency and low surface fracture in the residue of the fumed silica filled composite. This could also explain the better erosion performance of the composite as a result of the shielding effect of the coherent residue preventing progressive erosion. Moreover, the surface morphology outcomes of the dry-arc tested composites are consistent with those of the IPT, which validates the use of the dry-arc test as part of this framework. The surface roughness outcomes show a rougher surface waviness and higher values of R_a for the ground silica filled composite, which could further indicate the weakness and porosity of the residue, leading to an inferior performance under the IPT. The overall conclusion of the study suggests a significant role for the silica filler size in suppressing the erosion of SiR under DC voltage as a result of its influence on the eroded residue characteristics.

Author Contributions: Conceptualization, A.Y.A., R.A.G. and E.D.; methodology, A.Y.A., R.A.G. and E.D.; software, A.Y.A.; validation, R.A.G. and E.D.; formal analysis, A.Y.A., R.A.G. and E.D.; investigation, A.Y.A.; resources, R.A.G. and E.D.; data curation, A.Y.A.; writing—original draft preparation, A.Y.A.; writing—review and editing, A.Y.A., R.A.G. and E.D.; visualization, A.Y.A.; supervision, R.A.G. and E.D.; project administration, R.A.G. and E.D.; funding acquisition, R.A.G. All authors have read and agreed to the published version of the manuscript.

Funding: The authors of the paper would like to thank the Natural Sciences and Engineering Research of Canada (NSERC) and the National Research Council Canada (NRC) for their financial support.

Acknowledgments: The authors would like to thank Souheil-Antoine Tahan, Simon Laflamme, Mohammad Saadati and Joel Grignon for providing the permission, technical help and advice needed to complete the microscopic part of this study.

Conflicts of Interest: The authors declare no conflict of interest.

References

1. Sun, J.; Li, M.; Zhang, Z.; Xu, T.; He, J.; Wang, H.; Li, G.; Rensselaer Polytechnic Institute; State Grid Corporation of China; China Electric Power Research Institute. Renewable energy transmission by HVDC across the continent: System challenges and opportunities. *CSEE J. Power Energy Syst.* **2017**, *3*, 353–364. [CrossRef]
2. Fleeman, J.A.; Gutman, R.; Heyeck, M.; Bahrman, M.; Normark, B. EHV AC and HVDC transmission working together to integrate renewable power. In Proceedings of the 2009 CIGRE/IEEE PES Joint Symposium Integration of Wide-Scale Renewable Resources into the Power Delivery System, Calgary, AB, Canada, 29–31 July 2009; p. 1.
3. Meyer, L.; Jayaram, S.; Cherney, E.A. Thermal conductivity of filled silicone rubber and its relationship to erosion resistance in the inclined plane test. *IEEE Trans. Dielectr. Electr. Insul.* **2004**, *11*, 620–630. [CrossRef]
4. El-Hag, A.H.; Simon, L.C.; Jayaram, S.H.; Cherney, E.A. Erosion resistance of nano-filled silicone rubber. *IEEE Trans. Dielectr. Electr. Insul.* **2006**, *13*, 122–128. [CrossRef]
5. Nazir, M.T.; Phung, B.T.; Yu, S.; Zhang, Y.; Li, S. Tracking, erosion and thermal distribution of micro-AlN+ nano-SiO_2 co-filled silicone rubber for high-voltage outdoor insulation. *High Volt.* **2018**, *3*, 289–294. [CrossRef]
6. Ramirez, I.; Jarayam, S.; Cherney, E.A. Performance of silicone rubber nanocomposites in salt-fog, inclined plane, and laser ablation tests. *IEEE Trans. Dielectr. Electr. Insul.* **2010**, *17*, 206–213. [CrossRef]
7. Ansorge, S.; Schmuck, F.; Papailiou, K.O. Improved silicone rubbers for the use as housing material in composite insulators. *IEEE Trans. Dielectr. Electr. Insul.* **2012**, *19*, 209–217. [CrossRef]
8. Bruce, G.P.; Rowland, S.M.; Krivda, A. Performance of silicone rubber in DC inclined plane tracking tests. *IEEE Trans. Dielectr. Electr. Insul.* **2010**, *17*, 521–532. [CrossRef]
9. Cherney, E.A.; Gorur, R.S.; Krivda, A.; Jayaram, S.H.; Rowland, S.M.; Li, S.; Marzinotto, M.; Ghunem, R.A.; Ramirez, I. DC inclined-plane tracking and erosion test of insulating materials. *IEEE Trans. Dielectr. Electr. Insul.* **2015**, *22*, 211–217. [CrossRef]
10. Ghunem, R.A.; Jayaram, S.H.; Cherney, E.A. Suppression of silicone rubber erosion by alumina trihydrate and silica fillers from dry-band arcing under DC. *IEEE Trans. Dielectr. Electr. Insul.* **2015**, *22*, 14–20. [CrossRef]
11. Koné, D.; Ghunem, R.A.; Cissé, L.; Hadjadj, Y.; El-Hag, A.H. Effect of residue formed during the AC and DC dry-band arcing on silicone rubber filled with natural silica. *IEEE Trans. Dielectr. Electr. Insul.* **2019**, *26*, 1620–1626. [CrossRef]
12. Alqudsi, A.; Ghunem, R.; David, E. Analyzing the Role of Filler Interface on the Erosion Performance of Filled RTV Silicone Rubber under DC Dry-band Arcing. *IEEE Trans. Dielectr. Electr. Insul.*. (accepted; in press).
13. IEC 60587. *Electrical Insulating Materials Used Under Severe Ambient Conditions—Test Methods for Evaluating Resistance to Tracking and Erosion*; IEC: Geneva, Switzerland, 2007.
14. ASTM D 495-14. *Standard Test Method for High-Voltage, Low-Current, Dry Arc Resistance of Solid Electrical Insulation*; ASTM: West Conshohocken, PA, USA, 2018.
15. ASTM D7984-16. *Standard Test Method for Measurement of Thermal Effusivity of Fabrics Using a Modified Transient Plane Source (MTPS) Instrument*; ASTM: West Conshohocken, PA, USA, 2016.
16. ASME B46.1. *Surface Texture (Surface Roughness, Waviness, and Lay)*; ASME: New York, NY, USA, 2019.
17. Hshieh, F.-Y. Shielding effects of silica-ash layer on the combustion of silicones and their possible applications on the fire retardancy of organic polymers. *Fire Mater.* **1998**, *22*, 69–76. [CrossRef]
18. Timpe, D.C., Jr. Silicone Rubber Flame Resistance. In Proceedings of the Hose Manufacturer Conference, Cleveland, OH, USA, 11–12 June 2007.
19. Camino, G.; Lomakin, S.M.; Lageard, M. Thermal polydimethylsiloxane degradation. Part 2. The degradation mechanisms. *Polymer* **2002**, *43*, 2011–2015. [CrossRef]
20. Camino, G.; Lomakin, S.M.; Lazzari, M. Polydimethylsiloxane thermal degradation. Part 1. Kinetic aspects. *Polymer* **2001**, *42*, 2395–2402. [CrossRef]
21. Hermansson, A.; Hjertberg, T.; Sultan, B.-A. The flame retardant mechanism of polyolefins modified with chalk and silicone elastomer. *Fire Mater.* **2003**, *27*, 51–70. [CrossRef]

22. Delebecq, E.; Hamdani-Devarennes, S.; Raeke, J.; Cuesta, J.-M.L.; Ganachaud, F. High Residue Contents Indebted by Platinum and Silica Synergistic Action During the Pyrolysis of Silicone Formulations. *ACS Appl. Mater. Interfaces* **2011**, *3*, 869–880. [CrossRef] [PubMed]
23. Hamdani, S.; Longuet, C.; Perrin, D.; Lopez-Cuesta, J.; Ganachaud, F. Flame retardancy of silicone-based materials. *Polym. Degrad. Stabil.* **2009**, *94*, 465–495. [CrossRef]
24. Nazir, M.T.; Phung, B.T.; Hoffman, M. Performance of silicone rubber composites with SiO_2 micro/nano-filler under AC corona discharge. *IEEE Trans. Dielectr. Electr. Insul.* **2016**, *23*, 2804–2815. [CrossRef]
25. Kumagai, S.; Wang, X.; Yoshimura, N. Solid residue formation of RTV silicone rubber due to dry-band arcing and thermal decomposition. *IEEE Trans. Dielectr. Electr. Insul.* **1998**, *5*, 281–289. [CrossRef]
26. Deng, H.; Hackam, R.; Cherney, E.A. Role of the size of particles of alumina trihydrate filler on the life of RTV silicone rubber coating. *IEEE Trans. Power Deliv.* **1995**, *10*, 1012–1024. [CrossRef]
27. Kozako, M.; Higashikoji, M.; Tominaga, T.; Hikita, M.; Ueta, G.; Okabe, S.; Tanaka, T. Fabrication of silicone rubber nanocomposites and quantitative evaluation of dispersion state of nanofillers. *IEEE Trans. Dielectr. Electr. Insul.* **2012**, *19*, 1760–1767. [CrossRef]

 energies

Review

Application and Suitability of Polymeric Materials as Insulators in Electrical Equipment

SK Manirul Haque [1,*], Jorge Alfredo Ardila-Rey [2], Yunusa Umar [1], Abdullahi Abubakar Mas'ud [3,4,*], Firdaus Muhammad-Sukki [5], Binta Hadi Jume [6], Habibur Rahman [7] and Nurul Aini Bani [8]

1 Department of Chemical and Process Engineering Technology, Jubail Industrial College, Jubail 31961, Saudi Arabia; Umar_y@jic.edu.sa
2 Department of Electrical Engineering, Universidad Técnica Federico Santa María, Av. Vicuña Mackenna 3939, Santiago de Chile 8940000, Chile; jorge.ardila@usm.cl
3 Department of Electrical and Electronics Engineering Technology, Jubail Industrial College, Jubail 319261, Saudi Arabia
4 Prince Saud bin Thunayan Research Centre, Royal Commission for Jubail, Al Jubail 35718, Saudi Arabia
5 School of Engineering & the Built Environment, Merchiston Campus, Edinburgh Napier University, 10 Colinton Road, Edinburgh EH10 5DT, UK; f.muhammadsukki@napier.ac.uk
6 Chemistry Department, University of Hafr Batin, Al Jamiah, Hafar Al-Batin 39524, Saudi Arabia; albintu2@gmail.com
7 Department of General Studies, Jubail Industrial College, Jubail 31961, Saudi Arabia; habibur_r@jic.edu.sa
8 Razak Faculty of Technology and Informatics, Universiti Teknologi Malaysia, Jalan Sultan Yahya Petra, Kuala Lumpur 54100, Malaysia; nurulaini.kl@utm.my
* Correspondence: haque_m@jic.edu.sa (S.K.M.H.); abdullahi.masud@gmail.com (A.A.M.)

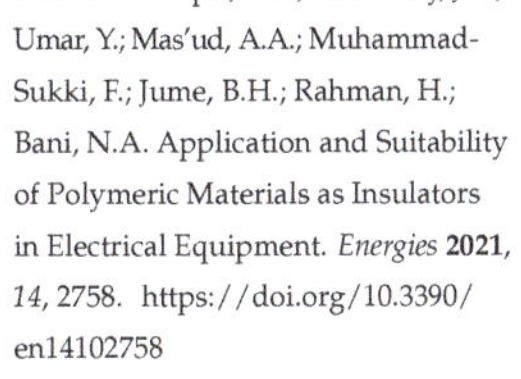

Citation: Haque, S.M.; Ardila-Rey, J.A.; Umar, Y.; Mas'ud, A.A.; Muhammad-Sukki, F.; Jume, B.H.; Rahman, H.; Bani, N.A. Application and Suitability of Polymeric Materials as Insulators in Electrical Equipment. *Energies* **2021**, *14*, 2758. https://doi.org/10.3390/en14102758

Academic Editors: Issouf Fofana, Stephan Brettschneider and Laurentiu Marius Dumitran

Received: 8 March 2021
Accepted: 7 May 2021
Published: 11 May 2021

Abstract: In this paper, the applications of thermoplastic, thermoset polymers, and a brief description of the functions of each subsystem are reviewed. The synthetic route and characteristics of polymeric materials are presented. The mechanical properties of polymers such as impact behavior, tensile test, bending test, and thermal properties like mold stress-relief distortion, generic thermal indices, relative thermal capability, and relative thermal index are mentioned. Furthermore, this paper covers the electrical behavior of polymers, mainly their dielectric strength. Different techniques for evaluating polymers' suitability applied for electrical insulation are covered, such as partial discharge and high current arc resistance to ignition. The polymeric materials and processes used for manufacturing cables at different voltage ranges are described, and their applications to high voltage DC systems (HVDC) are discussed. The evolution and limitations of polymeric materials for electrical application and their advantages and future trends are mentioned. However, to reduce the high cost of filler networks and improve their technical properties, new techniques need to be developed. To overcome limitations associated with the accuracy of the techniques used for quantifying residual stresses in polymers, new techniques such as indentation are used with higher force at the stressed location.

Keywords: polymeric material; thermoplastic; thermoset; elastomer; epoxy resin; electrical properties; mechanical properties; high-voltage applications; partial discharges

1. Introduction

Various types of composite and polymeric materials are suitable as insulators for electrical systems. The dielectric strength of polymers depends on the application and other external factors such as electrode size, shape, and nature of the outer surface, among others, and test conditions [1]. The studies suggested materials permittivity behaviors can play an essential factor and available polarizable aromatic rings like bromine and iodine that can enhance its dielectric constant. The applications and materials listed below (Table 1) with different values are used for capacitors insulation [2]. According to findings in [3], higher voltage stress, thinner insulation, and higher working temperature are the primary requirements for electrical equipment as insulating to possess higher dielectric strength and higher temperature ratings. Müller et al. [3] stated that composite materials are becoming an essential part of today's industry due to low weight advantages. An additional benefit of polymeric materials is the ease of processing, low relative permittivity, adhesive properties [4–6], corrosion resistance, high fatigue strength, outstanding performance, faster assembly, and favorable cost to traditional materials [7,8]. Consequently, composite materials offer the opportunity to provide the suitable product with the final application's required performance, thereby optimizing the price-performance ratio [9]. Additionally, nanocomposites are also characterized by distinctive advantages, including homogenous structure, no fiber rupture, optical transparency, improved or unchanged processability [7].

Table 1. Dielectric materials for capacitor insulations. Adapted from [2].

Materials	Dielectric Constant, $\acute{\varepsilon}$ (rt)
Isotactic polypropylene	2.28
Atactic polypropylene	2.16
Polyphenylene sulfide	3.5
Polyethylene terephthalate	3.3
Polycarbonate	3.0
Polyimide	2.78–3.48
Polyurea	5.18–6.19
Polyurethanes	3.84–6.09
Polyvinylidene fluoride	4.09–10.5

Polymeric materials and processes used for manufacturing cables at different voltage ranges are described, and their application to HVDC is discussed. The evolution and limitations of polymeric insulated power cables for various applications and their advantages and future trends are mentioned. These are extensively applicable with high-voltage (HV) systems as insulation material. Polymers can work at high temperatures under electrical strain due to their high breakdown toughness. This study contributes to the breakdown field as well as the electric strength of polymeric materials. Composite materials are conveniently used as insulators, with dielectric strength in the range of 10^6–10^9 V/m at room temperature [10]. The composite materials are convenient as insulators. It can be prepared by mingling two or more materials with distinct properties that simultaneously display unique properties. Natural fibers (cellulose, cotton, silk, wool) with sand, quartz, and natural resins extracted from plants with petroleum deposits (shellac, pitch, or linseed oil) were used to prepare the first composite material for the insulation system. The early stage of development for composite material looks on only at new materials with fewer design criteria. With time for action, the early development stage needs to focus on the mechanical, electrical properties and operating temperatures [11]. Inorganic materials are incorporated with micro and nanoscale to form a composite that procures awareness in power and voltage [7,12–17]. The alumina (Al_2O_3), silica (SiO_2), titanium oxide (TiO_2), glass and carbon fibers, carbon nanotube, graphene sheet combined as filler received particular attention concerning traditional composite material as it shows improved and enhanced electrical and mechanical properties than the single polymer [18–40]. Microcomposite polymer requires a large amount of filler than the nanocomposite and looks like the original polymer, and hence the nanocomposite behavior remains unchanged to density. Additionally, the length between two adjacent fillers is minimal compared to typical micro composites. The interaction between filler and polymer matrices depends on the fixed surface area, which is more with nanocomposite due to the ratio between nano and micro is 3:1 [7]. Nowadays, smart polymeric materials are a well-known composite for the new generation in biochemical sciences. It can enclose unique potential by using different compounds with distinct properties for biological and medicinal applications. These polymers are more sensitive to the environment, and as such, a small difference in the background will drastically produce changes in their physical properties.

The advantages of materials performance in the polymer composite field give rise signals to basic research and development unit to produce low-cost synthetic route with new composite materials that can pursue efficient energy. The HV insulation system's critical parameters are excellent adhesion to the substrate, grater glass transition temperature, lower ability for moisture absorption, and the polymer material's excessive thermal stability. The problem of polymer's high-dielectric properties can be solved by introducing voids into polymers, which help decrease their dielectric strength. The performance of polymers depends on size as well as the distribution of gaps in the structure.

Therefore, the present review article is adapting attention to developing the polymers' electrical features concerning the polymeric materials' thermal and electrical properties as insulators. Most commonly, epoxy resin and polyethylene are cross-linked with different sizes of particles used for insulating systems. These polymeric materials are applicable with HV systems power generators, transformers, cables, or in general, for electrical equipment. Additionally, this paper covers polymers' electrical behavior, mainly their dielectric strength, partial discharge in polymeric insulation, and high current arc resistance to ignition. Partial discharge has been recognized as a suitable technique to assess polymeric materials for insulation applications. Partial discharge refers to electrical discharges appearing in HV equipment insulation subjected to high voltage stress.

In Section 2, the types of polymers are described. Properties of polymers are depicted in Section 3. Section 4 describes specific techniques to determine the suitability of polymeric insulations for high voltage apparatus applications, while Section 5 demonstrates some real life examples of this type of insulation system for power equipment. Afterwards, Section 6 discusses the challenges and future directions. Finally, in Section 7, the conclusions from this review are presented.

2. Polymers

Polymers are high molecular weight and long-chain compounds formed by connecting many monomers (small units) through covalent bonding. Small companies' connections could lead, during polymerization, to different arrangements of polymer chains, namely linear, branched or cross-linked. The next subsections include the different types of polymers and how they are prepared and characterized.

2.1. Types of Polymers

2.1.1. Thermoplastic Polymers

These are mostly linear or branched polymers that soften and flow when heated, rapidly shape into complex products while in a melted state, and then hardens (solidify) when cooled. The hardening and softening process as a function of the material temperature is fully reversible [41]. Thus, most thermoplastics can be remolded many times without chemical structure effects, although chemical degradation may limit the number of cycles. The apparent advantage of thermoplastic polymers is that waste thermoplastics can be recycled, and a piece that is broken or rejected after molding can be re-grounded and remolded. The plastics used for drink bottles are typical thermoplastics that are widely recycled [42]. The largest scale volume thermoplastics include polyethylene, polypropylene, polystyrene, poly (vinyl chloride). Other thermoplastics include polyethylene terephthalate, polycarbonate, and polyamide (nylon).

Electricity is essential to our standard of living because it powers almost every aspect of our lives, but electricity is potentially lethal. On the other hand, plastics do not conduct electricity and are therefore used in various applications where their insulating properties are needed. Plastics are especially suited to housings for goods such as hairdryers, electric razors, and food mixers as they protect the user from the risk of electric shock. The thermoplastic polymers used in electric machines as insulators [43] are provided in the following Table 2.

Table 2. Application of thermoplastic with electrical system.

Material	Application
Polyethylene	Cable and wire insulation
Polypropylene	Kettles
Polyvinyl chloride	Cable and wire insulation, cable trucking
Polystyrene	Refrigerator trays/linings, TV cabinets
Polycarbonate	Telephones
Acrylonitrile butadiene styrene	Telephone handsets, keyboards, monitors, computer cases
Polyamide	Food processor bearings and adaptors
Ethylene-vinyl acetate	Freezer door strips, lean vacuum hoses, handle-grips
Polyesters	Business machine parts, coffee machines, toasters
Polyphenylene oxide	Coffee machines, TV housings

2.1.2. Thermoset Polymers

These polymers possess a cross-linked network structure formed exclusively by a covalent bond. Based on their cross-link, thermosets are stiff and brittle materials but are stable at elevated temperatures and resistant to solvents and other chemicals [44]. Thermosets do not melt upon heating because the cross-link prevents the chains from sliding past each other. When heated, thermoset material softens under certain temperatures, not melting, but further heating will cause decomposition by breaking down the chain's covalent bonds. Unlike thermoplastics, thermosets are shaped by placing them into a mold. The chemical reaction is initiated to cause cross-links that cause the material to harden and take a permanent shape. The process of crosslinking is called curing. Thus, thermoset materials become healed or set with thermal energy. The nature of the curing of a thermoset material is similar to baking a cake. The ingredients, which include polymer or monomer (that is capable of forming cross-links), colorants, curing agents, fillers, and other additives, are mixed and placed into the mold of the desired shape. The mixture is heated to crosslink and then cooled to facilitate removal from the mold. Thermoset is widely used to insulate electric wiring, while thermosets (which are thermally stable) are used for switches, circuit breakers, light fittings, and handles. Table 3 shows applications of common thermosets in electrical applications [45].

Table 3. Common thermoset polymer for electrical application.

Material	Application
Alkyd resins	Circuit breakers, switchgear
Amino resins	Lighting fixtures
Epoxy resins	Electrical components
Phenol formaldehyde	Fuse boxes, knobs, switches, handles
Urea-formaldehyde	Fuse boxes, knobs, switches

2.1.3. Elastomers

Elastomers are polymers that can be stretched usually too many times their original length but quickly return to the original shape without suffering permanent deformation when the stress is removed. Elastomer comes from the word "elastic" (describing the ability of a material to return to its original shape when the load is removed) and "mer" (from polymer, in which "poly" means many and "mer" means part). Upon elongation, the polymer chains assume a more ordered arrangement that corresponds to lower entropy and results in warming of the elastomer [9]. When the stress is removed, the elastomer contracts and cools at the same time. Figure 1 illustrates the concept of elastomers.

Thermoplastic elastomer (TPE) has gained recognition as the third generation of polymeric materials for high voltage (HV) insulators. The TPE electrifying aspects were observed to achieve few particular HV insulations, at least for the distribution class applications, especially in light contamination environments. Nowadays, TPEs, mainly silicone rubber (PDMS), ethylene-vinyl acetate copolymer (EVA), ethylene propylene diene rubber

(EPDM), styrene-butadiene rubber (SBR), chloroprene rubber, have become the material of choice as an insulator due to their performance and less expense [46–48].

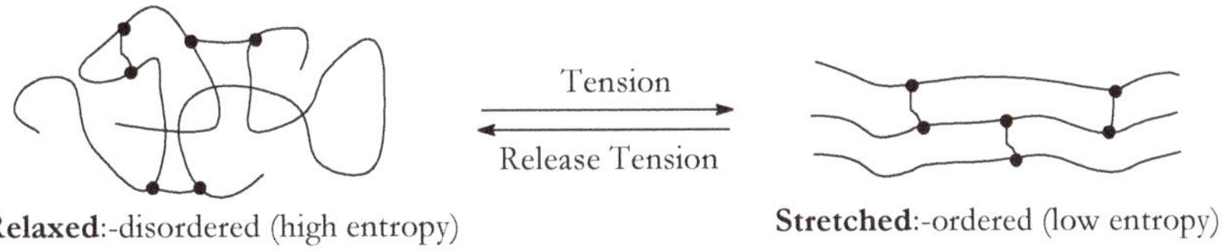

Tension
Release Tension
Relaxed:-disordered (high entropy)
Stretched:-ordered (low entropy)

Figure 1. Model of elastomer with a low degree of cross-linking under stress.

2.2. Preparation and Characterization of Polymeric Materials

Several techniques are used in producing the nanocomposites to have optimal dispersion of fiber in the matrix [49]. A well-known and established method of processing polymer nanocomposites is the sol-gel process, in-situ polymerization, solution mixing process, melt mixing process, and in-situ intercalative polymerization [3,6,13,50–52]. In addition, Thomas et al. [50], Müller et al. [3], Ilona et al. [6], and Shaoyun et al. [52] have extensively discussed the process of preparing the nanocomposite polymer.

2.2.1. Poly(9,10-phenanthrenequinone) for HV Electrode

Polyanthraquinone (P.A.Q.) can be produced directly with halogenated quinone used for low voltage system (Figure 2). Still, high voltage quinone- based polymers such as poly(9,10-phenanthrenequinone) (P.F.Q.) required indirect polymerization consisting of five steps of synthesis (Figure 3). Initially, a precursor 2,7-dibromo-9,10-phenanthrenequinone, as a monomer, was prepared by using phenanthraquinone and N-bromosuccinimide (N.B.S.) in the presence of concentrated sulfuric acid (H_2SO_4) followed by bromination [53]. Then reduced by tin (Sn) in acidic medium (Hydrochloric acid, HCl, and glacial acetic acid, CH_3COOH) and a monomer, 2,7-dibromo-9,10-dihydroxyphenanthrene was obtained [54]. Later, acetylation of this hydroxy compound monomer using acetic anhydride, pyridine, ethyl acetate produces the monomer (9,10-diacetoxy-2,7-dibromophenanthrene) [55]. Finally, a polymer poly(9,10-diacetoxyphenanthrene) was obtained using 1,5-cyclooctadiene, nickel complex Ni(C.O.D.)$_2$ and 2,2′-bipyridyl dissolved with D.M.F. after polymerized with bromophenanthrene, [56]. This polymer was treated with lithium aluminum hydride ($LiAlH_4$), anhydrous tetrahydrofuran (T.H.F.) in an acidic medium (HCl) resulted in less active poly(9,10-phenanthrenequinone), called PFQ_L. The final step, oxidization of less active PFQ_L to P.F.Q. (poly(9,10-phenanthrenequinone) utilizing 2,3-dichloro-5,6-dicyanobenzoquinone (D.D.Q.), as an oxidizing agent. The reduced graphene oxide (rGO) was prepared by incorporating graphene oxide with potassium permanganate and H_2SO_4, H_3PO_4, HCl, H_2O_2, and hydrazine hydrate [57]. The composite (rGO-PFQ) was produced by adding rGO during the polymerization reaction with 9,10-diacetoxy-2,7-dibromophenanthrene in the presence of Ni(C.O.D.)$_2$ and 2,2′-bipyridyl [58].

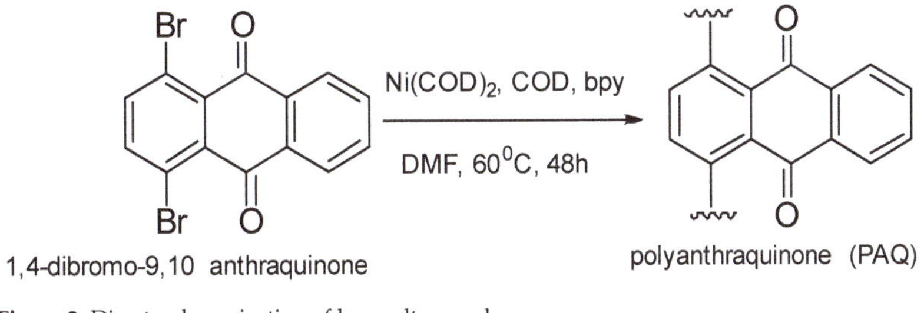

Br O
Br O
Ni(COD)$_2$, COD, bpy
DMF, 60^{0}C, 48h
O
O
1,4-dibromo-9,10 anthraquinone
polyanthraquinone (PAQ)

Figure 2. Direct polymerization of low voltage polymer.

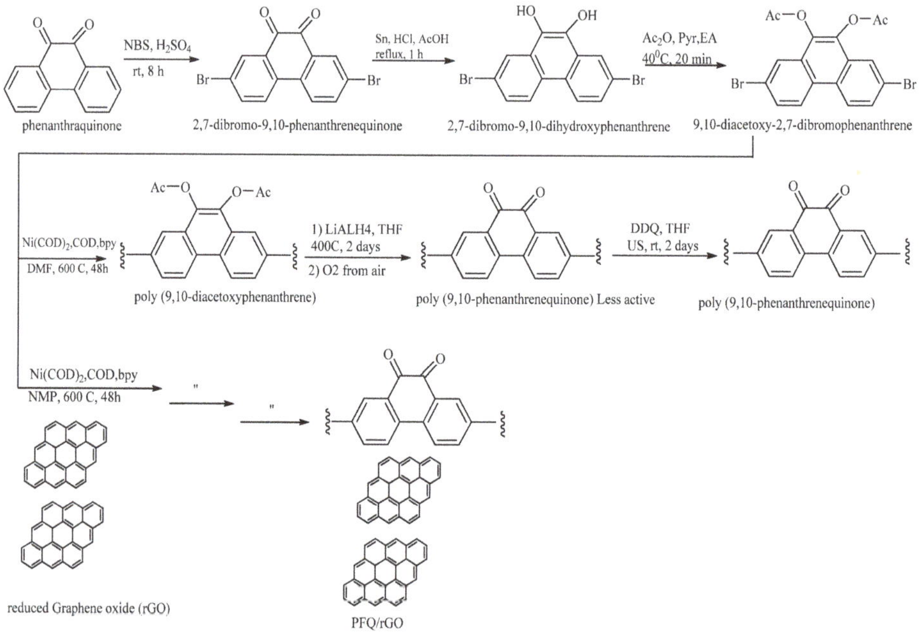

Figure 3. Indirect polymerization of HV polymer.

Low voltage quinone-based polymers like P.A.Q. can be directly polymerized with Ni(C.O.D.)$_2$. Whereas indirect multistep polymerization prevents the oxidation of Ni(C.O.D.)$_2$ while protecting quinone groups with acetyl groups helped to synthesize increased operating voltage polymers such as P.F.Q. Additionally, P.F.Q./rGO composite has a better porosity, enhancing electrical performance and rate capabilities compare with P.A.Q. polymers. Indirect polymerization of high redox potential ortho quinones also helps to synthesize higher operating voltage polymers such as P.F.Q. preferred over direct polymerization of para-quinones producing low voltage polymers.

2.2.2. Thermoplastic High Performance for Cable Insulation System

The polyethylene is cross-linked with different fillers' sizes to change the polymer's physical properties and make it capable of working with HV systems (cable system).

Polyethylene Materials

High-density polyethylene (HDPE) and low-density polyethylene (LDPE) are commodity polymers that are widely used for varieties of applications that include electrical insulations, packaging, household items and automotive parts. Some of the interesting properties of these polymers include chemical resistance, rigidity, stiffness and thermal stability. For electrical applications, these polymers are commonly used as insulators due to their exceptional electrical, mechanical, chemical and thermal properties. Polymer-based electric insulators exhibit high dielectric strength, high resistivity, low dielectric loss, and adequate mechanical properties. HDPE (density of 0.947 g/cm^3 and melt flow index of 42 g/10 min) and LDPE (density of 0.92 g/cm^3 and melt flow index of 25 g/10 min) are reported to have average dielectric strength of 70 and 79 kV/mm respectively. Though, LDPE exhibit relatively high electric strength, but its weak mechanical properties, thermal resistance and melt flow viscosity restricted its application in certain areas. However, these properties can be improved by mixing the LDPE with other polymers (polymer blend) or other materials (polymer composites) to produce material with optimum desirable properties [59].

The Dow material acting as the base resin was a non–commercial low-density polyethylene (LDPE) blended with high-density polyethylene (HDPE, Dow 40055E) to prepare new polymeric material that has the capacity to work with high temperature that can be applied as an insulation product with different types of cable. Small-scale material was synthesized by utilizing HDPE and LDPE (20:80) with the help of an extruder (Haake PTW 16/40D) [60]. However, for large scale, cross-linked low-density polyethylene (XLPE) and LDPE were produced as a reference for mini cables by incorporating the LDPE as a base material Berstorff ZE40UT extruder for blending [61]. The samples were prepared for mechanical (1.7 mm thickness) and breakdown (thickness 85 μm) testing using compression molding. The behavior of materials with temperature was studied with Mettler Toledo FP82 hot stage. The sample's thickness required for the mini cable material will be 4 mm and blended on the Troester extruder.

The crystallization temperature (Tc) isothermal for the LDPE/HDPE blended material in the range of 113–119 °C indicating the formation of lamellar crystals due to the presence of HDPE [61]. It also suggests that polymeric material's cooling temperature must be within the range of 0.5–10 °C/min, enhancing the breakdown strength and regenerating electrical effects [60]. When blended materials are cooled slowly, it would be challenging to retain mechanical properties that require high operating temperatures. Generally, stiff materials were formed than XLPE, which help to construct cable due to their operating condition, usually 30 °C higher than XLPE. The XLPE cable and mini cable breakdown data were studied and suggested maximum voltage below 400 kV; conversely, none of the extruded thermoplastic systems failed and shown both systems were extremely good electric materials for cable. The PE blended materials can be used for mini cables and XLPE based materials as reference for mini cables.

Polypropylene Materials

Polypropylene (PP) is a thermoplastic and commodity polymer used in a wide variety of applications. Its properties are similar to polyethylene, but it is slightly harder, stronger and more heat resistant. PP is produced via chain-growth polymerization from high purity propylene monomer that is obtained from the cracking of the petroleum hydrocarbons. Upon polymerization, PP can form three basic chain structures depending on the relative orientation of the methyl groups, namely isotactic PP (methyl groups are positioned at the same side with respect to the backbone of the polymer chain), syndiotactic (alternating methyl group arrangement) and atactic (irregular methyl group arrangement). The stereoregularity of these polymers is determined by the catalyst used to prepare it. In 1950s, Natta showed that the Ziegler organometallic type catalyst could be used to produce stereoregular PP with high crystallinity. Isotatic PP was synthesized by using heterogeneous Ziegler-Natta catalyst of a violet crystalline modified titanium (III) chloride with a co-catalyst or activator, usually organometallic compounds such as diethylaluminium chloride. However, this polymerization reaction simultaneously produces syndiotactic PP and atactic PP as minor products because these Ziegler-Natta based catalysts are multi-sites. Syndiotactic PP can be produced selectively using different catalysts such as homogenous Ziegler-Natta based catalyst and metallocene based catalyst. These different catalysts produce different microstructures of syndiotactic PP with different crystallization behavior and properties through different polymerization mechanisms. Atactic polypropylene can also be produced selectively using metallocene catalysts, atactic polypropylene produced this way has a considerably higher molecular weight [62,63].

For electrical applications, PP is commonly used as insulators due to their excellent electrical, mechanical, chemical and thermal properties. For high-voltage insulation, PP with average molecular of 250,000 g/mol, a density of 0.9 g/cm^3 and melt flow index of 12 g/10 min has been reported to have average dielectric strength of 55 kV/mm which is lower than the average values of 70 and 79 kV/mm for pure HDPE and LDPE respectively. However, the electric strength of PP was improved when blended with LDPE and HDPE samples in different ratios. A maximum values of 63 kV/mm for PP/LDPE blend and 67 kV/mm for PP/HDPE blend were reported [59].

Similarly, different ethylene ratios (9, 12, 15 mol%) blended with a polymer having various percentage weight and formed copolymer systems of propylene-ethylene (namely, VERSIFY™ 2200, 2300, and 2400; series no) for high voltage application. The Dow H358-02 was used as the isotactic polypropylene throughout the studies [64]. Some of the components need to be dissolved during the blending, and xylene was the best solvent. The blending process can be summarized in the below points:

- Solution blending;
- Melt blending;
- Extrusion.

The desired properties of the material with polypropylene achieved with the presence of space-filling morphologies. PEC blends' molecular structure was an essential parameter for its behavior, and LDPE used in polyethylene was a secondary choice. Materials were formed with various ratios and cooled with different rates, behave like an XLPE system. All materials are compatible with low temperatures. Considering the material's thickness, the cable performance with PP blend is the excellent and average thickness for E2200 (3.39 mm) and XLPE (4.34 mm) for mini cable insulation. The PE blended materials can be used for mini cables and XLPE based materials as references for mini wires.

Carbon Nanotube–Polyurethane Nanocomposite

The composite material produced with thermoplastic and carbon nanotubes (CNTs) can be used with a high voltage system. For this purpose, extruder temperature and screw speed were maintained at 215 °C and 300 rpm, respectively, for mixing CNTs (3 wt%) and polyurethane. The composites are in the form of granules. The thermoplastic polyurethane (TPU) materials are dried for 3 h, at 90 °C, followed by injection molding. Molding and melt temperature (40 and 220 °C) were controlled with a plate thickness of about 2 mm, but for extrusion, melt temperature was 185 °C and plate thickness 1.5 mm [12].

The acceptance criteria for any material for the industrial application have a large-scale melting range. The developed TPU can work with high temperature and confirmed by transmission electron microscopy (TEM) with injection molding, and extrusion samples gave across 10 m. It is homogeneous as well as elongation breaks about 560%. The size of dispersed particles of powder sample was obtained with the scanning electron microscope (SEM). It also confirmed the attachment of CNTs into the moiety of composite and not emerged through tubular protrusions by anchoring junctions with structural size about 20 μm after accumulation fragments with 100 mm plane surface level in the powder sample. The weathering was investigated by keeping composite test material and reference with latitude northern at 50° for 9 to 18 months and found that the filler will not degrade, only the matrix [65,66]. Analytical ultracentrifugation (AUC) and photoelectron spectroscopy (XPS) studies gave a strong indication of carbon nanotubes' presence in the nanocomposite materials. The internal diameter of pure material means the matrix of polyurethane and CNTs is smaller than the composite. Similar behavior can be seen in a composite having polyamide (PA) and silicon dioxide as well as with carbon nanotubes mixed polyoxymethylene (POM) and cement [66]. TGA recorded the decomposition of CNTs in the range of 500–650 °C [67]. Therefore, the composite material with CNTs has importance in electrical equipment, mainly in rollers and electromagnetic shielding.

2.2.3. Epoxy Resin of 9,9′-bis-(3,5-dibromo-4-hydroxyphenyl)anthrone-10 and Jute Composite

Thanki et al. [68] synthesized an epoxy resin from 9,9′-bis-(3,5-dibromo-4-hydroxyphenyl) anthrone-10 and epichlorohydrin by diluting them in isopropyl alcohol, and adding sodium hydroxide solution dropwise as a catalyst. After refluxing at 70 °C, they obtain a brown resin that was purified by extraction with chloroform [68].

At room temperature, synthesized EANBr was dissolved in tetrahydrofuran, and the curing agent was added (EPK 3251). The fabric jute was mixed with the resultant solution, and the remaining solvent evaporated. The mylar film was introduced in duplicate between developed sheets, and temperature and pressure were controlled, also silicon lubricant was used as mold release spray. The product was known as J–EANBr composite [68].

The Fourier transform infrared spectroscopy (FTIR) confirmed the presence of alkane (C–H, 1450 cm^{-1}), stretching (–OH, =C–H, –C=O, Ar C==C, Ar–O–R, C–O–H, C–Br) at 3532.56, 3070.21, 1664.48, 1592.61, 1254.19, 1071.1 and 631.2 cm^{-1} absorption frequencies respectively with EANBr sample. The nuclear magnetic resonance (NMR) studies indicated the formation of EANBr. The differential scanning calorimetry (D.S.C.) thermograms of EANBr and EANBrC are compared and found the broad endothermic transition of EANBrC (291.4 °C) and EANBr (265.3 °C) are expected due to tangible change and confirmed by no weight loss in the corresponding thermogravimetric analysis (TGA) thermogram. EANBr has a single step degradation reaction compared with J–EANBr, transition (291.4 °C) and two-step degradation and stable up to 310 °C [69]. EANBr and EANBrC are thermally steady up to 340 and 310 °C, respectively. As related to EPK 3251 cured EAN (360 °C), EPK 3251 cured EANBr (310 °C) has shown lower thermal stability. As well, the maximum weight loss (T_{max}) value for EANBrC (416.5 °C) is significantly more than EPK 3251 cured EAN (394.4 °C) but EANBrC (416.5 °C) has exhibited a higher value of T_{max} than EANBr (407.1 °C).

The mechanical and electrical properties of J-EANBr is better than J-EAN. In terms of tensile strength, almost similar with J-EAN with lower flexural strength. The electric strength is 40% lower in case of J-EANBr but volume resistivity is 29 times better than J-EAN, primarily expected to distinct degrees of cross linking and polarity, which impacts interfacial bond, due to annulment of partial charges because of OH groups of jute. The jute and EANBr have a better option as polymeric materials for electrical components due to their excellent properties with harsh environment, better hydrolytic activity, and thermal stability for low load bearing housing, electrical and electronic applications.

3. Properties of Polymers

3.1. Mechanical

Polymer chemical and physical properties are dependent on their molecular weight, chemical composition, and physical structures [70]. One of the properties of polymeric material is strong, which is the stress required to break the sample [6]. These strength properties are as follows: tensile, compressional, flexural, impact [71]. Factors affecting the strength of polymers are molecular weight, cross-linking, and crystallinity [71].

Young's modulus or tensile modulus is the induced stress divided by strain in the elastic region [71]. Ultimate elongation is defined as its ability to undergo deformation. Toughness measures the energy absorbed by the material to deform without fracturing.

The mechanical properties of the polymer are affected by temperature. Figure 4 shows a typical graph of stress versus strain and the effect of temperature, with an increase in the temperature, the elastic modulus, and tensile strength decrease, but the ductility increases [71].

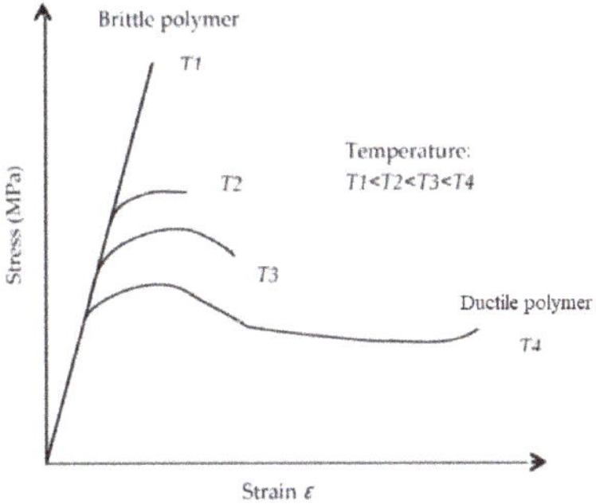

Figure 4. Effect of temperature on the mechanical properties of the polymer. Adapted from [71].

Viscoelasticity, two types of deformations exist, namely elastic and viscous. Elastic deformation is recoverable, while viscous deformation is a plastic deformation where the deformation is permanent upon removing the applied stress [71].

HV insulating systems' fabrication involves utilizing composite polymer; its application could be seen in electrical appliances. They are subjected to quivering wear and tear due to the rate of occurrence of magnetic force and shearing stress. To improve the mechanical performance of polymers, inorganic fillers are added to the polymers [72]. To enhance the strength and toughness is the need for polymer nanocomposites [73]. The composites' mechanical properties are strongly influenced by the filler's size and shape, the matrix properties, and the interfacial adhesion between the filler and matrix [74]. The basic principles of mechanical properties comprise tensile, compressive, bending, shear, and impact behavior is of importance [75]. The polymeric material is tested with standardized tests such as tensile strength, Izod impact strength, and softening point, but its application in a critical situation increases. Thus, the choice of material used depends upon a balance of stiffness, toughness, processability, and price in applying polymer [76].

3.1.1. Impact Behavior

Impact energy is known to reduce the static strength, reliability and improving tensile properties that simultaneously impact property. The basic stuff is impact toughness, which measures the needed energy to split a particular specimen. Figure 5a,b show supported beam load and cantilever beam load, respectively [77]. The determination of the Charpy impact strength of an unnotched (a_{cU}) (KJ/m^2) specimen is given in Equation (1).

$$a_{cU} = \frac{W_c}{b \times h} \tag{1}$$

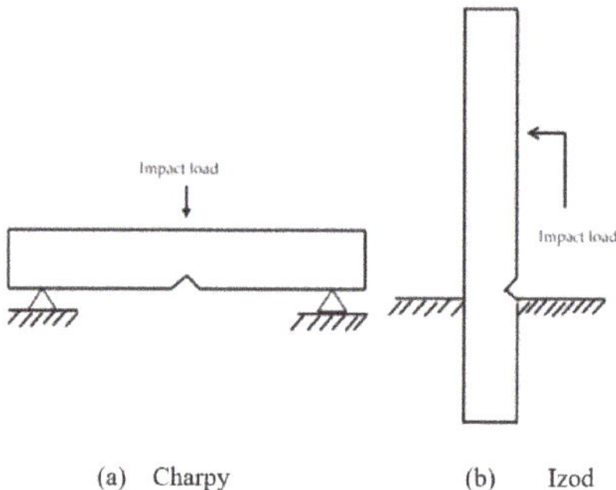

Figure 5. (**a**) Charpy (beam load) and (**b**) Izod (cantilever beam load). Adapted from [77].

Notched Charpy impact strength (a_{cN}) (KJ/m^2) can be determined by Equation (2), where *Wc* is the absorbed energy, *b* (m) is the with, and *h* (m) is the height.

$$a_{cN} = \frac{W_c}{b_N \times h} \tag{2}$$

The difference between Charpy impact strength acU and notched Charpy impact strength a_{cN} indicates how sensitive a material is to external notches, i.e., takes the problematic notch effect for the Charpy impact test into consideration and shows how effective fillers are. Thus, notch sensitivity can be calculated from the quotients of a_{cN} and a_{cU} indicated in Equation (3):

$$k_z = \frac{a_{cN}}{a_{cU}} \times 100\% \tag{3}$$

3.1.2. Tensile Test

Tensile tests involve the application of tensile force causing the elongation until the specimen breaks, and various loading conditions are essential and cannot be overemphasized [78]. The young modulus (E) (kN/m^2) can be estimated with Equation (4):

$$E = \frac{\sigma_2 - \sigma_1}{\varepsilon_2 - \varepsilon_1} \times 100\% \tag{4}$$

where $\sigma_2 - \sigma_1$ is a change in stress (kN/m^2) and $\varepsilon_2 - \varepsilon_1$ is a change in a strain which is dimensionless. A typical stress–strain curve is shown in Figure 6. Non-flexible polymers have high Young's modulus, while ductile polymers are elastic modulus with the ability to withstand extended elongation without fracturing [71]. In contrast, elastomers have a low stress–strain relationship with tough elastic texture [71].

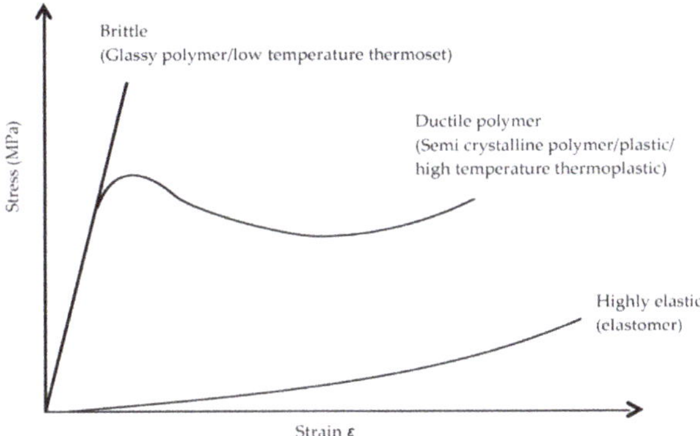

Figure 6. Stress–strain curves of various polymeric materials. Adapted from [77].

3.1.3. Bending Test

One of the most significant mechanical properties is the flexural strength and varies with specimen depth, temperature, and test span length [71]. The flexural strength (σ_{bh}), (MPa), known as the maximum stress at break, can be quantified using Equation (5):

$$\sigma_{bh} = \frac{\sigma_{f2} - \sigma_{f1}}{\varepsilon_{f2} - \varepsilon_{f1}} \times 100\% \tag{5}$$

The flexural or bending test is typically used in the quantification of the flexural strength. The flexural modulus (E_h), (MPa) can be estimated with Equation (6):

$$E_h = \frac{1}{4} \frac{L^3}{bh^3} \frac{\Delta F}{\Delta f} \tag{6}$$

where F is the breaking force in (N), L is the support distance in (mm), b is the width (mm) of the specimen in (mm), h is the thickness of the specimen in (mm), $\Delta F/\Delta f$ is the slope of the force-deflection curve. The load is applied at the specimen's center under standardized bending speed, temperature, and humidity. It is upon the most common type of loading encountered and essential in determining the polymer's characteristic values. The initial flexural modulus can be calculated using a slope of the line measuring Δf and ΔF as presented in Figure 7, depicting a typical force bending flexural test.

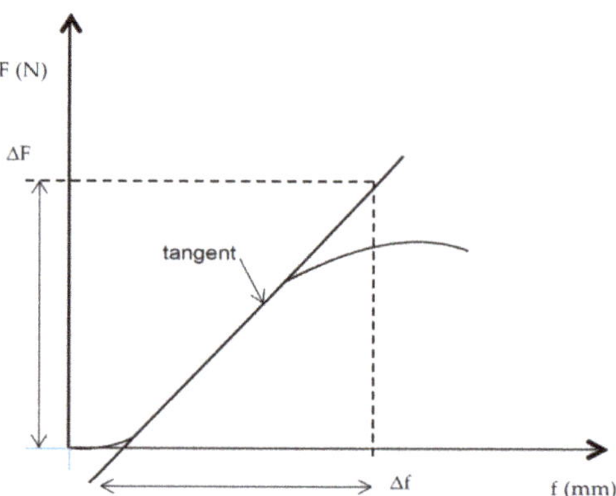

Figure 7. The initial slope of the bending force-deflection diagram. Adapted from [77].

Lim et al. [79] use a combination of the different fillers of aluminum oxide (Al_2O_3), Zinc oxide (ZnO), and organoclay. However, it was discovered that with 1.5% aluminum oxide, improved mechanical, burning rate, and dielectric breakdown compared to Zinc oxide and organoclay. Hedir et al. [80] irradiated the crosslinked polyethylene (XLDPE) with fluorescent lamps for 240 h. The results show a decline in resistivity, mechanical properties, and contact angle, increasing water retention and weight loss. Yasmin and Daniel [81] and Yang et al. [82] attributed that the tensile strength and elastic modulus can be increased by adding fillers at optimum content. Ray et al. [83] research on filler effect by using up to the upper limit and postulated that the stiffness and rigidity increase up to a certain level. Afterward, it harms the mechanical properties.

3.2. Thermal

Significant efforts have been made to enhance the thermal properties of various polymers, as well as to produce heat tolerant polymers, in order to qualify them for use in applications requiring efficient serviceability at high temperatures. The production of heat-resistant polymers began in the late 1950s with the goal of producing heat-resistant polymers that could satisfy the needs of the aerospace and electronics industries. Heat resistant polymers differ from ordinary polymers in that they can retain their desired properties at high temperatures. While there is no universal consensus about how to classify heat resistant polymers, the following is a helpful general classification [84]: (a) Heterocyclic polymers such as polyoxadiazoles, polybenzimidazoles; (b) condensation polymers such as polyhydrazides, polyesters, polyazomethines, and polyamides; (c) Condensation heterocyclic copolymers such as poly(amide-imide)s, poly(ether-imide)s, poly(amide-hydrazide) and poly(ester-imide)s, and (d) Ladder polymers including poly(benzimidazo benzophenanthroline) and polyquinoxaline.

3.2.1. Mold Stress-Relief Distortion

Products from injected molded polymers have residual stresses that may unfavorably affect the product performance, directly related to other processing conditions such as holding pressure, molding, and melting temperature [85]. Injection modeling in polymers is a common phenomenon. It has several merits such as low-cost mass production and good finishing's. However, due to the tricky deformations/bends or twist and pressure, residual stresses build up. The viscoelastic properties of polymers coupled with their high pressure during molding leads to complex situations such as orientation, stretching, and chain relaxation [85].

For most polymeric parts, residual stresses are quantified by destructive methods such as layer removal and hole drilling [86]. Although these techniques have problems related to their accuracy, their major drawback is reusing the measured parts. To overcome these limitations, new methods such as indentation are used. This technique involves applying higher force at the stressed location to align the same depth and add residual stresses at the indented locations.

3.2.2. Generic Thermal Indices

UL746B [87] and IEC60216 [88] were widely applied to determine relative thermal indices of polymers suitable for applications related to electrical applications. This is important to decide on the highest operating temperature considering the lifetime of the products. For polymeric materials, three essential properties can lead to their thermal categorization. These include electrical, mechanical/tensile, and impact properties.

The thermal index of polymers relies on the base polymers' property due to their deterioration, decomposition, and degradation caused by chemical reactions. In general, polymers' relative thermal index is vital in determining base polymers' long-standing thermal resistance [89].

3.2.3. Relative Thermal Capability

Nowadays, synthesis polymers are readily available and produced through the extrusion process. Due to the broader application of polymers, it is essential to know the relative thermal capability of polymers. It is vital to understand their thermal properties at molten temperatures, assembly, and the final product.

The thermal conductivity of polymer composites refers to both polymers and their fillers [90]. However, the importance of polymers' thermal conductivity has not been overemphasized concerning that of the fill. It is generally known that polymers' thermal conductivity is vital at low filler loadings due to the separation of the polymer matrix's thermally conductive fillers. To obtain high thermal conductivity in composite polymers, there is the need to create a continuous filler network. However, one major issue with network formation is high cost, reduced mechanical properties. The fundamental factors influencing polymers' thermal conductivity are their chain structure, crystallinity, and orientation domains [90]. On the other hand, polymers have thermal conductivities that are lower than ceramic or metals. Table 4 shows the thermal conductivities of renowned polymers, usually with low thermal conductivity [91].

Table 4. Thermal conductivities of polymers.

Polymer	Conductivity (W/m.k)
Epoxy resin	0.19
Polysulfone	0.22
Low-density polyethylene	0.3
High-density polyethylene	0.44
Polyethylene	0.11
Polycarbonate	0.2
Polystyrene	0.14
Polyimide, thermoplastic	0.11
polyphenylsulfone	0.35
Polyvinyl chloride	0.19
Nylon-6	0.25
Nylon-6.6	0.26
polymethylmethacrylate	0.21
Polyphenylene sulfide	0.3
Poly copolymer	0.33
Poly ethylene-vinyl acetate	0.34

3.2.4. Relative Thermal Index

The performance of two materials (developed new and reference) are compared to their thermal stability. Underwriters Laboratories (UL) standards are involved in projects with thermal aging of polymeric material. Properties of polymeric material can be changed by increasing the temperature, which will be considered the main factor in determining the particular material's thermal aging as per the procedure mentioned in UL 746B. The mathematical equation will give a relation between time and temperature, as presented in Equation (7):

$$\ln(t) = A + B/T \tag{7}$$

where, A = Constant, frequency factor; B = Energy constant for activation; T = Absolute temperature in Kelvin.

A comparison between newly produced polymers with control material will be made. The results established a relative thermal index based on the material's specific property, color, and thickness [87,88].

3.2.5. Temperature Excursions beyond the Maximum Temperature

The long-term aging data related to the product's stability, developed with a polymeric material, can establish a better thermal index. The temperature extrusions with short-term accepted for specific application for the material have enough thermal indices. The material will go to transition temperature immediately or anytime during the operation with electrical equipment. The heating appliance, initially thermostat will come automatically after highest temperature excursion but later on-air ambient temperature increases, the circuit will clear by thermostat and peak temperature decreases [92].

3.3. *Electrical*

3.3.1. Dielectric Strength

In certain cases, a material's dielectric strength is the determining factor in the design of the device in which it would be used. The common way to know the dielectric strength of a material is by applying a voltage until the material or insulator reaches a point where the electrical properties breakdown. The breakdown normally manifests itself as an electrical arc between the electrodes, resulting in a drastic reduction in resistance.

It is challenging to know the exact value of any material's dielectric strength because it may vary due to external factors and depending on the application [93]. Practically, it is challenging to realize the intrinsic dielectric strength of polymers because of external factors, which may lower polymers' dielectric strength. Most pure polymeric plastics such as PVC, PMM, PET, polycarbonate, polyethylene, and polypropylene have a dielectric strength range of 100 to 300 kV/cm. Some halogenated polymers such as Teflon show up to 700 kV/cm [94]. Table 5 provides the approximate values of dielectric strength of few commercial polymeric materials used in electrical equipment [94].

Table 5. Dielectric constant and dielectric strength of some polymers [94].

Compound	Dielectric Constant (1 MHz)	Dielectric Strength (kV/cm)
Poly(vinyl chloride)	2.9–3.1	140–200
Polyacrylonitrile	4.0–4.2	-
Poly(methyl methacrylate)	2.8–2.9	100–300
Polycarbonate	2.8–3.0	150–340
Poly(ethylene terephthalate)	3.0–3.5	150–200
Polytetrafluoroethylene (Telfon)	2.0–2.1	600–700
Polypropylene	2.2–2.3	230–250
Polyethylene	2.2–2.3	200–300
Silicone Oil	2.5	150
Fused Silica	-	250–400
Distilled Water	80	65–70
Air	1.0	15–30

Although it is difficult to understand the actual dielectric strength of polymeric material due to the above factors involved, many researchers have endeavored to increase the dielectric strength of polymeric materials proportional to the voltage for a specific thickness by adopting three primary methodologies are and filling nanoparticles in polymers, engineering filler-polymer interfaces, and coating film surfaces [95].

Factors affecting the dielectric strength of polymers include:

1. Environmental exposure: Certain ecological conditions such as severe exposure to chemicals, radiation, ozone, and oxidation weaken or break the chemical bond of polymers [93]. Most polymers fail prematurely due to moisture that creates conducting path within their layers, leading to treeing [96]. Performing an approach on polymers together with the presence of contaminants fast-tracks the breakdown process.
2. Electrode effects: It is a fact that the electrode properties may influence the breakdown strength of polymers depending on temperature [97]. Different electrode materials, sizes, and geometries [92] can also affect polymers' breakdown strength. The area of the electrodes is inversely proportional to the dielectric strength [97].
3. Temperature: One significant factor that affects the dielectric strength of polymers is temperature. However, for polymers subjected to high temperatures, there is the likelihood of oxidation and corona coupled with severe degradation and tracking of the material [93].
4. Voltage application: The rate of voltage change on polymers can also affect their dielectric strength. Fast application of voltage encourages electrical conduction, while slower voltage application promotes unavoidable degradation due to local heating, causing lower dielectric strength [93].
5. Frequency: The frequency of the applied voltage is another factor that influences the dielectric strength of polymers. The heat created in any dielectric is related to the applied frequency [93].
6. Specimen width: The thickness of a polymeric material is inversely proportional to its dielectric strength [96]. Increasing thickness of dielectric material creates a weaker path that may go a long way in causing breakdown [93]. Defects such as a cavity, metallic components, or contaminants within any polymer material provide an avenue for electrical discharges such as partial discharge or corona that may lead to severe degradation of the specimen and lower dielectric strength [96].

3.3.2. Dielectric Constant and Dissipation Factor

Generally, polymeric materials are used in two ways i.e., to isolate electrical network devices from one another and from the ground, and to act as a capacitor dielectric component. For polymeric materials to be used as insulating materials, the insulating material's capacitance should be as low as possible while retaining sufficient chemical and mechanical, chemical properties. As a consequence, it's preferable to use a material with a low relative permittivity. For the other application, a high dielectric constant value is desirable such that the capacitor dimensions should be reduced to a minimum value.

Polymeric materials used to provide insulation as well as capacitor dielectrics should have minimal dissipation factor to minimize the material heating. Since the dielectric loss increases linearly with frequency for a given loss index value, a low loss index value is recommended for high-frequency applications. When comparing materials with nearly the same dielectric constant or when using some material under such conditions that the dielectric constant remains basically constant, the dissipation factor is a quantity to consider [98].

3.3.3. Volume or Surface Resistivity

Volume resistivity may be used to help design a polymeric material for a specific application. Humidity or temperature resistivity will vary dramatically, and this must be taken into account when planning for operating environments of the polymeric material. Volume resistivity is commonly used in the study of an insulating material's uniformity, either in terms of processing or to detect conductive impurities that impair the material's durability but are difficult to detect using other methods. Ref [99] defines a commonly used research technique for determining electrical insulating materials' insulation resistance, volume resistance, and surface resistivity, as well as their conductivity. This technique only considers measurements taken when the DC voltage is applied.

In summary, the electrical breakdown strength of polymers decreases with increasing electrode area, the frequency of the voltage, the thickness of the sample, absorbed moisture, temperature, and reducing the rising rate of ramp voltage.

4. Assessment of Polymeric Insulations for HV Applications

To determine the suitability of polymeric insulations for high voltage apparatus applications, specific techniques can be deployed and discussed in this section.

4.1. Partial Discharges

PD has been established to evaluate different polymeric materials for application in HV apparatus, be it cable, electrical motors, transformers, etc. PD is defined as a localized electrical discharge that partially bridges the conductors' insulation and maybe closer or far away from a conductor [100,101]. In another context, the PD is regarded as an incomplete breakdown of the HV insulation system. PD activity mostly relies on the electric field strength applied to a specific area and its non-uniformity. For short gaps, PD possesses fast and slow-rise-time pulses that generate small electrical sparks or arcs or pseudo glow discharges, or pulseless glows [102,103]. Four main types of PD occur Corona, Surface discharges, cavities, and electrical trees (Figure 8).

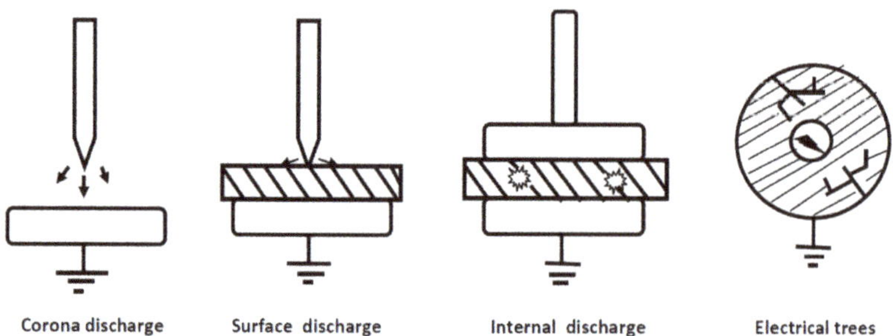

Figure 8. Types of Partial Discharge.

Internal discharge is a common phenomenon in voids within solid or liquid dielectrics [104]. These voids can embed in the insulation material during the manufacturing process of the apparatus. This form of discharge is extremely important and has a significant effect on polymeric insulation. A surface discharge, on the other hand, occurs at the HV insulating with a high tangential electric field [105]. Internal discharge is a common phenomenon in voids within solid or liquid dielectrics [104]. These voids can embed in the insulation material during the manufacturing process of the apparatus. This type of discharge, on the other hand, is less concentrated and hazardous than internal discharge. Corona discharge, on the other hand, occurs in air insulation and is usually harmless. Treeing is a form of discharge that is triggered by a sequence of internal discharges. Since the advent of polymer nanocomposites, various studies have been conducted on PD properties of such materials. Kozako et al. [106], for example, reported that only two wt% nanofiller is needed to increase the PD resistance of polyamide/layered silicate nanocomposites. Another investigated by Tanaka et al. [7] looked at PD resistance in epoxy/layered silicate nanocomposites. The authors concluded that by incorporating a limited amount of nanofillers into epoxy resins, PD tolerance of polymer nanocomposites could be greatly enhanced. Many experimental findings on the enhancement of partial discharge property of polymer nanocomposites have been reported. Henk et al. [107] studied the effects of nanoparticle amorphous silica dispersion in epoxy and polyethylene. Nano-silica was found to have a significant impact on thermoset partial discharge durability (epoxy and cross-link polyethylene). However, no effect was found with thermoplastics (low density polyethylene and medium density polyethylene).

4.2. High-Current Arc Resistance to Ignition (HAI)

The High-Ampere Arc Ignition (HAI) testing is a method that studies and assesses the electrical insulation flammability. This method is described in detail at the standard for safety UL 764A [108–110]. The HAI test subjects three specimens of the studied electrical insulation to electric arcs, recording the average number required to produce ignition in it, with a maximum of 200 (over this number is considered no ignition of the material). The test generates the electric arcs by using two round electrodes in contact with the studied electrical insulation; these electrodes have 3.2 mm of diameter; the difficulty is described in Figure 9. One of the electrodes is fixed and made of copper, where the other is stainless steel (303 alloys) movable rod; through the separation of these rods, an electric arc is generated. The electrodes are placed in a 45° plane of the studied electrical insulation specimen, and the fixed rod is sharpened to a 30° chisel point and the movable rod to a 60° conical point. The stainless steel rod movement is done using an air piston controlled by an electrical relay [108,109].

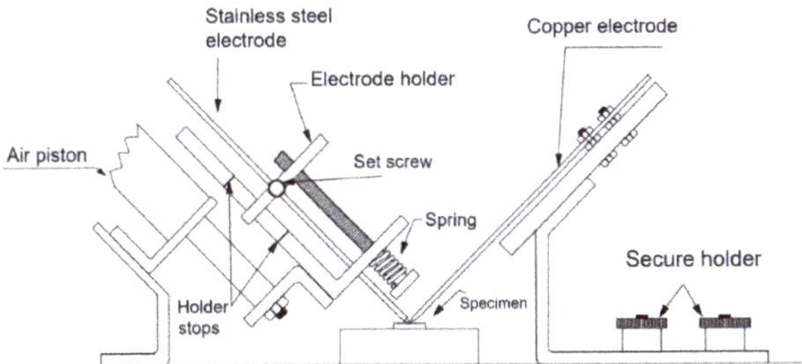

Figure 9. Electrode positioning and operating mechanism in HAI. Adapted from [109].

The testing method proposed by the standard UL 764A starts by applying a voltage of 240 V AC at a frequency of 60 Hz. An air core impedance is connected in series to the electrodes for yielding a short circuit current of 32.5 A, with a power factor of 0.5. The rate of electric arcs generated is 40 arcs/min, and the separation of the electrodes is an average of 254 $\pm$ 2.54 mm/s. There is no need to synchronize the electrodes' separation with a particular value of the sinusoidal current variation [108,109].

Several issues have been established for the replicability of this test, such as the identification of the ignition in the studied electrical insulation and the placement of the electrodes, that by not being described in detail, could lead to different kinds of arcs (random intensity or white flashes) [108,109]. To fix the latter issue, it was identified that the electrodes should return to their original position between cycles [108]. The identification problem was, in the beginning, attenuated by the use of a dark glass shield that allowed for a better distinction of the candle-like flame, which indicated the origin of the ignition in the studied specimen. However, this was improved by the use of a liquid crystal light attenuator [109]. The liquid crystal light attenuator allows for a better distinction of a low or high-energy arc, improving the test by providing the operator the crucial information regarding the test's continuity. The continuous series of soft energy arcs indicate that the electrodes should be moved to a different testing area of the studied specimen. Additionally, to enhance the distinction of a high from a low energy arc, the use of a digital arc energy meter is advised. This device records the magnitude of current and voltage for discrete intervals to calculate the electric arc's energy magnitude. These recommendations allow for a more descriptive signature between different materials taking into account its resistance to ignition [109].

The initial interest in making an HAI test in DC conditions originated from the automotive industry, which changed the traditional 12 V DC battery supply to a 36 V DC battery supply with a 42 V DC charging unit. This originated an interest in determining the ignition resistance of the automotive polymeric materials [111]. Initially, a primary circuit for studying the glass-fiber-reinforced plastics (GRP) for a DC supply was proposed, as shown in Figure 10. This test was based on the HAI test described by the standard UL 764A. This circuit uses a 200 V AC source; the capacitors generate a DC discharge between electrodes A and B, the release takes 0.5 to 12.8 ms depending on the capacitance values. The rate of the discharges is 20–40 arcs/min until the arc ignites the specimen. The use of a high-speed camera (4800 frames/s) allowed recording the length of the whole process, registering the arc discharge, arc ignition, and the combustion duration [112]. Similar studies have been realized in common polymers for the automotive industry [113–115].

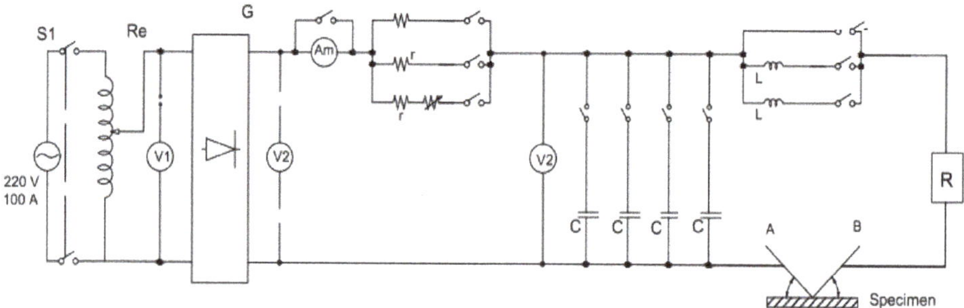

Figure 10. Basic circuit diagram of high current DC arc ignition testing apparatus formed by power switch (S1), regulator (Re), AC voltmeter (V1), rectifier (G), DC voltmeter (V2), DC ammeter (Am), charging resistance (r), capacitor (C), discharge resistance (R), fixed electrode (A), mobile electrode (B) and Inductance for current stabilization (L). Adapted from [112].

The interest in the DC testing method has increased in recent years due to the rise of renewable technologies like solar photovoltaics and wind power and DC microgrids' implementation [116]. This lead to proposing new DC-HAI testing, based on the original HAI testing described by the standard UL 764A, with the inclusion of some improvements, such as an automated control via a LabView-based program. The proposed testing setup is shown in Figure 11, where the movable electrode has three different positions: start, stable and safe. For this test, there are nine configuration parameters [116]:

- Output current
- Start arc delay
- Stable position delay
- Safe position delay
- Stop arc delay
- Start position delay
- Start position
- Stable position
- Safe position.

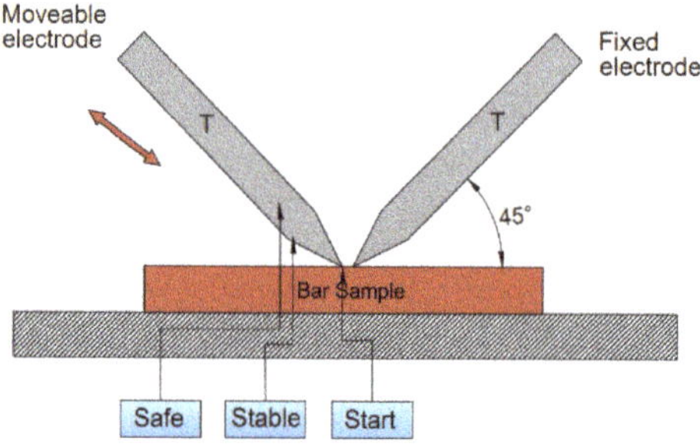

Figure 11. Resistance to high-ampere arc ignition (DC-HAI). Adapted from [117].

The safe position is the most considerable distance between the electrodes, allowing for a better judgment regarding the specimen's ignition. On the other hand, the stable position's purpose is to establish and arc stability [116]. The test realized by this proposed method resulted in a standard deviation ratio of 5% for DC arc that took 200 ms and 8% for an arc that has currents of 10 A; this proved to be consistent and repeatable [116].

5. Applications

5.1. Transformer Insulation

Epoxy resins, polyester, silicone, and imides are the most commonly used polymers for electric machines and dry transformers (see Figure 12) [6]. Dry-type transformers, known as epoxy-insulated transformers, are applied in areas requiring high fire protection, such as oil depots, high-rise buildings, and airports. Reinforced polyester with glass filaments is primarily used as a support for the protective system in oil transformers. These films increase the load due to breaking and provide strong resistance to infringement [118]. These tapes are suitable for leaded insulation wires, cable winding, and external protection. On the other hand, silicone transformer oil is primarily used as a coolant in high voltage power transformers. It has good heat capacity values, low viscosity, and high dielectric strength. Polyesteramideimides have long time endurance to 180–210 °C and have high heat resistance [119]. They are used as wire insulations in oil transformers.

Figure 12. Epoxy-resin cast dry type transformer.

5.2. Insulated Power Cables

In most power cables ranging from high to medium level, polymers are extensively applied as insulation materials. Such polymeric materials include polyethylene (PE), HDPE, LDPE, ethylene-propylene rubber (EPR), etc. Nevertheless, LDPE is more versatile and, until the 1960s, it was the most widely used polymeric material. Nowadays, cross-linked polyethylene (XLPE) has been preferred over paper because of its enhanced properties, such as efficiency and the ability to withstand high temperatures [120]. However, recently, XLPE cables have started to be replaced by HDPE in more advanced distribution systems because of their higher resistance to lightning strikes and water (see Figure 13) [121]. In some medium cables that require greater flexibility, EPR is used [120]. However, for low voltage applications, PVC is much more preferred because of its low manufacturing and durability cost. In new technologies such as the HVDC, polymeric cables are not used, and instead, oil impregnated paper is widely utilized. For such applications, polymeric cables are highly prone to partial discharges. Novel techniques such as the modifying the thermal resistivity and reducing the space have been proposed for the development of the polymer-insulated HVDC cables [122]. Other techniques being developed to improve polymeric cables under the HVDC include the development of nanocomposites, addition of inorganic particles [7,123,124]. All these additions can improve mechanical strength, thermal stability, and stringent dielectric breakdown. The research now focuses on applying nanoparticles to enhance the dielectric properties of nanocomposite [125]. For polyethylene and poly-vinyl chloride insulation, their rated voltages can be up to 275 kV and 3.3 kV respectively.

Figure 13. XPLE insulated power cable.

Nowadays, the focus is on developing insulating cables with unique properties capable of operating at high temperatures and electric stress levels. A recent research demonstrates the existence of a byproduct-free cross-linked copolymer blend that is seen as a potential solution to the widely used XLPE commonly employed for high voltage DC cable insulation [126]. The result show that the copolymer blend's loss tangent is three to four times lower than that of XLPE, with magnitudes of 0.12 at 70 °C and 0.01 at 50 °C [126]. The copolymer has shown good electrical properties and is free from cross-linking byproducts. As a result, this material is a promising option for HV components, such as HVDC cables, that require clean insulation materials.

5.3. *Electrical Encapsulation Materials*

Electrical encapsulation materials are required to properly operate transformers, motor coils, sensors, and solenoids [127]. Among the first electrical encapsulation materials were the resins, which had many advantages for protecting encapsulated electrical components [128], such as high insulation, low relative permittivity, low cost, and easy synthesis [129]. Thermoplastic resins have shown a better performance than thermosets by requiring thinner walls for them to be stronger than thermosets [127]. In addition, thermoplastics are produced in faster cycles, generate lesser scrap, and lack the environmental issues related to thermosets [127]. Nevertheless, when submitted to specific temperatures, resins' behavior could cause shrinkage or expansion of the material, generating focused stress and leading to premature failure [128]. A test is proposed to study this impulsive failure factor where cured resins samples are subjected to a wide range of temperatures. The embedment pressure is calculated; this study shows that the embedment pressure is higher for lower temperatures [128].

To manufacture most high-performance coils, thermoplastics such as PA 66, PBT polyester, and PET polyester polymer compositions are used for the encapsulation [127]. PA 612 could be applied as the encapsulation material for encapsulated sensors because it can withstand repeated thermal cycling more than polyesters [127]. Nowadays, to encapsulate high voltage multichip power assemblies, usually silicone gels are implemented; because of the high electrical insulation and softness, these are mainly used if there are bonding wires [130]. However, for temperatures above 250 °C, these materials have shown breaking symptoms. In cases requiring a higher temperature operation, silicone elastomer has proven to be a proper replacement [130].

The most common material for integrated circuits (IC) encapsulation is epoxy resins [131]. These properties differ among them because of the different formulations made to improve specific properties that enhance individual performances. These are related to thermal resistance, electric insulation, reliability in moisture, reliability temperature, and pressure [132]. Incorporating fillers such as alumina, boron nitride, alumina nitride, or other ceramic powder increases the thermal conductivity of the encapsulation's electrical insulation [129]. The inorganic filler reaches 65% to 90% of the total weight [131]. In recent years the incorporation of microscale and nanoscale insulating fillers in epoxy resins has been studied to increase the mechanical stress resistance and increase the dielectric strength [133–135]. Those investigations had shown that increasing the filler content increased the current conduction and volume conductivity, a conclusion that was not expected given the nature of the inorganic fillers [131]. This hint that the epoxy/filler intermolecular interaction could be directly related to the bulk transportation capabilities [131]. The issue with the epoxy/filler encapsulation is related to generating charge propagation over the surface of the ICs, due to the significant injection of electronic charge from the embedded bond wires [136]. The amount of filler in an epoxy molding compound varies the dielectric strength because at high temperatures for higher quantities of filler, the conductivity and electric field dependence increases [137]. This could be attributed to the consequent increment in the filler particles and the epoxy matrix [137].

5.4. Electrical and Electronic Plastics

Many electrical applications are requiring high graded polymer. At present, electrical apparatus commonly use plastic as an insulation system. Plastics have good dielectric strength, heat performance, and water resistance, making them ideal for electrical components. Several computer parts are made up of polymers. In general, conductive polymers are utilized in every computer element to ensure conductivity so that the device works well. On the other hand, plastics are used in making external parts of home appliances such as TV, toasters, juicers, and blenders. Currently, most electrical tools are made of plastics. Other plastic materials applications are relays, circuit breakers, transformer comments, cabling, and wires [138].

5.5. High Voltage Transmission Line Insulators

In the past, porcelain ceramic insulators were used in both transmission and distribution lines. In high-voltage transmission systems, polymer or composite insulators are becoming more common. Polymer Insulators, which are distinguished by their compact size, lightweight, high mechanical power, ease of installation, and low maintenance, have emerged as a new generation of high voltage transmission line insulators. The protective ribbed mold on polymer insulators is made of silicon organic rubber, which makes it different from other insulators [139].

Since its introduction in the early 1970s, electric utilities have gradually embraced polymer insulators as suitable substitutes for porcelain and glass insulators. Ethylene propylene polymers are used to make the insulators. EPR and Silicon Rubber are some of the polymers that are used as insulators. Among them, the EPR are one of the most weather-resistant synthetic polymers available [140]. It has superior ageing and color quality as well as excellent electrical, chemical, and mechanical properties. Heat, oxygen, ozone, and sunlight resistance is exceptional in all EPRs. Figure 14 shows a polymer suspension insulator used for high voltage line. Most polymer insulators are designed with a rated voltage ranging from 7.5 to 765 kV. For bushings in high voltage lines, certain type of polymers are used. These include polyoxymethylene, polyphenylene sulfide and ultra-high molecular weight polyethylene.

Nowadays, the focus is to develop new stress control techniques using advanced materials. Many researchers have recently looked at using field grading material to minimize electric field enhancement on high-voltage insulators in order to improve the design of the equipment [141]. Two main types of grading exist, i.e., capacitive grading and resistive grading. In the capacitive grading, a number of fillers may be added to the host matrix in order to boost the dielectric materials permittivity. In this situation, the electric field on the overhead insulators is redistributed [142]. In resistive grading, the idea is to have the electric field varying with the conductivity so as to have a non-linear conducting behavior. In this case, the base polymer is filled with an inorganic filler to achieve nonlinear characteristics. When the electric field strength reaches a withstand level, the nonlinear grading material turns out to be conductive, which tends to homogenize the electric field propagation within the bulk of the insulation, thereby eliminating the field enhancement effect [143].

Figure 14. Polymer suspension insulators.

6. Challenges and Future Directions

For over a century, insulating materials have been developed. There have been ongoing advances in insulating materials, from pure polymers to nanocomposites, and the mechanical, thermal, electrical properties of such materials have significantly improved. For instance, there have been increase in the electrical strength of materials at low temperature by nine folds, the withstand temperature of materials has increased 15 times. Furthermore, the thermal conductivity and the breakdown strength of materials increased by 30 and 3000 times respectively [144].

Despite all the aforementioned developments, there are still problems and challenges. For instance, the fundamental chemistry and physics underlying improved dielectric properties of polymer nanocomposites is not well understood, interface modifications is not well developed, insufficient nanoparticle dispersion and the repeatability of related experiments has been poor. As a result, there is refine individual materials while simultaneously improving composites' overall performance. Another challenge has to do with the biodegradable nature of conductive polymers, severely limiting their applications.

This research suggests new directions for polymers used as insulating materials: improving interface modification and manufacturing technology, investigating novel dispersion and surface modification techniques for nano materials, analyzing the formation of interfaces using computer calculations and analog simulations, designing multilayer technologies and components for application in a simpler and smaller items [7]. Furthermore, adding practical nanofillers to a polymer matrix, such as graphene, will significantly improve the conductivity of biodegradable polymers.

In terms of electrical applications, the focus should be on developing nanostructured materials for ultra-capacitors, electro-optic, discharge-resistant high-voltage equipment insulation, sensors, and actuators.

7. Conclusions

This paper extensively reviewed polymer materials, thermoplastics, and thermosets for application in electrical apparatus. Polyethylene has been the widely applied material for manufacturing cables at the medium and high voltage range due to its high electrical strength and low production costs. Using a cross-linking process, good thermal and mechanical characteristics can be achieved with XLPE, HDPE, and EPR. On the other hand, at low voltage for indoor applications, PVC is replaced with superior polymeric insulation due to safety and public health regulations. For HVDC applications, polymeric insulated cables have not been as successful as oil-impregnated paper cables because of some operational conditions that reduce reliability and increase the functioning costs.

On the other hand, smart polymeric materials were more useful in biological and medicinal applications due to their sensitivity to the environment. However, it has been observed that more research is required in the mechanical, electrical, and thermal stresses of polymers to increase the reliability and power density. Besides, residual stresses are a common phenomenon that affects the production of injected molded polymers. These stresses come up due to deformations/bends, twists, or pressure. The high pressure during molding also leads to complex situations such as chain reaction, stretching, and relaxation. These residuals are usually quantified using techniques with lower accuracy. To improve the accuracy, new techniques such as indentation are typically employed.

It is also important to note that for polymers, thermal conductivity refers to the polymers and their fillers. It is interesting to note that thermal conductivities are significant at low filler loadings due to the polymer matrix's thermally conductive fillers' disjointing. However, polymers' properties can be changed by increasing the temperature according to the procedure mentioned in UL 746B. Additionally, several factors affect the dielectric strength of polymers, which include: (1) environmental; (2) electrode effects; (3) temperature; and (4) voltage application and frequency.

However, current research concentrates on developing novel materials (e.g., nanofiller-added polymers) that possess additional capabilities such as improved mechanical strength and electrical erosion reduction. In general, PD has already been established to evaluate the quality of high voltage insulation systems. Future development trends for nanocomposites should concentrate on developing nanostructured materials for ultra-capacitors, electro-optic, discharge resistant high voltage equipment insulation, sensors and actuators.

Author Contributions: S.K.M.H. formulated the study and contributed with the abstract, introduction, polymer characteristics, and conclusion; J.A.A.-R. and A.A.M. involved the abstract, introduction, electrical properties, and conclusion; Y.U. and H.R. contributed the abstract, types, synthesis, thermal properties of the polymer, and conclusion; B.H.J. helped with introducing mechanical properties and N.A.B. with electrical properties and abstracts; F.M.-S. contributed the thermal properties, applications, conclusions as well as references. All the authors extensively reviewed the paper, provided areas of improvement, and added substantive information. All authors have read and agreed to the published version of the manuscript.

Funding: The authors would like to express their appreciation to Agencia Nacional de Investigación y Desarrollo (ANID) for the support received through the projects Fondecyt regular 1200055 and Fondef ID19I10165 and the UTFSM for the project PI_m_19_01. Nurul Aini Bani would like to thank the Ministry of Higher Education (MOHE), Malaysia via Universiti Teknologi Malaysia (UTM) (Research cost center no. Q.K130000.3556.06G43) for the support received.

Institutional Review Board Statement: Not applicable.

Informed Consent Statement: Not applicable.

Data Availability Statement: Not applicable.

Conflicts of Interest: The authors declare no conflict of interest. The funders had no role in the design of the study; in the collection, analyses, or interpretation of data; in the writing of the manuscript, or in the decision to publish the results.

References

1. Alfalahi, A.H.M. Study of factors affecting the dielectric strength of [EP/ZrO2-y2O3] Nanocomposites. *J. Eng. Appl. Sci.* **2018**, *13*, 484–488.
2. Nasreen, S.; Treich, G.M.; Baczkowski, M.L.; Mannodi-Kanakkithodi, A.K.; Cao, Y.; Ramprasad, R.; Sotzing, G. Polymer Dielectrics for Capacitor Application. In *Kirk-Othmer Encyclopedia of Chemical Technology*; John Wiley & Sons, Inc.: Hoboken, NJ, USA, 2017; pp. 1–29.
3. Müller, K.; Bugnicourt, E.; Latorre, M.; Jorda, M.; Echegoyen Sanz, Y.; Lagaron, J.; Miesbauer, O.; Bianchin, A.; Hankin, S.; Bölz, U.; et al. Review on the processing and properties of polymer nanocomposites and nanocoatings and their applications in the packaging, automotive and solar energy fields. *Nanomaterials* **2017**, *7*, 74. [CrossRef]
4. Bur, A.J. Dielectric properties of polymers at microwave frequencies: A review. *Polymer* **1985**, *26*, 963–977. [CrossRef]
5. Rimdusit, S.; Ishida, H. Development of new class of electronic packaging materials based on ternary systems of benzoxazine, epoxy, and phenolic resins. *Polymer* **2000**, *41*, 7941–7949. [CrossRef]
6. Pleşa, I.; Noţingher, P.; Schlögl, S.; Sumereder, C.; Muhr, M.; Pleşa, I.; Noţingher, P.V.; Schlögl, S.; Sumereder, C.; Muhr, M. Properties of polymer composites used in high-voltage applications. *Polymers* **2016**, *8*, 173. [CrossRef]
7. Tanaka, T.; Montanari, G.C.; Mulhaupt, R. Polymer nanocomposites as dielectrics and electrical insulation-perspectives for processing technologies, material characterization and future applications. *IEEE Trans. Dielectr. Electr. Insul.* **2004**, *11*, 763–784. [CrossRef]
8. Madi, A.; He, Y.; Jiang, L.; Yan, B. Surface tracking on polymeric insulators used in electrical transmission lines. *Indones. J. Electr. Eng. Comput. Sci.* **2016**, *3*, 639. [CrossRef]
9. Lalankere, G. Opportunities and challenges of employing composite materials in switchgear industry. In Proceedings of the 20th International Conference on Electricity Distribution (CIRED), Prague, Czech Republic, 8–11 June 2009; p. 644.
10. Britton, L.G. *Avoiding Static Ignition Hazards in Chemical Operations*; John Wiley & Sons, Inc.: Hoboken, NJ, USA, 1999; ISBN 9780470935408.
11. Stone, G.C.; Boulter, E.A.; Culbert, I.; Dhirani, H. Historical Development of Insulation Materials and Systems. In *Electrical Insulation for Rotating Machines*; Kartalopoulos, S., Ed.; John Wiley & Sons, Inc.: Hoboken, NJ, USA, 2004; pp. 73–94.
12. Wohlleben, W.; Meier, M.W.; Vogel, S.; Landsiedel, R.; Cox, G.; Hirth, S.; Tomović, Ž. Elastic CNT–polyurethane nanocomposite: Synthesis, performance and assessment of fragments released during use. *Nanoscale* **2013**, *5*, 369–380. [CrossRef]
13. Camargo, P.H.C.; Satyanarayana, K.G.; Wypych, F. Nanocomposites: Synthesis, structure, properties and new application opportunities. *Mater. Res.* **2009**, *12*, 1–39. [CrossRef]
14. Han, Z.; Garrett, R. Overview of Polymer Nanocomposites as dielectrics and electrical insulation materials for large high voltage rotating machines. *NSTI-Nanotech* **2008**, *2*, 727–732.
15. Keith Nelson, J. Overview of nanodielectrics: Insulating materials of the future. In Proceedings of the 2007 Electrical Insulation Conference and Electrical Manufacturing Expo, Nashville, TN, USA, 22–24 October 2007; pp. 229–235.
16. Matthews, F.L.; Rawlings, R.D. *Composite Materials: Engineering and Science*, 1st ed.; CRC Press: Cambridge, UK, 1999; ISBN 1855734737.
17. Sheer, M.L. Advanced composites: The leading edge in high performance motor and transformer insulation. In Proceedings of the 20th Electrical Electronics Insulation Conference, Boston, MA, USA, 7–10 October 1991; pp. 181–185.
18. Tanaka, T.; Imai, T. Advances in nanodielectric materials over the past 50 years. *IEEE Electr. Insul. Mag.* **2013**, *29*, 10–23. [CrossRef]
19. Rajendran Royan, N.R.; Sulong, A.B.; Sahari, J.; Suherman, H. Effect of Acid- and Ultraviolet/Ozonolysis-Treated MWCNTs on the electrical and mechanical properties of epoxy nanocomposites as bipolar plate applications. *J. Nanomater.* **2013**, *2013*, 717459. [CrossRef]
20. Druffel, T.; Geng, K.; Grulke, E. Mechanical comparison of a polymer nanocomposite to a ceramic thin-film anti-reflective filter. *Nanotechnology* **2006**, *17*, 3584–3590. [CrossRef]
21. Seena, V.; Fernandes, A.; Pant, P.; Mukherji, S.; Ramgopal Rao, V. Polymer nanocomposite nanomechanical cantilever sensors: Material characterization, device development and application in explosive vapour detection. *Nanotechnology* **2011**, *22*, 295501. [CrossRef]
22. Petrović, Z.S.; Cho, Y.J.; Javni, I.; Magonov, S.; Yerina, N.; Schaefer, D.W.; Ilavský, J.; Waddon, A. Effect of silica nanoparticles on morphology of segmented polyurethanes. *Polymer* **2004**, *45*, 4285–4295. [CrossRef]
23. Qian, Y.; Lindsay, C.I.; Macosko, C.; Stein, A. Synthesis and properties of vermiculite-reinforced polyurethane nanocomposites. *ACS Appl. Mater. Interfaces* **2011**, *3*, 3709–3717. [CrossRef] [PubMed]
24. Sánchez-Adsuar, M.S.; Linares-Solano, A.; Cazorla-Amorós, D.; Ibarra-Rueda, L. Influence of the nature and the content of carbon fiber on properties of thermoplastic polyurethane-carbon fiber composites. *J. Appl. Polym. Sci.* **2003**, *90*, 2676–2683. [CrossRef]
25. Koerner, H.; Price, G.; Pearce, N.A.; Alexander, M.; Vaia, R.A. Remotely actuated polymer nanocomposites—Stress-recovery of carbon-nanotube-filled thermoplastic elastomers. *Nat. Mater.* **2004**, *3*, 115–120. [CrossRef]
26. Kim, H.; Miura, Y.; Macosko, C.W. Graphene/Polyurethane nanocomposites for improved gas barrier and electrical conductivity. *Chem. Mater.* **2010**, *22*, 3441–3450. [CrossRef]
27. Khan, U.; May, P.; O'Neill, A.; Vilatela, J.J.; Windle, A.H.; Coleman, J.N. Tuning the mechanical properties of composites from elastomeric to rigid thermoplastic by controlled addition of carbon nanotubes. *Small* **2011**, *7*, 1579–1586. [CrossRef]

28. Zhang, R.; Dowden, A.; Deng, H.; Baxendale, M.; Peijs, T. Conductive network formation in the melt of carbon nanotube/thermoplastic polyurethane composite. *Compos. Sci. Technol.* **2009**, *69*, 1499–1504. [CrossRef]
29. Potschke, P.; Haussler, L.; Pegel, S.; Steinberger, R.; Scholz, G. Thermoplastic polyurethane filled with carbon nanotubes for electrical dissipative and conductive application. *KGK Kautschuk Gummi Kunststoffe* **2007**, *60*, 432–437.
30. De Girolamo Del Mauro, A.; Grimaldi, A.I.; La Ferrara, V.; Massera, E.; Miglietta, M.L.; Polichetti, T.; Di Francia, G. A simple optical model for the swelling evaluation in polymer nanocomposites. *J. Sens.* **2009**, *2009*, 703206. [CrossRef]
31. Chen, W.; Tao, X. Self-organizing alignment of carbon nanotubes in thermoplastic polyurethane. *Macromol. Rapid Commun.* **2005**, *26*, 1763–1767. [CrossRef]
32. Koerner, H.; Liu, W.; Alexander, M.; Mirau, P.; Dowty, H.; Vaia, R.A. Deformation–morphology correlations in electrically conductive carbon nanotube—thermoplastic polyurethane nanocomposites. *Polymer* **2005**, *46*, 4405–4420. [CrossRef]
33. BASF. Polyurethanes Elastollan Material Properties. Available online: http://www.polyurethanes.basf.de/pu/Elastollan/Elastollan_Materialeigenschaften (accessed on 7 March 2019).
34. Zhang, L.; Zhu, J.; Zhou, W.; Wang, J.; Wang, Y. Thermal and electrical conductivity enhancement of graphite nanoplatelets on form-stable polyethylene glycol/polymethyl methacrylate composite phase change materials. *Energy* **2012**, *39*, 294–302. [CrossRef]
35. Cong, H.; Radosz, M.; Towler, B.F.; Shen, Y. Polymer–inorganic nanocomposite membranes for gas separation. *Sep. Purif. Technol.* **2007**, *55*, 281–291. [CrossRef]
36. Wang, T.-L.; Yang, C.-H.; Shieh, Y.-T.; Yeh, A.-C.; Juan, L.-W.; Zeng, H.C. Synthesis of new nanocrystal-polymer nanocomposite as the electron acceptor in polymer bulk heterojunction solar cells. *Eur. Polym. J.* **2010**, *46*, 634–642. [CrossRef]
37. Ruiyi, L.; Qianfang, X.; Zaijun, L.; Xiulan, S.; Junkang, L. Electrochemical immunosensor for ultrasensitive detection of microcystin-LR based on graphene–gold nanocomposite/functional conducting polymer/gold nanoparticle/ionic liquid composite film with electrodeposition. *Biosens. Bioelectron.* **2013**, *44*, 235–240. [CrossRef]
38. Kaur, J.; Lee, J.H.; Shofner, M.L. Influence of polymer matrix crystallinity on nanocomposite morphology and properties. *Polymer* **2011**, *52*, 4337–4344. [CrossRef]
39. Kutvonen, A.; Rossi, G.; Ala-Nissila, T. Correlations between mechanical, structural, and dynamical properties of polymer nanocomposites. *Phys. Rev. E* **2012**, *85*, 41803. [CrossRef]
40. Kutvonen, A.; Rossi, G.; Puisto, S.R.; Rostedt, N.K.J.; Ala-Nissila, T. Influence of nanoparticle size, loading, and shape on the mechanical properties of polymer nanocomposites. *J. Chem. Phys.* **2012**, *137*, 214901. [CrossRef]
41. Ducháček, V. Recent developments for thermoplastic elastomers. *J. Macromol. Sci. Part B* **1998**, *37*, 275–282. [CrossRef]
42. Grigore, M.; Grigore, E.M. Methods of recycling, properties and applications of recycled thermoplastic polymers. *Recycling* **2017**, *2*, 24. [CrossRef]
43. Habib Mnaathr, S.; Saher, H.; Al-Hseenawy, N. The influences of thermoplastic polymers on the electrical efficiency if used as material insulation in cables. *Chem. Mater. Eng.* **2014**, *2*, 169–172. [CrossRef]
44. Ashby, M.F.; Shercliff, H.; Cebon, D. *Materials: Engineering, Science, Processing and Design*, 1st ed.; Elsevier Butterworth-Heinemann: Oxford, UK, 2007; ISBN 0750683910.
45. Hanna, D.; Goodman, S.H. *Handbook of Thermoset Plastics*, 3rd ed.; Elsevier Science: Amsterdam, The Netherlands, 2013; ISBN 9781455731091.
46. Ismail, N.H.; Mustapha, M. A review of thermoplastic elastomeric nanocomposites for high voltage insulation applications. *Polym. Eng. Sci.* **2018**, *58*, E36–E63. [CrossRef]
47. Ellingford, C.; Wemyss, A.M.; Zhang, R.; Prokes, I.; Pickford, T.; Bowen, C.; Coveney, V.A.; Wan, C. Understanding the enhancement and temperature-dependency of the self-healing and electromechanical properties of dielectric elastomers containing mixed pendant polar groups. *J. Mater. Chem. C* **2020**, *8*, 5426–5436. [CrossRef]
48. Wemyss, A.M.; Ellingford, C.; Morishita, Y.; Bowen, C.; Wan, C. Dynamic polymer networks: A new avenue towards sustainable and advanced soft machines. *Angew. Chem.* **2021**. [CrossRef]
49. Adeosun, S.O.; Lawal, G.I.; Balogun, S.A.; Akpan, E.I. Review of green polymer nanocomposites. *J. Miner. Mater. Charact. Eng.* **2012**, *11*, 385–416. [CrossRef]
50. Thomas, S.; Stephen, R.; Thomas, S.; Bandyopadhyay, S. Polymer nanocomposites: Preparation, properties and applications. *Rubber Fibers Plast. Int.* **2007**, *1*, 49–56.
51. Gacitúa, W.E.; Ballerini, A.A.; Zhang, J. Polymer nanocomposites: Synthetic and natural fillers a review. *Maderas Cienca Y Tecnol.* **2005**, *7*, 159–178. [CrossRef]
52. Fu, S.; Sun, Z.; Huang, P.; Li, Y.; Hu, N. Some basic aspects of polymer nanocomposites: A critical review. *Nano Mater. Sci.* **2019**, *1*, 2–30. [CrossRef]
53. Vo, T.H.; Shekhirev, M.; Kunkel, D.A.; Orange, F.; Guinel, M.J.-F.; Enders, A.; Sinitskii, A. Bottom-up solution synthesis of narrow nitrogen-doped graphene nanoribbons. *Chem. Commun.* **2014**, *50*, 4172–4174. [CrossRef] [PubMed]
54. Xianying, D.; Meilin, W.; Bencai, D.; Shujing, L.; Yuli, Z.; Xiangjun, Z.; Libing, G. 2,7-Dibromo-9-Hydroxyl Phenanthrene Derivatives and Preparation Method Thereof. China Patent CN102775279B, 5 March 2014.
55. Långvik, O.; Sandberg, T.; Wärnå, J.; Murzin, D.Y.; Leino, R. One-pot synthesis of (R)-2-acetoxy-1-indanone from 1,2-indanedione combining metal catalyzed hydrogenation and chemoenzymatic dynamic kinetic resolution. *Catal. Sci. Technol.* **2015**, *5*, 150–160. [CrossRef]

56. Yamamoto, T.; Etori, H. Poly(anthraquinone)s Having a .pi.-Conjugation System along the Main Chain. Synthesis by organometallic polycondensation, redox behavior, and optical properties. *Macromolecules* **1995**, *28*, 3371–3379. [CrossRef]
57. Marcano, D.C.; Kosynkin, D.V.; Berlin, J.M.; Sinitskii, A.; Sun, Z.; Slesarev, A.; Alemany, L.B.; Lu, W.; Tour, J.M. Improved synthesis of graphene oxide. *ACS Nano* **2010**, *4*, 4806–4814. [CrossRef] [PubMed]
58. Pirnat, K.; Bitenc, J.; Vizintin, A.; Krajnc, A.; Tchernychova, E. Indirect synthesis route toward cross-coupled polymers for high voltage organic positive electrodes. *Chem. Mater.* **2018**, *30*, 5726–5732. [CrossRef]
59. Ahmed Dabbak, S.; Illias, H.; Ang, B.; Abdul Latiff, N.; Makmud, M. Electrical properties of polyethylene/polypropylene compounds for high-voltage insulation. *Energies* **2018**, *11*, 1448. [CrossRef]
60. Hosier, I.L.; Vaughan, A.S.; Swingler, S.G. Structure–property relationships in polyethylene blends: The effect of morphology on electrical breakdown strength. *J. Mater. Sci.* **1997**, *32*, 4523–4531. [CrossRef]
61. Green, C.D.; Vaughan, A.S.; Stevens, G.C.; Sutton, S.J.; Geussens, T.; Fairhurst, M.J. Recyclable power cable comprising a blend of slow-crystallized polyethylenes. *IEEE Trans. Dielectr. Electr. Insul.* **2013**, *20*, 1–9. [CrossRef]
62. Rishina, L.A.; Kissin, Y.V.; Lalayan, S.S.; Krasheninnikov, V.G. Synthesis of atactic polypropylene: Propylene polymerization reactions with $TiCl_4$–$Al(C_2H_5)_2Cl/Mg(C_4H_9)_2$ catalyst. *J. Appl. Polym. Sci.* **2019**, *136*, 47692. [CrossRef]
63. De Rosa, C.; Auriemma, F. Structure and physical properties of syndiotactic polypropylene: A highly crystalline thermoplastic elastomer. *Prog. Polym. Sci.* **2006**, *31*, 145–237. [CrossRef]
64. Green, C.; Vaughan, A.; Stevens, G.; Pye, A.; Sutton, S.; Geussens, T.; Fairhurst, M. Thermoplastic cable insulation comprising a blend of isotactic polypropylene and a propylene-ethylene copolymer. *IEEE Trans. Dielectr. Electr. Insul.* **2015**, *22*, 639–648. [CrossRef]
65. Wohlleben, W.; Brill, S.; Meier, M.W.; Mertler, M.; Cox, G.; Hirth, S.; von Vacano, B.; Strauss, V.; Treumann, S.; Wiench, K.; et al. On the lifecycle of nanocomposites: Comparing released fragments and their in-vivo hazards from three release mechanisms and four nanocomposites. *Small* **2011**, *7*, 2384–2395. [CrossRef] [PubMed]
66. Nguyen, T.; Pellegrin, B.; Bernard, C.; Gu, X.; Gorham, J.M.; Stutzman, P.; Stanley, D.; Shapiro, A.; Byrd, E.; Hettenhouser, R.; et al. Fate of nanoparticles during life cycle of polymer nanocomposites. *J. Phys. Conf. Ser.* **2011**, *304*, 12060. [CrossRef]
67. Mansfield, E.; Kar, A.; Hooker, S.A. Applications of TGA in quality control of SWCNTs. *Anal. Bioanal. Chem.* **2010**, *396*, 1071–1077. [CrossRef]
68. Thanki, J.D.; Parsania, P.H. Physico-chemical study of modified mannich base cured epoxy resin of 9,9′-bis(4-Hydroxy Phenyl) Anthrone-10 and its jute and glass composites. *J. Polym. Mater.* **2016**, *33*, 101–109.
69. Thanki, J.D.; Patel, J.P.; Parsania, P.H. Synthesis and characterization of epoxy resin of 9,9′-bis-(3,5-dibromo-4-hydroxyphenyl) anthrone-10 and its jute composite. *Des. Monomers Polym.* **2018**, *21*, 9–17. [CrossRef]
70. Su, W.-F. *Principles of Polymer Design and Synthesis*; Lecture Notes in Chemistry; Springer: Berlin/Heidelberg, Germany, 2013; Volume 82, ISBN 978 3 642 38729 6.
71. Balani, K.; Verma, V.; Agarwal, A.; Narayan, R. Physical, Thermal, and Mechanical Properties of Polymers. In *Biosurfaces*; Balani, K., Verma, V., Agarwal, A., Narayan, R., Eds.; John Wiley & Sons, Inc.: Hoboken, NJ, USA, 2015; pp. 329–344.
72. Schadler, L.S. Polymer-Based and Polymer-Filled Nanocomposites. In *Nanocomposite Science and Technology*; Ajayan, P.M., Schadler, L.S., Braun, P.V., Eds.; Wiley-VCH Verlag GmbH & Co. KGaA: Weinheim, Germany, 2004; pp. 77–153.
73. Lach, R.; Antonova Gyurova, L.; Grellmann, W. Application of indentation fracture mechanics approach for determination of fracture toughness of brittle polymer systems. *Polym. Test.* **2007**, *26*, 51–59. [CrossRef]
74. Mohamad, N.; Muchtar, A.; Ghazali, M.J.; Mohd, D.; Azhari, C.H. The effect of filler on epoxidised natural rubber-alumina nanoparticles composites. *Eur. J. Sci. Res.* **2008**, *24*, 538–547.
75. Kimura, K. Multistress aging of machine insulation systems. In Proceedings of the 1995 Conference on Electrical Insulation and Dielectric Phenomena, Virginia Beach, VA, USA, 22–25 October 1995; pp. 205–210.
76. Crompton, T.R. *Polymer Reference Book*; Smithers Rapra Technology Ltd.: Shewsbury, Shropshire, UK, 2006; ISBN 1859574920.
77. Grellmann, W.; Seidler, S. *Polymer Testing*; Hanser Publications: München, Germany, 2013; ISBN 9781569905494.
78. Bonten, C. *Kunststofftechnik Einführung und Grundlagen*, 1st ed.; Carl Hanser Verlag: München, Germany, 2014; ISBN 3446440933.
79. Lim, K.S.; Mariatti, M.; Kamarol, M.; Ghani, A.B.A.; Shafi Halim, H.; Abu Bakar, A. Properties of nanofillers/crosslinked polyethylene composites for cable insulation. *J. Vinyl Addit. Technol.* **2019**, *25*, E147–E154. [CrossRef]
80. Hedir, A.; Bechouche, A.; Moudoud, M.; Teguar, M.; Lamrous, O.; Rondot, S. Experimental and predicted XLPE cable insulation properties under UV radiation. *Turkish J. Electr. Eng. Comput. Sci.* **2020**, *28*, 1763–1775. [CrossRef]
81. Yasmin, A.; Daniel, I.M. Mechanical and thermal properties of graphite platelet/epoxy composites. *Polymer* **2004**, *45*, 8211–8219. [CrossRef]
82. Yang, H.S.; Kim, H.J.; Son, J.; Park, H.J.; Lee, B.J.; Hwang, T.S. Rice-husk flour filled polypropylene composites; mechanical and morphological study. *Compos. Struct.* **2004**, *63*, 305–312. [CrossRef]
83. Ray, D.; Bhattacharya, D.; Mohanty, A.K.; Drzal, L.T.; Misra, M. Static and Dynamic Mechanical Properties of Vinylester Resin Matrix Composites Filled with Fly Ash. *Macromol. Mater. Eng.* **2006**, *291*, 784–792. [CrossRef]
84. Mehdipour-Ataei, S.; Tabatabaei-Yazdi, Z. Heat Resistant Polymers. In *Encyclopedia of Polymer Science and Technology*; John Wiley & Sons, Inc.: Hoboken, NJ, USA, 2015; pp. 1–31.
85. Guevara-Morales, A.; Figueroa-López, U. Residual stresses in injection molded products. *J. Mater. Sci.* **2014**, *49*, 4399–4415. [CrossRef]

86. Pak, S.Y.; Kim, S.Y.; Kim, S.H.; Youn, J.R. Measurement of residual stresses in polymeric parts by indentation method. *Polym. Test.* **2013**, *32*, 946–952. [CrossRef]
87. UL Standard for Polymeric Materials. In *Long Term Property Evaluations (UL—746B)*; Underwriters' Laboratories: Chicago, IL, USA, 2018.
88. IEC Electrical Insulating Materials. In *Thermal Endurance Properties—Part 1: Ageing Procedures and Evaluation of Test Results (IEC 60216-1:2013)*; International Electrotechnical Commission: Geneva, Switzerland, 2013; pp. 1–66.
89. PBI. Advanced Materials Co. Ltd. Heat Resistance. Available online: http://www.pbi-am.com/en/properties/heat-resistance (accessed on 8 March 2019).
90. Chen, H.; Ginzburg, V.V.; Yang, J.; Yang, Y.; Liu, W.; Huang, Y.; Du, L.; Chen, B. Thermal conductivity of polymer-based composites: Fundamentals and applications. *Prog. Polym. Sci.* **2016**, *59*, 41–85. [CrossRef]
91. Ngo, I.-L.; Jeon, S.; Byon, C. Thermal conductivity of transparent and flexible polymers containing fillers: A literature review. *Int. J. Heat Mass Transf.* **2016**, *98*, 219–226. [CrossRef]
92. UL Standard for Polymeric Materials. In *Use in Electrical Equipment Evaluations (UL—746C)*; Underwriters' Laboratories: Chicago, IL, USA, 2018.
93. Jafri, S.Z. *Experimental Investigation of Dielectric Strength of Polymer Films*; Department of Electrical and Computer Engineering, University of Windsor: Windsor, ON, Canada, 2001.
94. Polymer Science Dielectric Strength of Polymers. Available online: https://polymerdatabase.com/polymer%20physics/Dielectric%20Strength.html (accessed on 14 July 2020).
95. Tan, D.Q. The search for enhanced dielectric strength of polymer-based dielectrics: A focused review on polymer nanocomposites. *J. Appl. Polym. Sci.* **2020**, *137*, 49379. [CrossRef]
96. Dissado, L.A.; Fothergill, J.C. *Electrical Degradation and Breakdown in Polymers*; Stevens, G.C., Ed.; IET: London, UK, 1992; ISBN 0863411967.
97. Hikita, M.; Nagao, M.; Sawa, G.; Ieda, M. Dielectric breakdown and electrical conduction of poly(vinylidene-fluoride) in high temperature region. *J. Phys. D Appl. Phys.* **1980**, *13*, 661–666. [CrossRef]
98. Jansen, J.A. Characterization of Plastics in Failure Analysis. In *Failure Analysis and Prevention*; ASM International: Novelty, OH, USA, 2018; pp. 437–459.
99. ASTM D257-14 Standard Test Methods for DC Resistance or Conductance of Insulating Materials. In *Annual B. ASTM Standard*; ASTM International: Pennsylvania, PA, USA, 2014.
100. Illias, H.A.; Tunio, M.A.; Bakar, A.H.A.; Mokhlis, H.; Chen, G. Partial discharge phenomena within an artificial void in cable insulation geometry: Experimental validation and simulation. *IEEE Trans. Dielectr. Electr. Insul.* **2016**, *23*, 451–459. [CrossRef]
101. Kreuger, F.H. *Partial Discharge Detection in High-Voltage Equipment*; Butterworths-Heinemann Ltd.: Oxford, UK, 1989; ISBN 9780408020633.
102. Bartnikas, R. Partial discharges. Their mechanism, detection and measurement. *IEEE Trans. Dielectr. Electr. Insul.* **2002**, *9*, 763–808. [CrossRef]
103. Benzerouk, D. *Pulse Sequence Analysis and Pulse Shape Analysis: Methods to Analyze Partial Discharge Processes*; University of Siegen: Siegen, Germany, 2008.
104. Chen, G.; Mokhlis, H.; Abu Bakar, A.H.; Illias, H.A.; Tunio, M.A. Partial discharge behaviours within a void-dielectric system under square waveform applied voltage stress. *IET Sci. Meas. Technol.* **2014**, *8*, 81–88.
105. Edin, H. Partial Discharges Studied with Variable Frequency of the Applied Voltage. Ph.D Thesis, Department of Electrical Engineering, KTH Royal Institute of Technology, Stockhom, Sweeden, 2001.
106. Kozako, M.; Fuse, N.; Ohki, Y.; Okamoto, T.; Tanaka, T. Surface degradation of polyamide nanocomposites caused by partial discharges using IEC (b) electrodes. *IEEE Trans. Dielectr. Electr. Insul.* **2004**, *11*, 833–839. [CrossRef]
107. Henk, P.O.; Kortsen, T.W.; Kvarts, T.; Saeidi, A. Increasing the PDendurance of epoxy and XLPE insulation by nanoparticle silica dispersion in the polymer. In Proceedings of the Nordic Insulation Symposium 2001, Stockholm, Sweeden, 11–13 June 2001.
108. Middendorf, W. Successful Performance of the High Current Arc Ignition Test. *IEEE Trans. Electr. Insul.* **1982**, *EI-17*, 429–433. [CrossRef]
109. Middendorf, W.H.; Kirshteyn, M. The high current arc ignition test. *IEEE Electr. Insul. Mag.* **1988**, *4*, 21–25. [CrossRef]
110. UL Standard for Polymeric Materials. In *Short Term Property Evaluations (UL—746A)*; Underwriters' Laboratories: Chicago, IL, USA, 2012.
111. Wagner, R.; Stimitz, J. *Properties of Plastic Materials for Use in Automotive Applications—Study of Arc Track Properties of Plastic Materials when Subjected to DC Voltages Ranging from 12 VDC—150 VDC: Addendum Report*; Underwriters' Laboratories: Chicago, IL, USA, 2004.
112. Watanabe, T.; Sato, M.; Kuwajima, H. High current DC arc ignition testing of GRP. *Composites* **1982**, *13*, 24–28. [CrossRef]
113. Zuercher, J.; Hetzmannseder, E. Testing Arc Faults in Automotive 42 V Applications. In Proceedings of the 42 Volt Automotive Systems Conference, Chicago, IL, USA, 17–18 September 2001.
114. Hetymannseder, E.; Zuercher, H.; Hastings, J. Method for Realistic Evaluation of Arc Faults Detection Performance. In Proceedings of the 21st International Conference on Electrical Contacts, Zurich, Switzerland, 9–12 September 2002; pp. 296–302.
115. Hastings, J.K.; Zuercher, J.C.; Hetzmannseder, E. Electrical Arcing and Material Ignition Levels. In Proceedings of the Society for Automotive Engineering (SAE) 2004 World Congress, Detroit, MI, USA, 8–11 March 2004.

116. Jiang, H.; Brazis, P.; Navarro, N. DC high-energy arcing ignition (HAI) resistance for polymeric materials: Part I: Consistency and repeatability of DC-HAI system. In Proceedings of the 2012 IEEE Symposium on Product Compliance Engineering Proceedings, Portland, OR, USA, 5–7 November 2012; pp. 1–4.
117. Stone, G.C. Advancements during the past quarter century in on-line monitoring of motor and generator winding insulation. *IEEE Trans. Dielectr. Electr. Insul.* **2002**, *9*, 746–751. [CrossRef]
118. North West Laboratory Insulation. Materials for Transformers Winding. Available online: http://ferrite.ru/products/electrical_tape_en/ (accessed on 8 March 2019).
119. Kirby, A.J. *Polyimides: Materials, Processing and Applications*; RAPRA Technology Limited Shawbury: RAPRA Review Reports; Pergamon Press: Oxford, UK, 1992; ISBN 9780080419664.
120. Moore, G.F. *Electric Cables Handbook*, 3rd ed.; Blackwell Science: Oxford, UK, 1997; ISBN 9780632040759.
121. Souza, R.E.; Gomes, R.M.; Lima, G.S.; Silveira, F.H.; De Conti, A.; Visacro, S. Analysis of the impulse breakdown behavior of covered cables used in compact distribution lines. *Electr. Power Syst. Res.* **2018**, *159*, 24–30. [CrossRef]
122. Hanley, T.L.; Burford, R.P.; Fleming, R.J.; Barber, K.W. A general review of polymeric insulation for use in HVDC cables. *IEEE Electr. Insul. Mag.* **2003**, *19*, 13–24. [CrossRef]
123. Tanaka, T. Dielectric nanocomposites with insulating properties. *IEEE Trans. Dielectr. Electr. Insul.* **2005**, *12*, 914–928. [CrossRef]
124. Abdel-Gawad, N.M.K.; El Dein, A.Z.; Mansour, D.-E.A.; Ahmed, H.M.; Darwish, M.M.F.; Lehtonen, M. Multiple enhancement of PVC cable insulation using functionalized SiO2 nanoparticles based nanocomposites. *Electr. Power Syst. Res.* **2018**, *163*, 612–625. [CrossRef]
125. Singha, S.; Thomas, M. Dielectric properties of epoxy nanocomposites. *IEEE Trans. Dielectr. Electr. Insul.* **2008**, *15*, 12–23. [CrossRef]
126. Kumara, S.; Xu, X.; Hammarström, T.; Ouyang, Y.; Pourrahimi, A.M.; Müller, C.; Serdyuk, Y.V. Electrical characterization of a new crosslinked copolymer blend for DC cable insulation. *Energies* **2020**, *13*, 1434. [CrossRef]
127. Wichmann, M.W.; Boyer, T.D.; Keller, C.T. Coil encapsulation with thermoplastic resins. In Proceedings of the Electrical Insulation Conference and Electrical Manufacturing and Coil Winding Conference, Rosemont, IL, USA, 25 September 1997; pp. 525–528.
128. Melick, D.; Browning, D.; Shelton, K. Evaluation of electrical material properties: Embedment stress testing on electrical encapsulation resins. In Proceedings of the Electrical/Electronics Insulation Conference, Chicago, IL, USA, 4–7 October 1993; pp. 639–641.
129. Vanga-Bouanga, C.; Savoie, S.; Frechette, M.F.; David, E. Modification of polyethylene properties by encapsulation of inorganic material. *IEEE Trans. Dielectr. Electr. Insul.* **2014**, *21*, 1–4. [CrossRef]
130. Locatelli, M.-L.; Khazaka, R.; Diaham, S.; Pham, C.-D.; Bechara, M.; Dinculescu, S.; Bidan, P. Evaluation of encapsulation materials for high-temperature power device packaging. *IEEE Trans. Power Electron.* **2014**, *29*, 2281–2288. [CrossRef]
131. Imperiale, I.; Reggiani, S.; Pavarese, G.; Gnani, E.; Gnudi, A.; Baccarani, G.; Ahn, W.; Alam, M.A.; Varghese, D.; Hernandez-Luna, A.; et al. Role of the Insulating fillers in the encapsulation material on the lateral charge spreading in HV-ICs. *IEEE Trans. Electron. Devices* **2017**, *64*, 1209–1216. [CrossRef]
132. George, J.; Compagno, T.; Rodgers, K.; Waldron, F.; Barrett, J. Reliability of plastic-encapsulated electronic components in supersaturated steam environments. *IEEE Trans. Components, Packag. Manuf. Technol.* **2015**, *5*, 1423–1431. [CrossRef]
133. Imai, T.; Sawa, F.; Nakano, T.; Ozaki, T.; Shimizu, T.; Kozako, M.; Tanaka, T. Effects of nano- and micro-filler mixture on electrical insulation properties of epoxy based composites. *IEEE Trans. Dielectr. Electr. Insul.* **2006**, *13*, 319–326. [CrossRef]
134. Castellon, J.; Nguyen, H.; Agnel, S.; Toureille, A.; Frechette, M.; Savoie, S.; Krivda, A.; Schmidt, L.E. Electrical properties analysis of micro and nano composite epoxy resin materials. *IEEE Trans. Dielectr. Electr. Insul.* **2011**, *18*, 651–658. [CrossRef]
135. Choudhury, M.; Mohanty, S.; Nayak, S.K.; Aphale, R. Preparation and characterization of electrically and thermally conductive polymeric nanocomposites. *J. Miner. Mater. Charact. Eng.* **2012**, *11*, 744–756. [CrossRef]
136. van der Pol, J.A.; Rongen, R.T.H.; Bruggers, H.J. Model and design rules for eliminating surface potential induced failures in high voltage integrated circuits. *Microelectron. Reliab.* **2000**, *40*, 1267–1272. [CrossRef]
137. Cornigli, D.; Reggiani, S.; Gnudi, A.; Gnani, E.; Baccarani, G.; Fabiani, D.; Varghese, D.; Tuncer, E.; Krishnan, S.; Nguyen, L. Characterization of dielectric properties and conductivity in encapsulation materials with high insulating filler contents. *IEEE Trans. Dielectr. Electr. Insul.* **2018**, *25*, 2421–2428. [CrossRef]
138. Nationwide Plastics. Plastics for Electrical Insulation. Available online: https://www.nationwideplastics.net/industries/electrical-insulation/ (accessed on 7 March 2019).
139. Vose, F.C.; Nichols, F.S. A polymer insulator for high-voltage transmission lines. *Electr. Eng.* **1963**, *82*, 684–688. [CrossRef]
140. Pigini, A. Polymeric Materials for HV Insulators. Available online: https://www.inmr.com/polymeric-materials-insulators/ (accessed on 12 April 2021).
141. Al-Gheilani, A.; Rowe, W.; Li, Y.; Wong, K.L. Stress Control Methods on a High Voltage Insulator: A Review. In *Energy Procedia*; Elsevier: Amsterdam, The Netherlands, 2017; Volume 110, pp. 95–100.
142. Qasim, S.A.; Gupta, N. Functionally graded material composites for effective stress control in insulators. In Proceedings of the IEEE International Conference on Properties and Applications of Dielectric Materials; Institute of Electrical and Electronics Engineers Inc.: Sydney, Australia, 2015; pp. 232–235.

143. Wang, F.; Zhang, P.; Gao, M.; Zhao, X.; Gao, J. Research on the non-linear conductivity characteristics of nano-Sic silicone rubber composites. In Proceedings of the Annual Report—Conference on Electrical Insulation and Dielectric Phenomena (CEIDP), Chenzhen, China, 20–23 October 2013; pp. 535–538.
144. Li, S.; Yu, S.; Feng, Y. Progress in and prospects for electrical insulating materials. *High Volt.* **2016**, *1*, 122–129. [CrossRef]

Article

Investigating the Effect of Several Model Configurations on the Transient Response of Gas-Insulated Substation during Fault Events Using an Electromagnetic Field Theory Approach

Muresan Alexandru [1], Levente Czumbil [1], Roberto Andolfato [2], Hassan Nouri [3] and Dan Doru Micu [1,*]

1 Department of Electrotechnics and Measurements, The Numerical Methods Research Laboratory (LCMN), Technical University of Cluj-Napoca, 400000 Cluj-Napoca, Romania; alexandru.muresan@ethm.utcluj.ro (M.A.); levente.czumbil@ethm.utcluj.ro (L.C.)
2 SINT Srl, 36061 Bassano del Grappa, Veneto, Italy; r.andolfato@sintingegneria.it
3 Power Systems Research Laboratory, University of West England (UWE), Bristol BS16 1QY, UK; Hassan.Nouri@uwe.ac.uk
* Correspondence: dan.micu@ethm.utcluj.ro

Received: 28 August 2020; Accepted: 20 November 2020; Published: 27 November 2020

Abstract: Assessment of very fast transient overvoltage (VFTO) requires good knowledge of the behavior of gas-insulated substation when subjected to very high frequencies. The international standards and guidelines generically present only recommendations regarding the VFTO suppression without a technical and mathematical background. Therefore, to provide an accurate image regarding the critical locations across a gas-insulated substation (GIS) from a transient response point of view, a suitable modeling technique has to be identified and developed for the substation. The paper aimed to provide an accurate assessment of the GIS holistic transient response through an electromagnetic field theory (EMF) approach. This modeling technique has always been a difficult task when it came to gas-insulated substations. However, recent studies have shown that through suitable Computer-aided design models, representing the GIS metallic ensemble, accurate results can be obtained. The paper investigated several simplifications of the computational domain considering different gas-insulated substation configurations in order to identify a suitable modeling approach without any unnecessary computational effort. The analysis was performed by adopting the partial equivalent element circuit (PEEC) approach embedded into XGSLab software package. Obtained results could provide useful hints for grounding grid designers regarding the proper development and implementation of transient ground potential rise (TGPR) mitigation techniques across a gas-insulated substation.

Keywords: gas-insulated substations; VFTO; EMF modeling; transient ground potential rise

1. Introduction

Typically, the gas-insulated substation (GIS) requires 10–25% of the area allocated to a conventional air-insulated substation due to its superior dielectric characteristics of the insulation material, sulphur hexafluoride (SF6), compared to air. SF6 gas is widely used in high-voltage energy applications also because of its high electric arc quenching capacity, which is chemically inert, nontoxic, and nonflammable [1]. Due to dielectric material strength reasons, local values of the electric field inside GIS enclosure, there are two types of GIS topology: three-phase system encapsulated in a single coaxial pipe for nominal voltages below 145 kV and each phase conductor individually encapsulated in a coaxial pipe for nominal voltages above 145 kV [2]. Besides the occurrence of the widely known and studied short circuit at power frequency fault, gas-insulated substations are subjected to severe transient regimes generated by the switching events. As a result of the relatively low speed of the disconnector switch contacts' movement during switching events, 0.6 cm/s [3], there is dielectric breakdown phenomena occurrence followed by the appearance of electric arc in the contact cavity causing the generation of transient overvoltage characterized by very-high-frequency (Very Fast Transient Overvoltage, VFTO) [4].

The amplitude and waveform of the very fast transient overvoltage (VFTO) depends on the configuration of the GIS substation, the trapped charge stored by the GIS equipment [5], the resistance of the electric arc formed between contacts [6], the capacity at the transformer terminals [7], and the gas pressure [7]. Measurements and digital simulations' amplitudes of transient overvoltage were determined to be between 0.5 p.u. and 2.05 p.u. [8–10]. During switching events in GIS, the transient electromagnetic wave travels toward the metal enclosure through sulphur hexafluoride (SF6)-air bushings and the metallic flanges located between each particular metallic enclosure [11,12]. Moreover, when the difference of the potential between the inner wall of the enclosure and the phase conductor exceeds the breakdown voltage of the dielectric material an electric arc is initiated, generating a short circuit between both metallic structures. When the voltage increases over the breakdown voltage of the insulating arrangement, an arc discharge takes place [13]. This is characterized by a heavy flow of current through the gas between the electrodes and the high dissipation of energy in the form of heat. The breakdown voltage of pure SF6 gas (below 1 bar gas pressure) is around 240 kV when a 20-mm gap between experimental electrodes is considered [14].

When a short circuit between the phase conductor and the inner wall of the enclosure occurs, due to voltage breakdown phenomena, an important amount of energy is leaked toward the metallic enclosure. Usually, GIS module enclosure is connected from place to place through grounding leads to the earthing system of the substation, thereby generating dangerous levels of transient ground potential rise (TGPR). In order to accurately compute and assess the TGPR occurring in GIS, the three-dimensional configuration of the metallic shell as well as grounding grid conductors need to be considered.

There are several locally applied mitigation techniques (like ferrite rings, shunt resistors, etc.) proposed in the literature [15–17]. However, these cannot be generally adopted due to mechanical constraints and they usually require high additional financial efforts. Moreover, the international standards and guidelines provide recommendations regarding the VFTO suppression without a technical and mathematical background [1]. An accurate assessment of the holistic transient response of the substation can highlight the critical locations across the GIS and its vicinity. As a result, suitable mitigation techniques can be applied only where they are necessary.

There are two main approaches widely used when GIS analysis needs to be employed: circuit theory and electromagnetic field theory approaches. The former is the predominant method adopted and used in industry and literature for simulation and investigation of the transient regimes in GIS. The circuit theory approach is based on the representation of encapsulated ensemble by equivalent electrical circuits and distributed parameters: propagation speed, equivalent capacitance, equivalent inductance, and characteristic impedance providing solutions in the frequency and time domain [18–24]. The circuit formulation methodology is based on Kirchhoff's laws embedded in different software interfaces providing numerical solutions for an energy system operating at a known voltage level, based only on the transverse mode of propagation of the electromagnetic wave, TEM, meaning that the method neglects the electric and magnetic field in the direction of propagation. [25]. The method neglects the effects of propagation losses (skin effect, etc.), which result in a lower damping coefficient for very-high-frequency components associated with transient regimes. An important drawback of the method consists of the impossibility to consider the three-dimensional configuration of the GIS metallic enclosure, as well as of the grounding grid conductors during the computational process. Considering such complex metallic structures possessed by a gas-insulated substation configuration (e.g., enclosure, structure resistance, grounding grid conductors) located in a relatively small air volume, electromagnetic couplings will occur between different metallic subsystems, which cannot be quantified in the final solution provided by circuit theory approach.

On the other hand, the electromagnetic field theory approach can be applied toward GIS three-dimensional geometries regardless of the complexity of the model to be studied [25]. Full-wave numerical electromagnetic analysis (NEA) methods are defined as methodologies allowing the direct numerical solution of Maxwell's equations in both frequency and time domain [25]. These methods are becoming the most promising approaches to study complex transient phenomena that cannot be straightforwardly solved by means of circuit theory or transmission line approaches (e.g., by using electromagnetic transient program (EMTP)). Computational electromagnetics are getting increasing attention not only in the research fields but also in the industrial fields as well as in gas-insulated substation analysis. A method solving Maxwell's equation directly can be classified into a differential equation–(DE) and an integral equation–(IE) based method [25]. The most commonly used numerical approaches in solving Maxwell equations, applicable to gas-insulated substation analysis, are method of moments (MoM) finite element method (FEM), finite differences time domain approach (FDTD) and partial equivalent element circuit (PEEC) method, which will be further applied and discussed. In the vast majority of the case studies available in the literature, MoM, the IE-based method approach applied to gas-insulated substation analysis, is focused on safety assessment during steady state, unbalanced condition, and power frequency faults [26–28]. Although the mentioned studies provide a comprehensive safety assessment analysis, the proposed Computer-aided design based (CAD) models representing the GIS metallic enclosure and adjacent inner phase conductors are not properly validated.

Current studies present FEM numerical approach as a viable solution for dielectric design, regarding GIS insulation material, in order to ensure that the electric field and its gradient are within acceptable limits, below critical values imposed by the manufacturer [2]. Furthermore, studies containing partial discharge detection in dielectric media (dielectric strength analysis) alongside mechanical stress analysis and voltage breakdown between disconnecting switch (DS) contacts are available in public literature [29–33]. FEM software packages can handle geometries regardless of the required accuracy (DS contacts' chamber). However, the limitation of the method arises when large geometries need to be included in the computational domain. FDTD, similar to FEM, is based on the DE form of Maxwell equations, resulting in the necessity to consider the entire volume space containing the geometry under study as computational domain, which implies a high amount of digital resources [34–39]. By adopting suitable boundary conditions (perfect matching layer [36], perfect electrical conductor [36]), the area of interest can be extended, considering several simplifying assumptions. Nevertheless, the computational effort required by considering the entire GIS metallic enclosure during the analysis represents an important drawback of the FDTD approach. The PEEC method is based on an integral equation description of the geometry that is interpreted in terms of circuit elements: partial inductances and partial capacitances, a partial potential, which represents the intermediate elements between electromagnetic field approach and interpretation of Maxwell's equations in circuit domain [40]. According to [41,42], the PEEC method shows satisfactory accuracy in comparison with experimental results and with simulation results calculated by the MoM and the FDTD method considering several types of electrodes with different diameters and positions with respect to the soil surface and lengths even when more complex structures are analyzed. Several PEEC numerical approaches applied to transient phenomena can be found in literature [43–47]. However, the extension of application toward gas-insulated substation transient analysis remains a challenging task due to high complexity of metallic structures involved in the computational domain.

The aim of the following study was to provide a clear understanding of the transient ground potential rise, across the metallic structures located inside GIS building, during switching operations considering different gas-insulated substation configurations (one GIS configuration contains a certain number of GIS modules) in order to identify the suitable modeling technique achieving efficient computational efforts. Through the proposed analysis methods, a parametric analysis was employed in order to accurately quantify the transient response of the system when an additional GIS bus section is connected to the model, by adopting a PEEC approach. For numerical modeling and simulation, the XGSLab software (XGSLab 9.4.1.5 version, SINT Ingegneria, Bassano del Grappa, Italy) package was used [48].

- Based on a rigorous state-of-the-art analysis, the authors wanted to highlight the drawbacks of the numerical methods applied toward GIS transient behavior analysis due to several important reasons. The paper aimed to address two major drawbacks of the electromagnetic modeling techniques:
- The lack of a validated digital avatar of the three-dimensional GIS metallic ensemble and adjacent metallic structures;

The computational effort limitations arising when the DE-based methods are employed by presenting the PEEC approach application on GIS transient behavior analysis thus allowing to consider of the entire substation during the simulation process.

Up to now, due to the fact that there is no available digital model representing the three-dimensional GIS enclosure, the transient response of the grounding system in the presence of the metallic enclosure, connected in several locations to the grid, hence creating closed current loops affecting the fault energy flowing throughout the substation, could not be assessed.

General Description of the System

The investigated system is a 110-kV substation with three-phase GIS modules placed in a dedicated building. The substation has a double bus configuration containing four GIS modules, further noted from BUS1 to BUS4, and a bus coupler (BC) arranged in a horizontal layout. The two-bus system of the double bus configuration is placed vertically, one on the top of the other (see Figure 1).

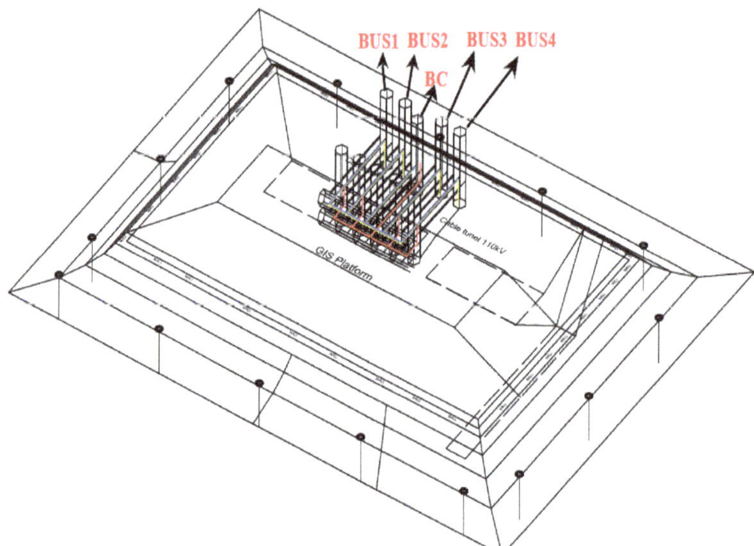

Figure 1. General layout of the substation, initial configuration.

Inside the coaxial pipe, the phase conductors are located in a radial configuration, with respect to the geometric origin of the enclosure.

The grounding grid layout can be described as follows: A copper strip contour (40 × 5 mm) is located on the GIS platform surroundings to which each GIS module is connected through two copper grounding leads. On the inner wall of the GIS building an additional copper strip contour is installed at h = 0.3 m, which ensures the conduction paths toward the external grounding system. Three copper conductor contours are buried outside the GIS building at different depths with respect to the soil surface and different distances from the GIS building wall. Taking into account the very fast time distribution of the transient regime, the auxiliary equipment located near the GIS building (autotransformers, lightning protection systems, etc.) were not considered during the computational process. The geometrical characteristics and the implemented materials, adopted during the computational process, associated with the system components are presented in Table 1.

Table 1. Geometric characteristics of the model components.

Model Component	Diameter d [mm]	Conductor Material
Phase conductor	30	Copper
aluminum Bar	40	aluminum
Copper strip	28.65	Copper
Copper conductor	12.36	Copper
Vertical rods	20	Copper

2. Modeling Concept

2.1. CAD Approach

The GIS metallic enclosure was modeled through six parallel aluminum bars interconnected in such a manner that the resultant geometry constitutes the GIS shell equivalent structure (hexagon-shape geometry) (see Figure 2). The thickness of the aluminum bars was considered to be equal with the distance between the inner and outer walls of the GIS enclosure, accounting also for the thickness of both metallic walls. The metallic flanges located at each GIS equipment enclosure ends were modeled through aluminum conductors with a similar diameter as the aluminum bars, in order to ensure the galvanic connection between every two consecutive bars.

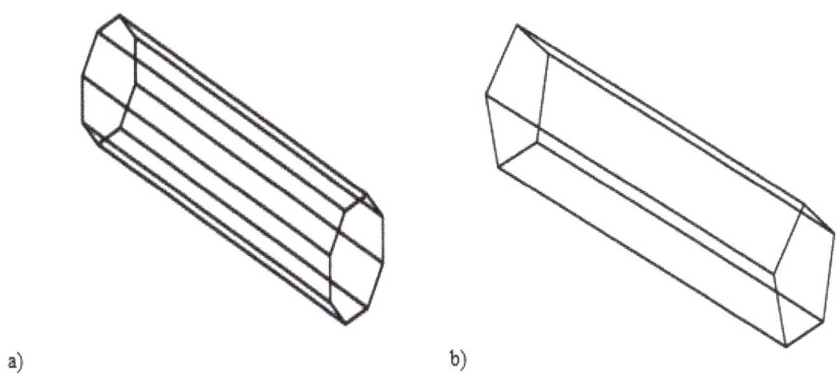

Figure 2. CAD representation of GIS pipe (**a**) octagon (**b**) pentagon.

To understand and to endorse the adopted hexagon-based geometry approach, three simplified CAD geometries (pentagon, hexagon, and octagon layouts) were designed and tested in similar modeling conditions for a single GIS bus section. The endorsement and validation of the geometric approach were performed considering lightning surge scenario, considering a voltage breakdown fault. The purpose of the study was to quantify how the different number of parallel aluminum bars representing the solid metallic pipe will impact the transient response of the enclosure during voltage breakdown fault, taking into account very-high-frequency transients.

From a geometric point of view, each CAD model was built following similar design procedures. The radius of geometric circle of the enclosure (located at a half distance between the inner and the outer GIS walls), r = 0.248 m, was taken as reference, with all three investigated geometries being inscribed in a circle with the corresponding circle.

The average errors computed analyzing the numerical values of maximum and minimum amplitudes measured for the first through fourth periods of the time domain waveform were between acceptable limits, between 2–5%. Due to the fact that the spatial distribution of the aluminum bars surrounding the phase conductor was slightly different for each particular model, dissimilarities between numerical values were to be expected. However, the neglectable differences between recorded TGPR values considering different models were obtained as a result of conductive coupling, which dominates the transient response of the system during voltage breakdown fault. According to graphical as well as numerical results (see Figure 3), there were negligible differences between the primary parameters of the TGPR when different enclosures were considered during the computational process. Due to geometric symmetry reasons as well as from a computational time point of view, the hexagon-based geometry was developed and further used for the real GIS model analysis.

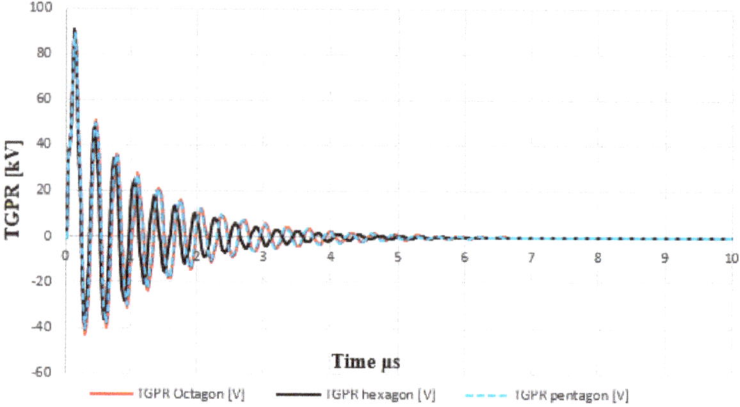

Figure 3. Transient ground potential rise during lightning surge scenario, pentagon-, hexagon-, and octagon-based geometries 10 µs.

Figure 4 illustrates the equivalent CAD model representing the metallic enclosure of a single GIS Bus configuration.

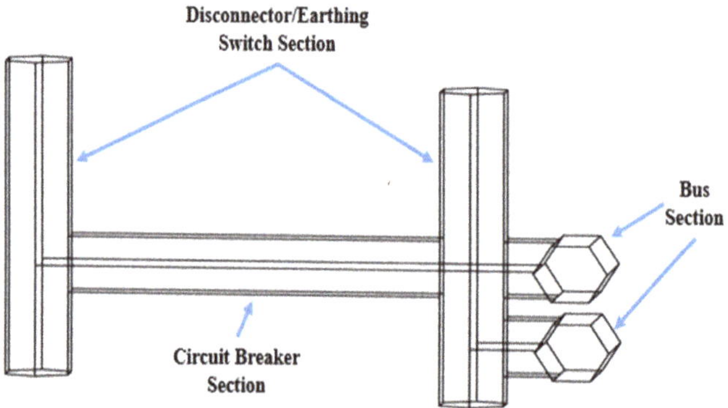

Figure 4. CAD model representing one GIS bus section enclosure.

2.2. Adopted Electromagnetic Field Theory Approach: Partial Equivalent Element Circuit Method

The partial element equivalent circuit (PEEC) method is derived from Maxwell's equations and provides a full-wave solution to them [41]. The PEEC method interprets Maxwell's equations to a circuit domain. The theoretical derivation of the PEEC method for the thin wire structures proposed by Yutthagowith and Ametani in [42] starts from a total electric field on a wire surface:

$$\bar{s}\cdot\overline{E^i}(\bar{r}) = \bar{s}\left(\overline{E^i}(\bar{r}) + \overline{E^s}(\bar{r})\right) = \bar{s}\cdot\frac{\bar{J}(\bar{r})}{\sigma} \tag{1}$$

Whole system equations in the frequency domain can be written in a matrix form corresponding to a modified nodal analysis (MNA) formulation, as shown in Equation (2):

$$\begin{vmatrix} j\omega P^{-1} + Y_a & A^T \\ -A & R + j\omega L \end{vmatrix} \cdot \begin{vmatrix} \phi \\ I \end{vmatrix} = \begin{vmatrix} I_S \\ U_S \end{vmatrix} \tag{2}$$

where A is an incident matrix that expresses the cell connectivity, R is a matrix of series resistances of current cells, L is a matrix of partial inductances of current cells including the retardation effect, P is a matrix of partial potential coefficients of the potential cells including the retardation effect, Y is a vector of potentials on the potential cells, I is a vector of currents along with current cells, U_S is a vector of voltage sources, I_S is a vector of external current sources, and Y_a is an additional admittance matrix of linear and nonlinear elements [25].

The purpose of the computer-based solvers is to accurately expend analytical methods elaborated for basic structures to real-case scenarios considering metallic towers, different geometries of the grounding grid, gas-insulated substation metallic enclosure, and the electromagnetic interaction between such structures. Although, initially, the PEEC numerical approach was intended to be applied on inductivity analysis across integrated circuit boards, improvements and modifications have been made in recent years in order to be successfully applied on transient regime analysis. As was presented in previous sections, PEEC approach has several advantages compared to DE form-based methods, which require the discretization of the entire computational domain's volume, thus imposing limitations regarding the analysis of complex geometries. Moreover, the outcome provided by PEEC approach does not require post-processing procedures in order to obtain the potential and currents along a circuit as FDTD, MoM, and FEM methods. Regarding the application toward gas-insulated substation transient behavior analysis, there are barely a few papers throughout public literature focused on this topic.

The substation is energized by two 220/110-kV autotransformers, which are modeled through equivalent impedances, $Z = 0.1\ \Omega$. Pure resistive loads were assumed during the computational process, modeled as transversal impedances to Earth connected at each particular phase conductor $Z = 1\ \Omega$. The VFTO source was modeled through an ideal electromotive force (EMF) generator situated at the tripped disconnector switch location. Beside the transient source, each metallic element became an individual electromagnetic source, which generates a particular electric and magnetic field, affecting the metallic structures located in nearby surroundings, the electromagnetic chain reaction. The transient waveform is described by a double exponential function, presented in [12] with the following parameters $\alpha = 2.31049 \cdot 10^5 \text{s}^{-1}$, $\beta = 8.17350372 \cdot 10^9 \text{s}^{-1}$, and $V_m = 1$ p.u. in order to obtain per unit results considering fault severity. The following relation describes the time domain distribution of the voltage source in the XGSLab software package:

$$V(t) = \frac{V_m}{k}\left(e^{\frac{-t}{\tau_1}} - e^{\frac{-t}{\tau_2}}\right) \tag{3}$$

where

$$\tau_1 = \frac{1}{\beta} \text{ and } \tau_2 = \frac{1}{\alpha}$$

with:

k = correction factor,
V_m = maximum amplitude of the voltage waveform,
τ_1 = time to peak parameter, and
τ_2 = time to half value.

The proposed analysis was performed assuming a phase-to-enclosure fault inside of BUS4 (see Figure 1 for BUS4 location inside GIS).

2.3. Assessment Methodology

To correctly quantify the contribution of each particular metallic section at the holistic transient behavior of the substation, a parametric analysis was employed as follows: The first stage of the computational process will contain the initial GIS configuration (5 GIS bus sections, see Figure 1), taking into account phase-to-enclosure fault inside BUS4 enclosure, modeled through a galvanic connection between the phase conductor and the inner wall of the enclosure. The results obtained in this first analysis stage will represent the baseline for the investigation carried out in this paper.

With each future analysis step an additional bus section will be removed from the model until the single GIS bus section configuration is achieved (see Figures 5 and 6 for analysis of step 2 and step 4). Minimizing the number of elements contained by the model will reduce the computational effort bringing improvements of the required computational time. With each GIS bus section extracted from the computational domain, the number of elements contained by the model decrease by approximately 16% (considering that the complete configuration contained five sections and the grounding grid structure remained constant during the simulations).

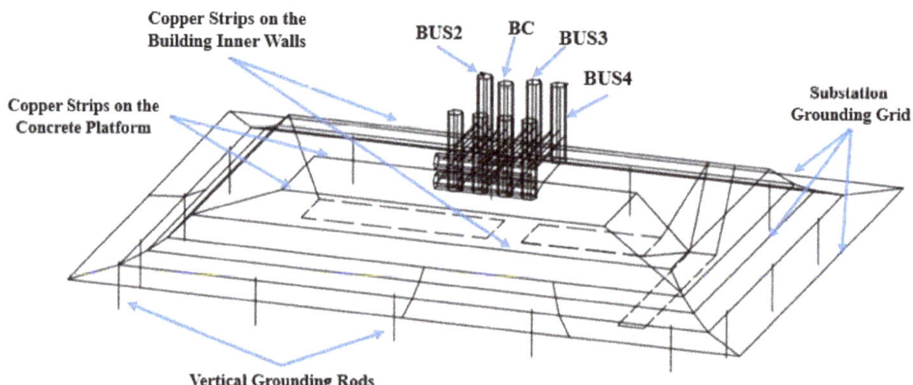

Figure 5. CAD representation of analysis stage 2: GIS model with four bus sections' modules.

Through overlapped graphical representations and numerical comparisons of the TGPR waveform computed for different GIS configurations, the transient response of the substation during voltage breakdown fault is assessed. This analysis will determine if the computed values of transient ground potential rise generated throughout the substation during phase-to-enclosure fault considering a single GIS bus section (the last stage of the investigation, see Figure 7) provide the worst-case scenario, within acceptable deviation range from the full GIS configuration model. As a result, a suitable modeling technique will be achieved and presented, and, therefore, the computational domain could be reduced for further analysis and investigations. Moreover, the transient response of the grounding grid in the presence of the metallic enclosure will be assessed and discussed. Due to the fact that there was no available 3D model representing the entire GIS metallic ensemble, quantifying the impact of the enclosure on the grounding grid performance when very-high-frequency transients flowed throughout the system was an impossible task. The response of the grid was expected not to be uniform across the substation, although it is based on the approximative symmetrical geometric shape of the earthing system due to the presence of the metallic enclosure. It is well known that during the very fast transient regime the effective area of the grounding grid is significantly reduced and, therefore, the grounding grid subsystems responsible for the fault energy clearance will be identified.

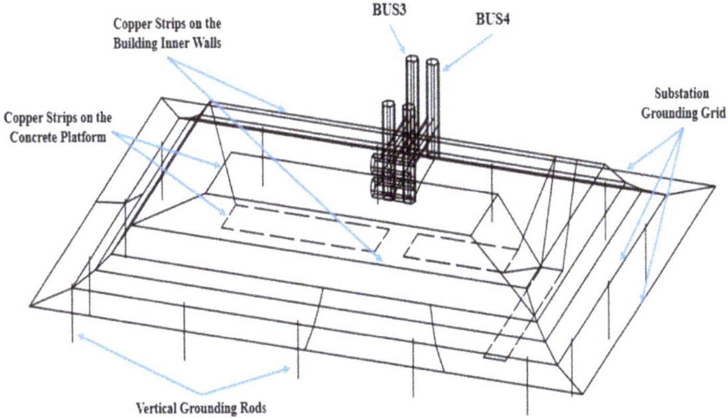

Figure 6. CAD representation of analysis stage 4: GIS model with two bus sections' modules.

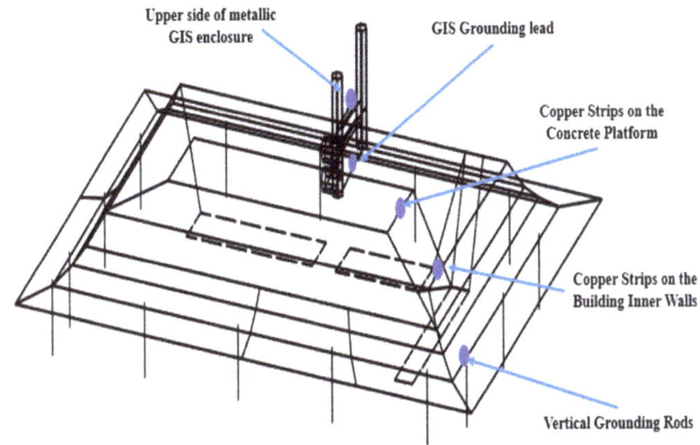

Figure 7. CAD representation of the final analysis stage: GIS model with a single bus section.

Figure 7 illustrates the CAD model considering the final analysis stage, single bus GIS configuration (containing only the faulted bus). The purple dots, highlighted in Figure 7, represent the established analysis locations across the substations, as follows: upper side of the BUS4 (faulted bus) enclosure, grounding lead that connects BUS4 enclosure with the earthing system, copper strip located on the GIS platform, copper strip located on the inner wall of the GIS building, and vertical rod.

Quantifying the total discharge current by the enclosure toward the grounding grid considering such complex electromagnetic interactions within several metallic structures requires a detailed analysis of the contribution of each particular bus section. The entire structure of the grounding grid was considered during the computational process. Hence, the impact of adding or subtracting particular GIS bus sections from the model was analyzed.

3. Simulations and Results

The simulation was performed using an I7-7700 CPU, 3.60 GHz, personal computer with 16.0 GB of RAM and required 4 h of computational time for an 800-μs simulation period for the analyzed five GIS buses' configuration.

In the following section, several simplifications of the computational domain are proposed, tested, and performed in order to try to reduce computation time. Moreover, reducing the number of elements contained by the model will simplify the mathematical formulation associated with the electromagnetic field problem under study. When GIS arrangements contain four, three, two, and single modules then the computational effort has been reduced accordingly.

It has to be mentioned that when large matrix systems describing the model are conceived by the electromagnetic algorithm implemented in XGSLab software package, a well-known problem, known as an ill-conditioned system, might arise due to the low-frequency breakdown numerical instability. An ill-conditioned problem defines a condition of a small change in the inputs (the right-hand side of the equations' system) that leads to a large change in the output without real correlation with the physical phenomenon. However, using an appropriate numerical solver and suitable mathematical techniques, an accurate solution can be provided even for matrix systems containing a large number of elements and unknowns [49]. When low frequencies were considered during the simulation process, an artificial phase shift of the transient electromagnetic wave was observed, caused by the numerical instabilities associated with the low-frequency breakdown [50]. In order to overcome the low-frequency breakdown challenge, frequencies between 0–1000 Hz were not considered during the computational process. Considering the very fast nature of the physical phenomenon, the output of the method was not affected by the low-frequency spectrum elimination.

Figure 8 shows characteristics of TGPR on the grounding lead associated with BUS4 under influence of faulted bus enclosure with the earthing system considering five, four, and one GIS bus section's(s') configuration. The analysis was performed on the grounding lead that connects the faulted bus enclosure with the earthing system for different gas-insulated substation configurations.

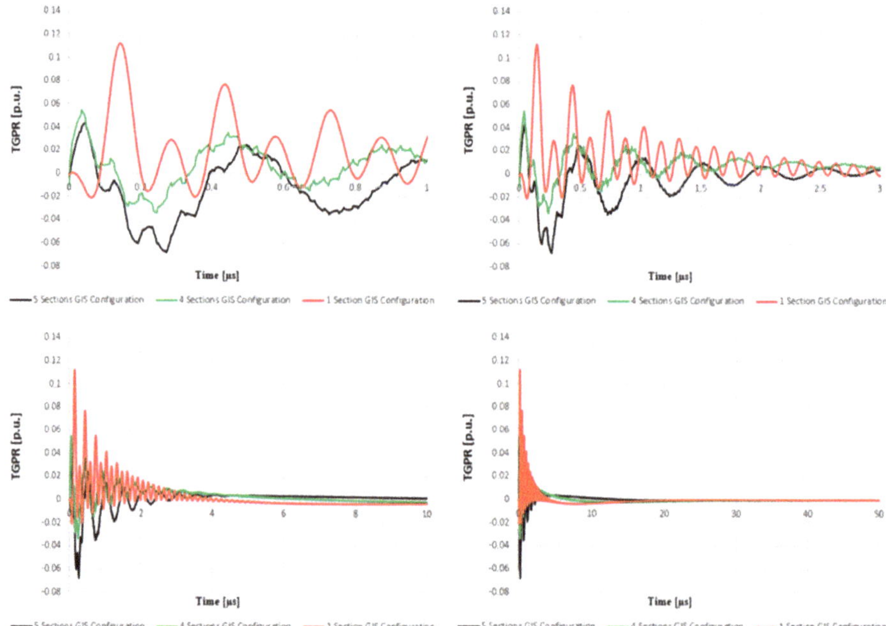

Figure 8. Transient ground potential rise on the grounding lead associated with BUS4.

Considering additional GIS bus sections into the computational domain is equivalent to generating multiple parallelism conditions between each pair of metallic elements contained by the model (aluminum bar–aluminum bar, aluminum bar–copper strip) through which electromagnetic couplings are developed. While single GIS bus module configuration (only the faulted bus section) was analyzed, the transient response computed at the level of grounding lead was more severe (approximately double), due to the fact that the fault energy was not cleared throughout a complex metallic structure as in the case of four and five GIS bus sections' configuration. It can be noted that the computed voltage waveforms were characterized by similar harmonics when four and five GIS modules were considered during the computational domain.

When the analysis was performed on the upper side of BUS4 metallic enclosure, similar behavior to that of the TGPR was observed, as in the previous case: Gradually extracting additional GIS bus section from the model caused an amplifying effect on the oscillatory character of the voltage waveform combined with the reduction of the maximum amplitude (see also Figures 9 and 10). A steeper character 0.045 (time to peak) µs can be observed when a single bus is considered in comparison with 0.14 µs when four and five buses are considered, respectively. Similar harmonics and time-to-peak parameters describing the transient voltage waveforms computed during the computational process considering five and four GIS bus sections, respectively, could be observed (see Figure 7).

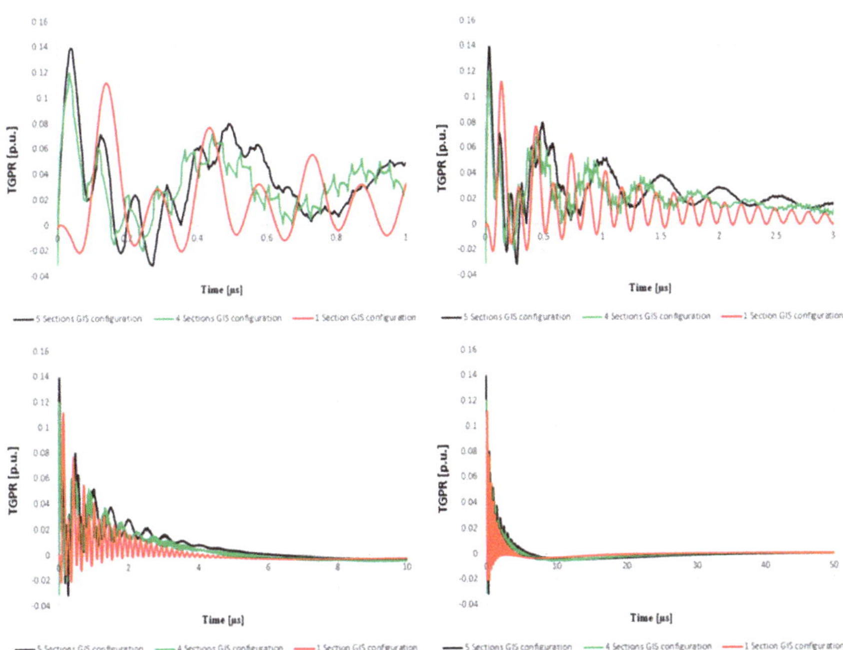

Figure 9. Transient ground potential rise on the upper side of metallic enclosure (BUS4).

Figure 10 illustrates the absolute maximum TGPR amplitude computed on the upper side of the metallic enclosure while considering different GIS configurations during the computational process. Besides the parallelism conditions generated by multiple GIS sections contained in the computational domain, the proportions of equivalent inductance, capacitance, and resistance into the circuit were modified. From the harmonic point of view, the multiple parallel GIS modules behaved as a filter (see Figures 8, 9 and 11).

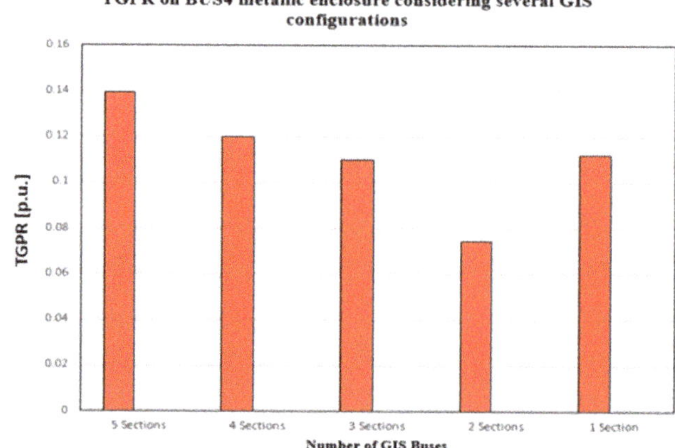

Figure 10. Comparison between TGPR computed considering five GIS configurations.

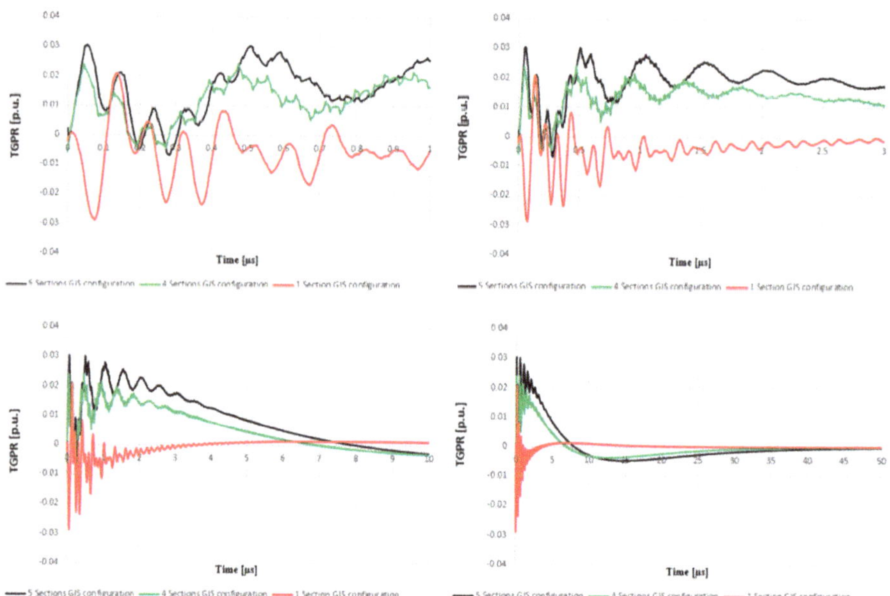

Figure 11. Transient ground potential rise across copper strip located on GIS platform, considering different GIS buses' configuration.

The amplification effect influencing the oscillatory character of the transient overvoltage waveform provided by gradually extracting bus sections from the model could be observed also across the copper strip ring located on GIS concrete platform surroundings (Figure 11). However, as the distance from the transient source increased, as well as from the aluminum bars' configuration, the impact of electromagnetic couplings developed between particular metallic enclosure elements did not affect in a similar manner the maximum TGPR values recorded as in previous locations.

Let us assume that the first significant frequency period can be determined within the first 0.2 μs (nonperiodic signal) in all the cases presented in Figure 11. The time-to-peak parameters related to TGPR waveforms computed while five, four, and one GIS bus section(s) were 0.0525 μs for first two cases and 0.0775 μs, respectively, however, with different polarities. The multiple reflections of the electromagnetic wave travelling within aluminum bars' configuration, considering multiple GIS bus sections contained by the model, generated relatively steeper voltage waveform in comparison with a single GIS bus configuration.

Figure 12 illustrates the transient overvoltage computed on the inner wall of the GIS building, analysis location, highlighted in Figure 5, considering different gas-insulated substation configurations.

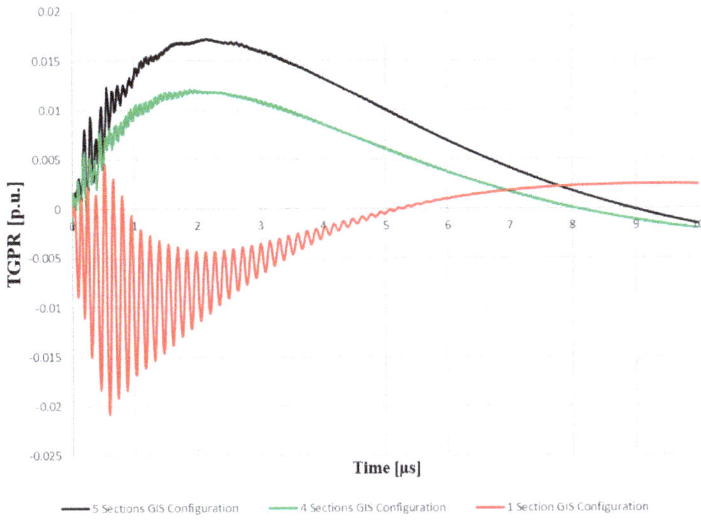

Figure 12. TGPR on the copper strip located on the inner wall of GIS building.

Considering the five GIS bus sections' configuration, the maximum amplitude describing TGPR computed across copper strip located on the inner wall of GIS building was attenuated with 87%, if compared with the maximum amplitude computed throughout the substation, which was considered as a reference value (on the metallic enclosure of BUS4).

When the transient electromagnetic wave reached the grounding grid, conductors located outside GIS building, vertical rods, and copper conductor, the oscillatory character was fully damped, as depicted in Figure 13, regardless of the GIS configuration considered during the computational process. Similar maximum amplitudes of the transient ground potential rise were computed on the vertical rod in comparison with those obtained at the previous analysis location; hence, the efficiency of the grounding grid decreased when the distance from the transient source increased. Comparing the TGPR computed, taking into account different GIS configurations, the worst-case scenario was obtained when five GIS buses were included in the computational domain and similar transient response of the vertical rods was computed when single and four GIS buses' configurations were considered. However, in order to achieve an accurate assessment regarding the holistic behavior of the substation, the number of GIS sections in the model must be limited in the computational process.

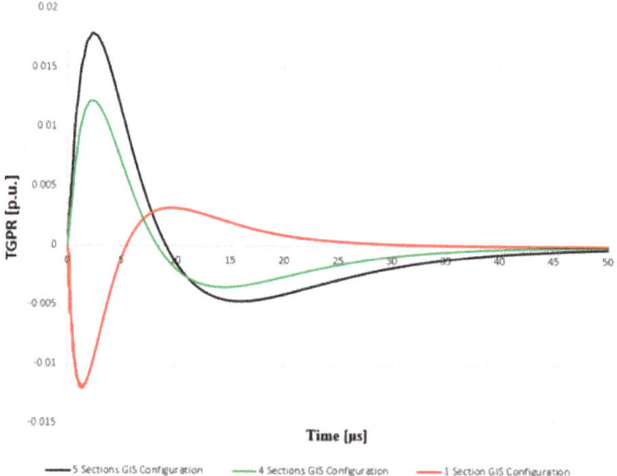

Figure 13. TGPR, one particular vertical rod, comparison considering several GIS configurations.

4. Discussions

In general, if the assessment of safety procedures is successfully achieved for the worst-case scenarios, as a conservative measure, it can be said that the substation is fully protected. However, this could result in an overdesign of the required protection system and additional unnecessary financial costs. In order to achieve a safety level of the substation, the main characteristics of the protection system must overcome the transient overvoltage generated across the substation that is described by frequency, amplitude, and time-to-peak parameter. Designing a suitable safety protection system requires that the computational domain must contain the on-site physical configuration.

It is important to mention that the TGPR waveform illustrated throughout the current paper contained incident, reflected, and refracted components. The XGSLab software package computed the electromagnetic interaction between each pair of two metallic elements contained by the model. Therefore, the outcome of the method contained the combined effects of inductive, capacitive, and conductive couplings, fully taken into account.

Based on the conclusions from the state-of-the-art section (Introduction), a 3D model representing the entire metallic ensemble was required to accurately assess the holistic transient response of a gas-insulated substation and adjacent metallic structures during voltage breakdown fault, considering the very fast transient regime. The main concept of the proposed modeling technique was to represent the metallic enclosure (coaxial pipe) through several parallel aluminum bars. A first step of the proposed analysis method was to quantify how different numbers of parallel aluminum bars representing the solid metallic pipe will impact the transient response of the enclosure during voltage breakdown fault, taking into account very-high-frequency transients. According to the computed results (see Figure 3) there were negligible differences between the primary parameters of the TGPR when different enclosures were considered during the computational process, considering certain operating conditions.

Keeping in mind the hexagonal geometric model proposed, identified as most suitable for large GIS modeling, the transient ground potential rise was computed at several locations across the substations. According to the graphical and numerical results presented throughout the current paper, the spatial distribution of the TGPR across the grounding grid, especially inside GIS building, was not uniform, although the geometric arrangement of the earthing system was relatively symmetric with respect to the GIS platform, due to the presence of the metallic enclosure. As the distance from the transient source increased, as well as from the aluminum bars' configuration, the impact of electromagnetic couplings developed between particular metallic enclosures did not affect in a similar manner the maximum TGPR values recorded at various analysis locations. The maximum attenuation coefficients were recorded when the TGPR was analyzed starting from the upper side of the metallic enclosure up the GIS platform (69% when five-section configuration was considered, see Table 2). When the observation point moved toward the extremities of the grounding grid (namely, on the vertical rods) negligible TGPR attenuation rates were obtained. Therefore, it can be concluded that during very fast transients flowing throughout the substation the grounding grid effective area was located near the transient source, more precisely, on the GIS concrete platform.

Table 2. Maximum absolute amplitudes describing the TGPR waveforms at different locations.

	TGPR on BUS4's Metallic Enclosure				
	5 Sections	4 Sections	3 Sections	2 Sections	1 Section
Max Potential [p.u.]	0.13927	0.12002	0.11002	0.07458	0.11208
	TGPR on the BUS4's Grounding Lead				
	5 Sections	4 Sections	3 Sections	2 Sections	1 Section
Max Potential [p.u.]	0.04369	0.05475	0.05475	0.04396	0.11208
Attenuation [%]	69%	54%	50%	41%	0%
	TGPR Copper Strip Located on GIS Platform				
	5 Sections	4 Sections	3 Sections	2 Sections	1 Section
Max Potential [p.u.]	0.03037	0.02355	0.02102	0.01707	0.02897
Attenuation [%]	30%	57%	62%	61%	74%
	TGPR on Copper Strip Located on the Inner Wall of GIS Building				
	5 Sections	4 Sections	3 Sections	2 Sections	1 Section
Max Potential [p.u.]	0.01723	0.01209	0.01209	0.00605	0.02086
Attenuation [%]	43%	49%	42%	65%	28%
	TGPR on Vertical Rod				
	5 Sections	4 Sections	3 Sections	2 Sections	1 Section
Max Potential [p.u.]	0.01785	0.01218	0.01119	0.0021	0.01202
Attenuation [%]	−4%	−1%	7%	65%	42%

When the parametric study is employed several discussions can be initiated regarding the quantification of each GIS bus section into the overall transient response of the system. Through this type of analysis, the proper modeling technique is established. Moreover, through the number of GIS buses' variation during the simulations, the stability and modularity of the proposed modeling technique are achieved. The outcome of the method in all simulation conditions is strictly related to physical phenomenon rather than pure numerical dependencies. According to the TGPR comparison presented in a previous section (see Figures 9, 11 and 12), the oscillatory character of the wave is damped within the traveling path due to resistive, inductive, and capacitive behavior of the metallic ensemble. Therefore, the frequency attenuation coefficient is directly proportional with the length of propagation path and equivalent with the number of elements contained by the model. According to the maximum overvoltage amplitudes presented in Table 2, when three, four, and five GIS sections are contained by the model, the area of the grounding grid clearing the transient energy is located inside the GIS building. Moreover, the copper strip situated beneath GIS metallic enclosure (copper strip ring and the grounding leads) will dominate the damping effect provided by the grounding grid during voltage breakdown (see Table 2).

When a single GIS bus configuration is considered during the computational process, the transient response of the system behaves differently, if compared with multiple sections' scenario. Similar TGPR maximum amplitudes are computed across the metallic enclosure and the grounding leads when a single bus section configuration is considered. Having multiple GIS sections in the model is equivalent to considering multiple existing electromagnetic couplings across the substation due to multiple parallelism conditions provided by the aluminum bars' ensemble, which greatly affect the overall transient response of the substation.

Different attenuation profiles of the transient electromagnetic wave phenomena are observed when different GIS configurations are considered during the computational process. According to the previous presented statements, it must be mentioned that, under no circumstance, should the simplification of the model in order to reduce the computational time be applied; whereas significant differences occurring between the topology of the computed time domain electromagnetic waveform in all considered configurations were observed (see Figures 9, 11 and 12). Taking into account the different TGPR waveforms' topologies computed while single and multiple bus arrangements are considered, in order to design a suitable safety protection system, the computational domain must contain the on-site physical configuration.

5. Conclusions

Several important objectives were accomplished during the current study, which contributes to a better understanding of the holistic transient behavior of the gas-insulated substation.

The numerical instabilities due to the low-frequency breakdown were overpassed by neglecting the first frequencies' decades from the spectrum, between 0–1000 Hz during the computational process. By means of fast Fourier transform, the numerical algorithm embedded in the XGSLab software package computed the harmonic components of the transient source waveform, and the solution of the associated mathematical problem was obtained in the frequency domain. Applying the inverse fast Fourier transform, the time domain solution was computed. Considering the very fast nature of the physical phenomenon, the harmonic components with characteristic frequencies below 1000 Hz did not affect the overall transient behavior of the system under study.

A novel electromagnetic field-based model was developed, representing a real gas-insulated substation configuration. Based on the TGPR assessment at various locations across the substations, the proper modeling technique was established.

The effective area of the grounding grid during the voltage breakdown fault was identified. Based on the numerical and graphical results, a pattern regarding the fault energy flowing throughout the substation was established, which will contribute to a better understanding of the holistic transient behavior of a typical GIS substation.

The electromagnetic couplings developed inside the GIS building between the metallic enclosure and certain components of the grounding grid affected the behavior of the latter near the GIS platform.

Author Contributions: Conceptualization M.A., L.C., and D.D.M.; methodology M.A. and L.C.; software R.A.; validation M.A. and L.C.; formal analysis D.D.M. and L.C.; investigation A.M. and L.C.; writing—original draft preparation M.A.; writing—A.M., L.C., and D.D.M; visualization D.D.M. and H.N.; supervision D.D.M. and H.N. All authors have read and agreed to the published version of the manuscript.

Funding: This research was funded the Technical University of Cluj-Napoca through its Ph.D. scholarship program. The APC was funded by European Union's Horizon 2020 research & innovation program under grant agreement Nr. 815301 and Ministry of Education and Research, "International Campus" project, Grant number: CNFIS-FDI-2020-0340.

Conflicts of Interest: The authors declare no conflict of interest.

References

1. IEEE Guide for Specifications for High-Voltage Gas-Insulated Substations Rated 52 kV and Above. In *IEEE Std C37.123-2016 (Revision of IEEE Std C37.123-1996)*; IEEE: New York, NY, USA, 2016; pp. 1–60. [CrossRef]
2. Lackovic, V. E-045 Gas Insulated Substation Definitions and Basics. Continuing Education and Development Inc., Stony Point, NY 10980. 2017. Available online: https://pdhstar.com/wp-content/uploads/2019/01/E-045-Gas-Insulated-Substation-Definitions-and-Basics-FINAL.pdf (accessed on 1 November 2020).

3. Ram, A.R. Estimation of Very Fast Transient Overvoltages on bushing connected in GIS. *Int. J. Eng. Sci. Technol.* **2012**, *4*, 457–472.
4. Kho, D.T.A.; Smith, K.S. Analysis of Very Fast Transient Overvoltages in a proposed 275 kV Gas Insulated Substation. In Proceedings of the International Conference on Power Systems Transients (IPST2011), Delft, The Netherlands, 14–17 June 2011.
5. Boggs, S.A.; Chu, F.Y.; Fujimoto, N.; Krenicky, A.; Plessl, A.; Schlicht, D. Disconnect Switch Induced Transients and Trapped Charge-Insulated Substations. *IEEE Power Eng. Rev.* **1982**, *2*, 21. [CrossRef]
6. Zhan, H.; Duan, S.; Li, C.; Yao, L.; Zhao, L. A Novel Arc Model for Very Fast Transient Overvoltage Simulation in a 252-kV Gas-Insulated Switchgear. *IEEE Trans. Plasma Sci.* **2014**, *42*, 3423–3429. [CrossRef]
7. Zhongyuan, Z.; Gui-Shu, L.; Xiang, C. Modeling of potential transformers in gas insulated substation for the very fast transient simulation. In Proceedings of the 2003 IEEE International Symposium on Electromagnetic Compatibility (EMC '03), Istanbul, Turkey, 11–16 May 2003; Volume 1, pp. 308–311.
8. Liu, B.; Tong, Y.; Deng, X.; Feng, X. Measuring of very fast transient overvoltage and very fast transient current generated by disconnector operating. In Proceedings of the 2014 International Conference on Power System Technology, Chengdu, China, 20–22 October 2014; pp. 1349–1354. [CrossRef]
9. Kuczek, T.; Florkowski, M. *Modeling of Overvoltages in Gas Insulated Substations*; ABB Corporate Research Center: Krakow, Poland, 2012.
10. Kondalu, M.; Subramanyam, P.S. Estimation of Re-striking Transient Over voltages in a 132 KV Gas insulated Substation. *Int. Electr. Eng. J. IEEJ* **2012**, *3*, 757–763.
11. Cai, Y.; Guan, Y.; Liu, W. Measurement of transient enclosure voltage in a 220 kV gas insulated substation. In Proceedings of the 2015 IEEE International Instrumentation and Measurement Technology Conference (I2MTC) Proceedings, Pisa, Italy, 11–14 May 2015; pp. 115–120.
12. Ruan, W.; Dawlibi, F.P.; Ma, J. Study of Transient Grounding Potential Rise in Gas Insulated Substation during Fault Conditions Using Electromagnetic Field and Circuit Theory Approaches. 1999, pp. 123–130. Available online: https://sestech.com/pdf/User2000_A.pdf (accessed on 1 November 2020).
13. Hong, A. Dielectric Strength of Air. In *The Physics Factbook*; 2000. Available online: https://hypertextbook.com/facts/2000/RachelChu.shtml (accessed on 1 November 2020).
14. Nechmi, H.E.; Beroual, A.; Girodet, A.; Vinson, P. Fluoronitriles/CO_2 gas mixture as promising substitute to SF6 for insulation in high voltage applications. *IEEE Trans. Dielectr. Electr. Insul.* **2016**, *23*, 2587–2593. [CrossRef]
15. Li, Q.; Wu, M. Simulation Method for the Applications of Ferromagnetic Materials in Suppressing High-Frequency Transients Within GIS. *IEEE Trans. Power Deliv.* **2007**, *22*, 1628–1632. [CrossRef]
16. Pathak, N.; Bhatti, T.S. Ibraheem Study of very fast transient overvoltages and mitigation techniques of a gas insulated substation. In Proceedings of the 2015 International Conference on Circuits, Power and Computing Technologies [ICCPCT-2015], Nagercoil, India, 19–20 March 2015; pp. 1–6.
17. Reddy, P.R.; Amarnath, J. Simulation of mitigation methods for VFTO' sand VFTC's in Gas Insulated Substations. In Proceedings of the 2013 15th International Conference on Advanced Computing Technologies (ICACT), Rajampet, India, 21 September 2013; pp. 1–4.
18. Bamne, S.; Kinhekar, N. Estimation and Analysis of VFTO at Various Locations in 132kV GIS Substation. In Proceedings of the 2018 International Conference on Computation of Power, Energy, Information and Communication (ICCPEIC), Chennai, India, 28–29 March 2018; pp. 399–403. [CrossRef]
19. Zhao, L.; Ye, L.; Wang, S.; Yang, Y.; Jiang, P.; Zou, X. Research on Very Fast Transient Overvoltage during Switching of Disconnector in 550kV GIS. In Proceedings of the 2018 IEEE 3rd International Conference on Integrated Circuits and Microsystems (ICICM), Shanghai, China, 24–26 November 2018; pp. 114–118. [CrossRef]
20. Zhuo, W.; Weiquan, W.; Qiang, W. Research of suppressing VFTO for 500kV GIS substation based on EMTP/ATP. In Proceedings of the 2011 International Conference on E-Business and E-Government (ICEE), Shanghai, China, 6–8 May 2011; pp. 1–4.
21. Na, W.; Li-ping, F.; Yu, L. The effect of apparatus model on Very Fast Transient Over-voltage. In Proceedings of the 2011 International Conference on Electric Information and Control Engineering, Wuhan, China, 25–27 March 2011; pp. 833–835.
22. Tavakoli, A.; Gholami, A.; Nouri, H.; Negnevitsky, M. Comparison Between Suppressing Approaches of Very Fast Transients in Gas-Insulated Substations (GIS). *IEEE Trans. Power Deliv.* **2012**, *28*, 303–310. [CrossRef]

23. Na, W.; Yu, L. Influence of Model of Circuit-breaker on Very Fast Transient Over-voltage. *Phys. Procedia* **2012**, *24*, 283–289. [CrossRef]
24. Haseeb, M.A.; Thomas, M.J. Computation of very fast transient overvoltages (VFTO) in a 1000 kV gas insulated substation. In Proceedings of the 2017 IEEE PES Asia-Pacific Power and Energy Engineering Conference (APPEEC), Bangalore, India, 8–10 November 2017; pp. 1–6.
25. Conference Internationale Des Grands Reseaux electriques (CIGRE) Working Group C4.501. Guideline for Numerical Electromagnetic Analysis Method and its Application to Surge Phenomena. 2013. Available online: https://infoscience.epfl.ch/record/187423/files/CigreTB_543.pdf (accessed on 1 November 2020).
26. Liu, J.; Dawalibi, F.P.; Brian, F. Majerowicz. Gas Insulated Substation Grounding System Design Using the Electromagnetic Field Method. In Proceedings of the 2012 China International Conference on Electricity Distribution, Shanghai, China, 10–14 September 2012.
27. Dawalibi, F.P.; Joyal, M.; Liu, J.; Li, Y. Realistic Integrated Grounding and Electromagnetic Interference Analysis Accounting for GIS, Cables and Transformers During Normal and Fault Conditions. In Proceedings of the 2013 IEEE PES Asia-Pacific Power and Energy Engineering Conference (APPEEC), Kowloon, China, 8–11 December 2013.
28. Zhang, J.; Qian, F.; Guo, B.; Li, Y.; Dawalibi, F. Grounding of Urban GIS Substation Connected to Commercial Buildings and Metallic Infrastructures. *Int. J. Mater. Mech. Manuf.* **2015**, *3*, 191–196. [CrossRef]
29. Chen, L.; Zhang, J.; Zeng, K.; Jiao, P. An edge-based smoothed finite element method for adaptive analysis. *Struct. Eng. Mech.* **2011**, *39*, 767–793. [CrossRef]
30. Chen, J.S.; Wu, C.T.; Belytschko, T. Regularization of material instabilities by mesh free approximations with intrinsic length scales. *Int. J. Numer. Meth. Eng.* **2000**, *47*, 1303–1322. [CrossRef]
31. Smajic, J.; Holaus, W.; Kostovic, J.; Riechert, U. 3D Full-Maxwell Simulations of Very Fast Transients in GIS. *IEEE Trans. Magn.* **2011**, *47*, 1514–1517. [CrossRef]
32. Smajic, J.; Shoory, A.; Burow, S.; Holaus, W.; Riechert, U.; Tenbohlen, S. Simulation-Based Design of HF Resonators for Damping Very Fast Transients in GIS. *IEEE Trans. Power Deliv.* **2014**, *29*, 2528–2533. [CrossRef]
33. James, J.; Albano, M.; Clark, D.; Guo, D.; Haddad, A. (Manu) Analysis of Very Fast Transients Using Black Box Macromodels in ATP-EMTP. *Energies* **2020**, *13*, 698. [CrossRef]
34. Tanaka, H.; Ohki, H.; Nagaoka, N.; Takeuchi, M.; Tanahashi, D.; Okada, N.; Baba, Y. Finite-difference time-domain simulation of partial discharges in a gas insulated switchgear. *High Volt.* **2016**, *1*, 52–56. [CrossRef]
35. Hikita, M.; Ohtsuka, S.; Okabe, S.; Wada, J.; Hoshino, T.; Maruyama, S. Influence of disconnecting part on propagation properties of PD-induced electromagnetic wave in model gis. *IEEE Trans. Dielectr. Electr. Insul.* **2010**, *17*, 1731–1767. [CrossRef]
36. Rao, M.M.; Thomas, M.J.; Singh, B.P. Electromagnetic Field Emission from Gas-to-Air Bushing in a GIS During Switching Operations. *IEEE Trans. Electromagn. Compat.* **2007**, *49*, 313–321. [CrossRef]
37. Nishigouchi, K.; Kozako, M.; Hikita, M.; Hoshino, T.; Maruyama, S.; Nakajima, T. Waveform estimation of particle discharge currents in straight 154 kV GIS using electromagnetic wave propagation simulation. *IEEE Trans. Dielectr. Electr. Insul.* **2013**, *20*, 2239–2245. [CrossRef]
38. Hikita, M.; Ohtsuka, S.; Wada, J.; Okabe, S.; Hoshino, T.; Maruyama, S. Study of partial discharge radiated electromagnetic wave propagation characteristics in an actual 154 kV model GIS. *IEEE Trans. Dielectr. Electr. Insul.* **2012**, *19*, 8–17. [CrossRef]
39. Yao, R.; Zhang, Y.; Si, G.; Wu, C.; Yang, N.; Wang, Y. Simulation Analysis on the Propagation Characteristics of Electromagnetic Wave in T-branch GIS Based on FDTD. In Proceedings of the 2015 IEEE 15th Mediterranean Microwave Symposium (MMS), Lecce, UK, 30 November–2 December 2015; pp. 1–4. [CrossRef]
40. Ruehli, A. Equivalent Circuit Models for Three-Dimensional Multiconductor Systems. *IEEE Trans. Microw. Theory Tech.* **1974**, *22*, 216–221. [CrossRef]
41. Yutthagowith, P.; Ametani, A.; Nagaoka, N.; Baba, Y. Application of a Partial Element Equivalent Circuit Method to Lightning Surge Analyses. In Proceedings of the 7th Asia-Pacific International Conference on Lightning, Chengdu, China, 1–4 November 2011.
42. Yutthagowith, P.; Ametani, A.; Nagaoka, N.; Baba, Y. Application of the Partial Element Equivalent Circuit Method to Analysis of Transient Potential Rises in Grounding Systems. *IEEE Trans. Electromagn. Compat.* **2011**, *53*, 726–736. [CrossRef]

43. Antonini, G.; Cristina, S.; Orlandi, A. PEEC modeling of high voltage towers under direct and nearby lightning strike. In Proceedings of the 10th International Symposium on High Voltage Engineering, Montreal, QC, Canada, 24–30 August1997; pp. 187–203.
44. Antonini, G.; Cristina, S.; Orlandi, A. PEEC modeling of lightning protection systems and coupling coaxial cable. *IEEE Trans. Electromagn. Compat.* **1998**, *40*, 481–491. [CrossRef]
45. Yutthagowith, P.; Ametani, A.; Nagaoka, N.; Baba, Y. Lightning-Induced Voltage Over Lossy Ground by a Hybrid Electromagnetic Circuit Model Method with Cooray–Rubinstein Formula. *IEEE Trans. Electromagn. Compat.* **2009**, *51*, 975–985. [CrossRef]
46. Wang, S.; He, J.; Zhang, B.; Zeng, R.; Yu, Z. A Time-Domain Multiport Model of Thin-Wire System for Lightning Transient Simulation. *IEEE Trans. Electromagn. Compat.* **2010**, *52*, 128–135. [CrossRef]
47. Alexandru, M.; Czumbil, L.; Polycarpou, A.; Nouri, H.; Andolfato, R.; Micu, D.D. Novel Electromagnetic Field Theory approach applied to Gas Insulated Substation Transient Behaviour during Lightning Surge Impulse. In Proceedings of the MedPower 2020, Online, 9–12 November 2020.
48. XGSLab User Guide. Available online: https://www.scribd.com/document/405051708/XGSLab-UserGuide-pdf (accessed on 1 November 2020).
49. Gope, D.; Ruehli, A.; Jandhyala, V. Solving Low-Frequency EM-CKT Problems Using the PEEC Method. *IEEE Trans. Adv. Packag.* **2007**, *30*, 313–320. [CrossRef]
50. Gür, U.M.; Ergül, O. Low-frequency breakdown of the potential integral equations and its remedy. In Proceedings of the 2017 Progress in Electromagnetics Research Symposium—Fall (PIERS—FALL), Singapore, 19–22 November 2017; pp. 676–682.

Publisher's Note: MDPI stays neutral with regard to jurisdictional claims in published maps and institutional affiliations.

Article

Partial Discharge Behaviour of a Protrusion in Gas-Insulated Systems under DC Voltage Stress

Thomas Götz *,†, Hannah Kirchner † and Karsten Backhaus

Chair of High Voltage and High Current Engineering, Faculty of Electrical and Computer Engineering, Institute of Electrical Power Systems and High Voltage Engineering (IEEH), Technische Universität Dresden, 01069 Dresden, Germany; kirchner.hannah@yahoo.de (H.K.); karsten.backhaus@tu-dresden.de (K.B.)

* Correspondence: Thomas.Goetz1@tu-dresden.de

† These authors contributed equally to this work.

Received: 20 May 2020; Accepted: 5 June 2020; Published: 16 June 2020

Abstract: High reliability, independence from environmental conditions, and the compact design of gas-insulated systems will lead to a wide application in future high voltage direct current (HVDC) transmission systems. Reliable operation of these assets can be ensured by applying meaningful and robust partial discharge diagnosis during development tests, acceptance tests, or operation. Therefore, the discharge behavior must be well understood. This paper aims to contribute to this understanding by investigating the partial discharge behavior of a distorted weakly inhomogeneous electrode arrangement in sulfur hexafluoride (SF_6) and synthetic air under high DC voltage stress. In order to get a better understanding, the partial discharge current is measured under the variation of the insulation gas pressure, the gas type, the electric field strength, and the voltage polarity. Derived from this, a classification of the different discharge types is performed. As a result, four different discharge types can be categorized depending on the experimental parameters: discharge impulses, discharge impulses with superimposed pulseless discharges, discharge impulses with superimposed pulseless discharges, and subsequent smaller discharges and pulseless discharges. Concluding suggestions for partial discharge measurements under DC voltage stress are given: recommendations for the necessary measurement time, the applied voltage and polarity, and useful measurement techniques.

Keywords: partial discharge; protrusion; gas-insulated system; HVDC; SF_6; synthetic air

---

1. Introduction

Gas-insulated systems (GIS) have been the state-of-the-art in high voltage alternating current (HVAC) transmission grids since the 1960s [1]. These systems have major advantages compared to air-insulated systems (AIS) like their space-saving design, their independence from environmental conditions, and their higher reliability. These advantages are beneficial for high voltage direct current (HVDC) transmission systems as well. Especially due to the growing importance of renewable energy sources and their integration into the existing power grid, longer transmission lines have to be built. This is economically feasible only by using HVDC technologies. The advantageous space-saving design of GIS can be used, for example, to reduce the size of offshore converter-platforms, to ensure a reliable power transmission in densely populated areas, and to allow building high voltage infrastructure close to protected landscapes due to their low visual impact. Therefore, gas-insulated HVDC systems are a space-saving solution for HVDC substations.

Due to the transition of the electric field, the directed movement of charge carriers, and the accumulation of charge carriers on gas-solid interfaces, the development of gas-insulated HVDC systems is challenging [2,3]. Reliable operation can be ensured, using partial discharge (PD) measurements during type tests, factory acceptance tests, on-site tests, and operation (monitoring).

For AC applications, electrical PD measurements using the phase-resolved partial discharge pattern (PRPD) have been well known since the 1960s for various types of defects [4]. Due to the missing phase relation, the procedures and analysis tools used under AC voltage stress cannot be applied directly for DC equipment. Additionally, the directed movement of charge carriers, space-charges in the gas, and accumulated charges at the gas-solid interfaces lead to a significantly different partial discharge behavior under DC voltage stress [5–12]. One major challenge, reported in the literature, is the occurrence of pulseless discharges, which cannot be detected using the conventional methods according to IEC 60270 [13] and ultra-high frequency (UHF) measurements [5,10,11]. Summarizing these challenges, the PD experts stated in the literature that a major problem during PD measurements and analysis under DC voltage stress is the lack of experience [14]. In addition, the physical PD behavior under DC voltage stress is not as well understood as under AC voltage stress, since space charges can accumulate over a long time period. However, a reliable PD diagnosis requires the knowledge of the discharge behavior. Hence, the aim of this investigation is to bridge the gap of knowledge for one typical defect in gas-insulated systems: a protrusion. Therefore, the partial discharge current of a fixed needle in a weakly inhomogeneous electric field is measured. The measurements reveal four different discharge types that may occur depending on the electric field strength, gas pressure, and voltage polarity.

The insulating gas sulfur-hexafluoride (SF_6) has been used as a dielectric in gas-insulated systems since the early 1960s. Due to the high global warming potential (GWP) of this gas and the related political will to reduce the emission of fluorinated gases, alternative gases with a low GWP come to the fore of manufacturers and customers [15–18]. These gases can be pure natural gases or mixtures of natural gases, like synthetic air. Furthermore, these alternatives can be gas-mixtures with highly electron affine components in order to improve their dielectric behavior. In this contribution, the discharge behavior of SF_6 is compared with the behavior of pressurized synthetic air as one example for an alternative gas.

2. Physical Fundamentals

Comprehensive knowledge about the physical fundamentals of the discharge formation is necessary, in order to understand the different discharge phenomena observed in this investigation.

A necessary requirement for the discharge inception is the presence of a starting electron. It can be emitted from the cathode [19] or by electron detachment of gas molecules. This detachment mainly occurs due to cosmic radiation [20,21]. The starting electron will be accelerated in the stationary electric field. As soon as the kinetic energy, absorbed from the electric field, is high enough, ionization due to the collision with other molecules or ions takes place. Newly generated electrons are again accelerated in the electric field. The number of ionization processes per electron and per unit of distance is described by the ionization coefficient α. At the same time, η electrons are attached to molecules or ions. If more electrons are generated than attached to ions or molecules ($\alpha > \eta$), an electron avalanche develops. Hence, the main parameter characterizing insulating gases is the effective ionization coefficient $\bar{\alpha} = \alpha - \eta$. If the effective ionization coefficient becomes $\bar{\alpha} > 0$, the number of free electrons in the gas increases like an avalanche. The ionization coefficient depends on the type of insulating gas [22], mainly determined by its electron affinity. For example, the electron affinity of oxygen is lower than that of SF_6. Nitrogen has no ability to attach free electrons. Hence, the electrical strength of synthetic air is lower than that of SF_6. The effective ionization coefficient $\bar{\alpha}$ can be calculated according to Equation (1) for SF_6 and according to Equations (2) and (3) for atmospheric air depending on the electric field strength E and the gas pressure [21,23]. Since the dielectric strength of an insulating gas depends on its gas density, $\bar{\alpha}$ is related to $p_{20\,°C}$, the gas pressure at 20 °C, in order to specify this density dependence. It is assumed that the discharge behavior of pressurized synthetic air and pressurized atmospheric air is comparable.

$$\frac{\bar{\alpha}_{SF_6}}{p_{20\,°C}} = 28\,\frac{1}{kV} \cdot \left(\frac{E}{p_{20\,°C}} - 89\,\frac{kV}{mm \cdot MPa}\right) \tag{1}$$

$$\frac{\bar{\alpha}_{air}}{p_{20\,°C}} = 0.22 \cdot \frac{mm \cdot MPa}{kV^2} \cdot \left(\frac{E}{p_{20\,°C}} - 24.4\,\frac{kV}{mm \cdot MPa}\right)^2 \tag{2}$$

$$\left\{\frac{E}{p_{20\,°C}}\middle|\, 24.4\,\frac{kV}{mm \cdot MPa} < \frac{E}{p_{20\,°C}} \leq 50\,\frac{kV}{mm \cdot MPa}\right\}$$

$$\frac{\bar{\alpha}_{air}}{p_{20\,°C}} = 0.5 \cdot \frac{(mm \cdot MPa)^{0.75}}{kV^{1.75}} \cdot \left(\frac{E}{p_{20\,°C}} - 24.4\,\frac{kV}{mm \cdot MPa}\right)^{1.75} \tag{3}$$

$$\left\{\frac{E}{p_{20\,°C}}\middle|\, 50\,\frac{kV}{mm \cdot MPa} < \frac{E}{p_{20\,°C}} \leq 120\,\frac{kV}{mm \cdot MPa}\right\}$$

According to the physics, the ionization coefficient of SF_6 is considerably lower than that of synthetic air (Figure 1). This leads to the assumption that the inception and growth of the PD avalanches will show remarkable differences between the insulating gases. Due to the significantly higher slope of the effective ionization coefficient of SF_6 compared to air in the zero crossing ($\bar{\alpha} \geq 0$), the discharge behavior of SF_6 is expected to be much stronger, dependent on small changes of the electric field strength.

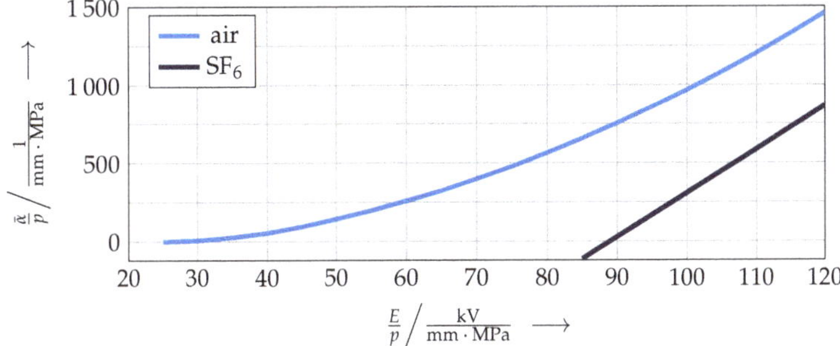

Figure 1. Effective ionization coefficient of air and SF_6 in dependence of electrical field strength and gas pressure (according to Equations (1)–(3)).

The charge carriers, electrons and ions, generate drift in the electric field with a certain velocity, which is inversely proportional to the gas pressure [20]. Due to their higher mass, the mobility μ_{ion} of positive and negative ions is significantly lower than the mobility μ_e of electrons (Table 1).

Table 1. Mobility of ions and electrons in SF_6 and air at approximately 0.1 MPa.

	μ_e in $\frac{cm^2}{V \cdot s}$	μ_{ion} in $\frac{cm^2}{V \cdot s}$	
SF_6	≈ 200	0.42 ... 1.0	[21,24–26]
Air	≈ 500	1.0 ... 2.5	[26–29]

The differences in the drift velocities of electrons and ions lead to a concentration of electrons in the avalanches' head, whereas the generated ions can be considered as remaining at their position. If the number of electrons in the avalanches' head exceeds the critical number of 10^8, the electric field strength of the avalanche, in addition to the background field, is sufficiently high to initiate photoionization, and thus, additional electron avalanches in the vicinity of the first discharge channel are started. This discharge process is well known as streamer discharge [30,31]. If streamer inception takes place, the electrical field strength exceeds the gas density dependent dielectric strength of the insulating gas in a certain region, the so-called critical volume.

In SF_6, the discharge inception is equivalent to the streamer inception [21]. In addition to this single streamer discharge, it is reported in the literature that one PD impulse with a high magnitude can be followed by several impulses with a lower magnitude [32,33]. These subsequent PD events are generated in the channel of the first streamer [32].

The charge carriers, generated by the streamers, can form a stable space charge, which significantly influences the electric field strength in this region. If the number of charges generated is equivalent to the number of charges drifting off from the space charge region, a constant pulseless direct current can be measured, known as glow discharge. It is evident that a streamer discharge might be superimposed on this glow discharge, if the space charge region becomes instable [21].

The described charge carrier movement in the electric field can be measured as a current (Figure 2). This current consists of a fast rising electron current I_e, representing the growth of the electron avalanche, and a slow ion current I_{ion}, representing the ion drift. According to the literature [33,34], the electron current is significantly higher than the ion current.

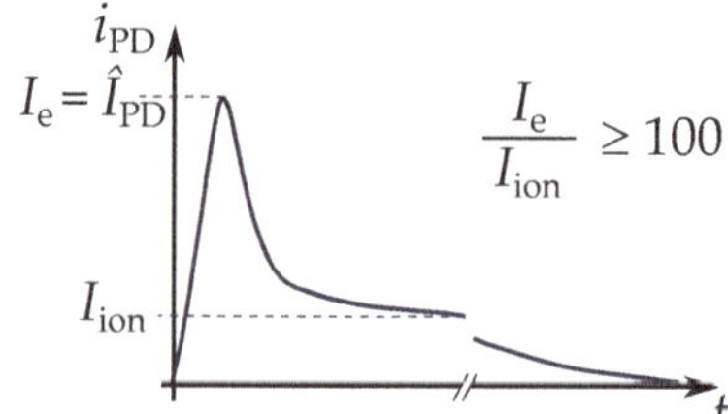

Figure 2. Partial discharge current consisting of electron current I_e and ion current I_{ion} [33].

To the authors' knowledge, there is no comprehensive study of the partial discharge behavior of a protrusion in gas-insulated systems under DC voltage stress with respect to PD current measurements. For this reason, in this contribution, the electron current I_e and the ion current I_{ion} are measured depending on the polarity, the applied voltage, and the gas pressure. It is expected that a classification of the occurring PD types can be derived from the measurements considering the current amplitudes and the time differences between subsequent discharge impulses. This classification would provide a basis for a meaningful interpretation of PD measurements in gas-insulated DC systems. A comparison of the results obtained in SF_6 with measurements in pressurized synthetic air will give an outlook for future research focusing on alternative insulating gases under DC voltage stress.

3. Experimental Setup

The experimental setup was located in a completely shielded room in order to achieve a high signal-to-noise-ratio (SNR), necessary for the measurement of low currents.

3.1. Generation of High DC Voltage with Low Ripple

Investigating the discharge physics under DC voltage stress requires a DC voltage with low ripple. This is necessary because a ripple factor δU of a few percent can lead to a phase dependent concentration of PD impulses in the voltage maximum and would therefore influence the PD behavior significantly [35].

For this reason, a symmetric Greinacher voltage doubler circuit (Figure 3) was used to generate high DC voltages [4]. It was fed with a power frequency $f = 50\,\text{Hz}$ and loaded by a current I. In contrast to common Greinacher circuits with a ripple of δU (Equation (4)), the smoothing capacitors C_s were charged every half-cycle, leading to a lower ripple δU_{symm} according to Equation (5).

$$\delta U = \frac{I}{2 \cdot f \cdot C_s} \tag{4}$$

$$\delta U_{symm} = \frac{I}{4 \cdot f \cdot C_s} \tag{5}$$

Hence, the ripple could be reduced by a factor of two. In addition, a high smoothing capacitance C_s was used to further reduce the voltage ripple. In contrast to usual realizations of the symmetric Greinacher circuit [4], two high voltage transformers with a primary voltage shifted by 180° were used in this investigation.

Due to the maximum reverse voltage of the rectifiers used, the maximum output voltage of this voltage doubler circuit was limited to $U_{DC\,max} = \pm 250\,\text{kV}$. The measured voltage ripple was approximately 250 V at its maximum ($\delta U_{symm} \leq 0.1\%$). The voltage measurement was performed using a calibrated ohmic voltage divider with a Highvolt MU17 peak voltmeter. A resistor R_d and an inductance L_d were placed in between the DC voltage supply and the test object C_p in order to limit the current in case of a breakdown.

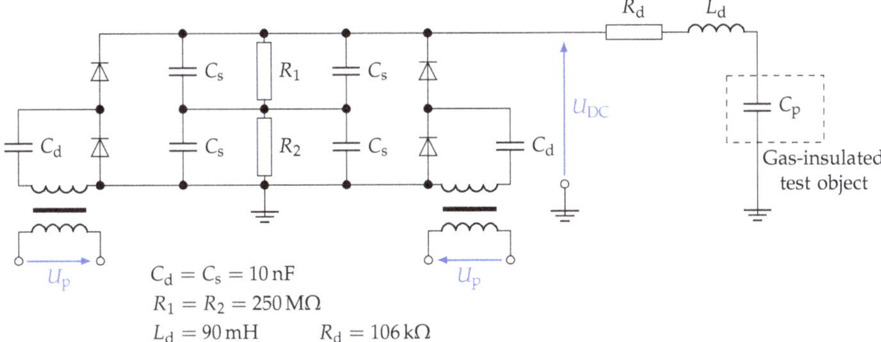

Figure 3. Circuit diagram of a symmetric Greinacher voltage doubler circuit for the generation of high DC voltage with low ripple.

3.2. Electrode Arrangement and Test Vessel

The high DC voltage supply was connected to the gas-insulated test vessel using a SF_6-air bushing (Figure 4a). The test vessel used was a commercially available part of a 420 kV GIS with a sandblasted encapsulation. It allowed investigations up to an absolute gas pressure of 0.7 MPa.

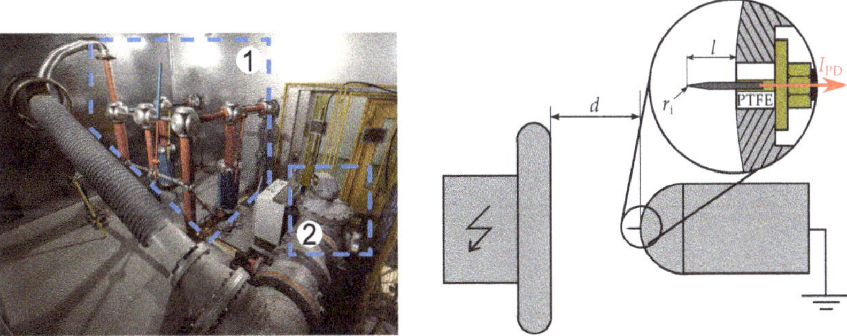

(a) Photo of the exerimental setup with ① symmetric Greinacher circuit and ② test vessel.

(b) Weakly inhomogeneous electrode arrangement with protruding needle.

Figure 4. Test setup and electrode arrangement.

In order to model the electric field of a real gas-insulated system, a weakly inhomogeneous electrode arrangement made of aluminum was placed inside the test vessel (Figure 4b). The gap distance between the half-sphere and the plate could be varied between $d = (0...100)\,\mathrm{mm}$. Hence, the degree of homogeneity η could be varied from 0.88 to 0.37. For the investigations presented in this paper, the gap distance was fixed at $d = 60\,\mathrm{mm}$. This resulted in $\eta \approx 0.52$, which was close to real applications [33]. In order to investigate the PD behavior of the insulating gases, a protrusion was placed in the middle of the half-sphere. The used needles were made of 100Cr6 steel and had a tip radius of $r_\mathrm{i} \approx 22\,\mathrm{\mu m}$. The length of the needle was $l = 5\,\mathrm{mm}$. To measure the partial discharge current I_PD, the needle was separated from the grounded sphere using an insulating PTFE ring and directly connected to an SMA adapter. The polarities mentioned in this investigation refer always to the polarity of the protrusion.

3.3. Measurement Setup

Since the expected PD current consisted of a pulseless current with superimposed impulse currents, the magnitude and frequency content of the whole signal were very broad. Hence, it was reasonable to use two different measurement setups.

3.3.1. Measurement of Impulse Currents

The measurements were performed using a high bandwidth oscilloscope-type Teledyne LeCroy WavePro 735 ZiA with an analogue bandwidth of 3.5 GHz, a maximum sample rate of 20 GS/s (using four channels), and a memory of 128 MSa per channel to allow long-term, high bandwidth measurements.

In order to get a deeper understanding of the partial discharge peak current $\hat{I}_\mathrm{PD}$ and the time differences between subsequent impulses, a high frequency current measurement was necessary. Hence, the needle electrode was directly connected to the oscilloscope using a low loss coaxial cable and a terminating resistor $R_\mathrm{meas} = 50\,\Omega$ in parallel to the input impedance of the oscilloscope used $R_\mathrm{osci} = 1\,\mathrm{M\Omega}\,||\,C_\mathrm{osci} = 16\,\mathrm{pF}$. Even though the used measurement device had a high analogue bandwidth, the measurement circuit (Figure 5a) led to a frequency dependent measurement impedance, especially due to the capacitance C_cable of the coaxial cable used (Figure 5b).

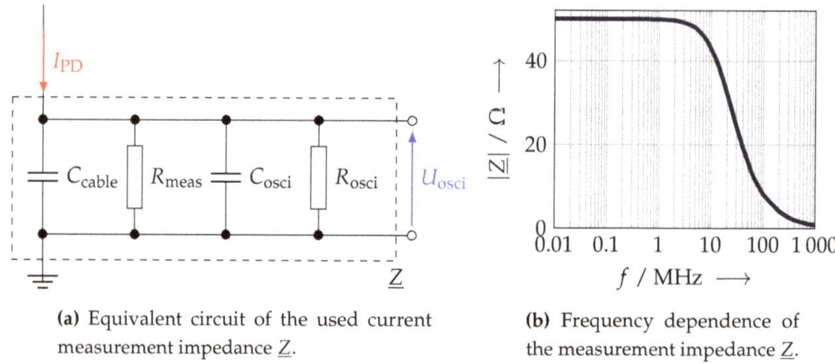

(a) Equivalent circuit of the used current measurement impedance $\underline{Z}$.

(b) Frequency dependence of the measurement impedance $\underline{Z}$.

Figure 5. Frequency dependence of the measurement circuit used.

Since the partial discharge current was calculated using Equation (6), the frequency dependence of the measurement circuit was not taken into account. This led to an amplitude error in the presented partial discharge peak currents, if their frequency content exceeded several megahertz. Due to the unknown frequency content of the partial discharge current, this approach seemed appropriate and was already used in other different publications [34,36]. Furthermore, one has to mention that the electrode arrangement influenced the measured PD current as well, mainly due to the stray capacitances between the protruding needle, the ground/high-voltage electrode, and the encapsulation of the test vessel [37]. Because the influence of these capacitances could not be easily quantified, they were neglected in this investigation.

Each occurring PD impulse triggered a 1 µs long sequence. The minimum time difference between two subsequent sequences was $\Delta t_{\mathrm{min}} = 1\,\mu\mathrm{s}$ according to the manufacturer. If a PD occurred in the dead time between two sequence recordings, it could not be detected.

$$I_{\mathrm{PD}} = \frac{U_{\mathrm{osci}}}{50\,\Omega} \tag{6}$$

Not only the partial discharge current amplitudes were investigated, but also the time differences between subsequent impulses. Since it was expected that they could occur with a rather high time difference of several milliseconds to seconds, the sequence mode of the oscilloscope was used (Figure 6) [38].

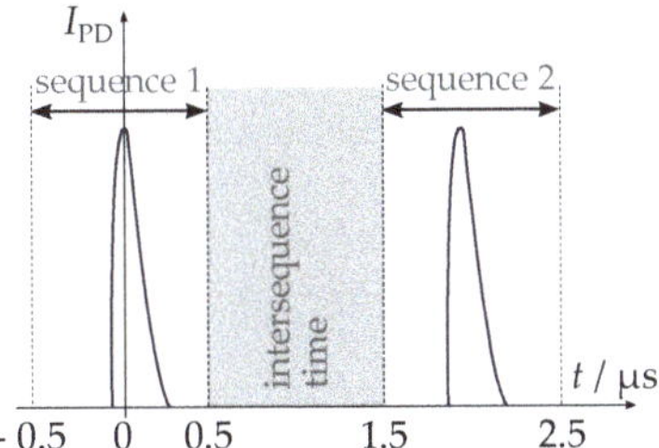

Figure 6. Sequence sampling mode.

3.3.2. Measurement of Pulseless Currents

Even though the pulseless PD current could be measured with the oscilloscope as well, it was, in terms of the expected high amplitude differences between the impulse current and the pulseless current (Figure 2), reasonable to measure the smaller pulseless currents separately. Therefore, a transimpedance amplifier was connected to the needle electrode instead of the oscilloscope. Two low-pass filters with a cut-off frequency of approximately 4 Hz and two amplifiers allowed precise measurements of the pulseless PD current with a high SNR in the range from 0.1 nA to 100 µA. The data logger was connected via Bluetooth to the measurement computer and transmitted the mean value of the measured direct current every second. The operational readiness of this measurement device was already proven by other investigations [3,39].

4. Test Execution and Data Evaluation

4.1. Test Execution

In order to understand the partial discharge behavior in dependence of the electric field strength, the investigations were performed at different voltage levels, starting from the inception voltage. Due to the missing general accepted definition of the inception voltage [38,40], the voltage with the first measurable partial discharge current impulse was determined as inception voltage U_{i} in this investigation (cf. Section 5.1). The investigations were performed at multiples of this voltage at absolute gas pressures of $p_{\mathrm{SF_6}} = 0.1\,\mathrm{MPa}$, 0.5 MPa and 0.7 MPa. The dew point of the SF_6 used was lower than −35 °C with a purity of at least 99%. The experiments in synthetic air were carried out at an absolute gas pressure of $p_{\mathrm{syn.\ air}} = 0.5\,\mathrm{MPa}$. The gas consisted of 20.5% oxygen (O_2) in nitrogen (N_2) according to the manufacturer. Its moisture content was less than 2.0 $\mathrm{ppm_{mol}}$.

Studying the impulse currents, the number of recorded sequences (each with a duration of 1 µs) depended on the time difference between subsequent impulses. For the measurements at inception voltage, five-hundred sequences were recorded and analyzed; if the applied voltage was higher than the inception voltage, 3000 to 4000 sequences were examined. An exception was the investigations at positive polarities of the protrusion at inception voltage: at higher gas pressures of $p_{SF_6} = 0.5\,\text{MPa}$ and $p_{SF_6} = 0.7\,\text{MPa}$, only ten sequences could be recorded due to the long time difference between two impulses.

The studies of the pulseless partial discharge current were carried out with a stepped voltage rising test. At each step, the voltage was kept constant for 5 min, and the mean value of the PD current measured within this period was evaluated.

The needle electrodes used were changed after every test execution in order to avoid the influence of changing tip radii on the PD behavior. More precisely, one needle was used for the investigations at one polarity and one pressure value, but different voltage levels.

4.2. Data Evaluation

In the following, the evaluation of the data acquired using the oscilloscope is discussed, to make the presentation of the results more comprehensible. The recorded sequences with 20,000 data points each were evaluated individually in MATLAB® to determine the impulse current amplitude $\hat{I}_{PD}$ and the time difference between subsequent sequences Δt (Figure 7). In this investigation, the time differences between two sequences were a measure for the time differences between subsequent impulses, since the length of one sequence of 1 µs was in most cases negligible with respect to the time difference $\Delta t >> 1\,\mu s$. If one sequence contained more than one partial discharge impulse, this would be evaluated separately.

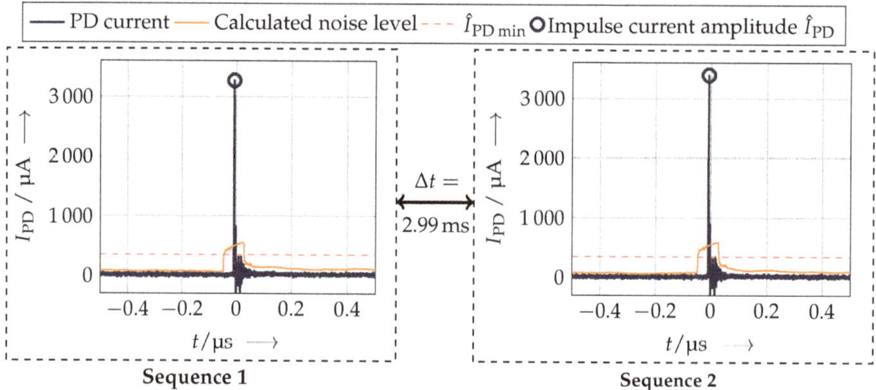

Figure 7. Sample sequences for the data evaluation (negative protrusion, $p_{SF_6} = 0.1\,\text{MPa}$, $U = 2 \cdot U_i$).

To achieve a meaningful data evaluation, the following definitions were made and applied to every recorded sequence: A noise level was calculated with a moving average out of 1500 data points multiplied by a factor of five. A PD current impulse needed to be larger than this calculated noise level and higher than a defined minimum impulse amplitude $\hat{I}_{PD\,min}$ to be evaluated as a PD event (Table 2).

Table 2. Parameters for the data evaluation.

	p/MPa	SF$_6$ $\lvert\hat{I}_{PD\,min}\rvert$/µA at U_i	$2 \cdot U_i$	$3 \cdot U_i$	$\Delta t_{imp\,min}$/ns	Synthetic Air $\lvert\hat{I}_{PD\,min}\rvert$/µA at U_i	$2 \cdot U_i$	$3 \cdot U_i$	$\Delta t_{imp\,min}$/ns
positive protrusion	0.1	350	350	- *	65	-	-	-	-
	0.5	100	350	- *	15	350	350	350	15
	0.7	100	- *	- *	15	-	-	-	-
negative protrusion	0.1	30	350	- *	65	-	-	-	-
	0.5	30	30	30	65	350	350	350	15
	0.7	35	35	- *	65	-	-	-	-

* No measurements possible because the measurement voltage was too close to breakdown voltage.

The parameter $\hat{I}_{PD\,min}$ was determined depending on the measured impulse current amplitudes, since the noise level was dependent on the chosen measurement range of the oscilloscope. If one sequence contained more than one single impulse, the definition of a minimum time difference between subsequent impulses in one sequence $\Delta t_{imp\,min}$ was necessary, to avoid the oscillations following an impulse being evaluated as another PD impulse.

5. Results

Following, the measurement results are presented. As a summary, a classification of the occurred partial discharge phenomena is derived that helps to improve the understanding and interpretation of PD measurement at a protrusion under DC voltage stress. Therefore, the impulse current amplitude $\hat{I}_{PD}$, the time difference between two subsequent sequences Δt, and the pulseless partial discharge current I_{mean} were analyzed in dependence of the gas pressure and the applied voltage.

5.1. Inception Voltage (Voltage Rising Test)

In order to determine the inception of current impulses and the pulseless PD current in dependence of the polarity of the protrusion and the gas pressure, a voltage rising test (VRT) was performed (Figure 8). The rate of voltage increase was set to $0.5\,\mathrm{kV} \cdot \mathrm{s}^{-1}$.

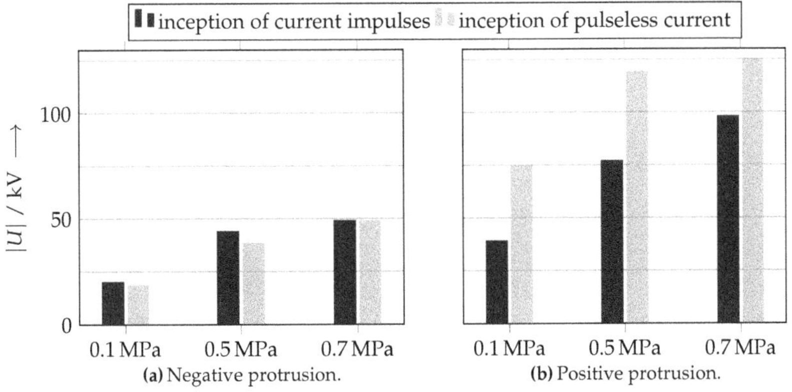

(a) Negative protrusion. **(b)** Positive protrusion.

Figure 8. Inception of current impulses and pulseless currents in dependence of gas pressure and the polarity of the protrusion in SF$_6$.

The measurements determined different voltages for the inception of impulse currents and pulseless PD currents. At a negative protrusion, these voltages were very similar. Only slight differences occurred due to measurement uncertainties and the statistical behavior of the discharge inception. At a positive protrusion, the inception of pulseless PD currents took place at higher voltages than the inception of current impulses. That led to the conclusion that a pulseless partial discharge current, probably a superposition of many small electron avalanches, was superimposed on the PD impulses and emerged in dependence of the polarity and the electric field strength.

The inception voltage of the current impulses at a positive protrusion was almost twice the inception voltage determined for a negative protrusion. These differences could be explained with the different supply of starting electrons in dependence of the polarity of the protrusion. Due to field emission processes from the cathode, more electrons were present in front of the negative protrusion [19]. The starting electrons at the positive protrusion had to be generated by collision detachment of negative ions close to the needle electrode [38]. Another fact influencing the growth of a PD impulse in the vicinity of a protrusion was the direction of movement of electrons and ions. Electrons at a positive protrusion were accelerated in the direction of the protrusion, whereas electrons at a negative protrusion were accelerated in the direction of the oppositely charged electrode. This underlined the polarity dependencies. The changing gas pressure had no influence on the described polarity effects.

An increased gas pressure led to an increased inception voltage for both polarities, which corresponded to the assumptions derived in the Physical Fundamentals Section. The effective ionization coefficient $\bar{\alpha}$ decreased with an increasing gas pressure (Figure 1).

In the following sections, the presented results always refer to the inception of PD current impulses for each polarity-pressure combination, which was thus determined as the inception voltage U_{i} in this investigation.

5.2. Partial Discharge Impulse Current

In the following, the investigations of the PD current impulses, in particular the impulse current amplitude $\hat{I}_{\mathrm{PD}}$ and the time difference between subsequent sequences Δt, are described. The investigation of the positive protrusion could only be carried out with a limited voltage, since higher voltages were too close to the breakdown voltage with a high risk of damaging the measurement equipment.

5.2.1. Amplitude of Partial Discharge Impulses

The amplitude of the partial discharge impulses depended on the polarity of the protrusion, the gas pressure, and the electric field strength (Figure 9).

In principle, a higher gas pressure led to a lower amplitude of the impulse current. The amplitude of the impulse current was in the same range for both polarities. For negative polarity of the protrusion and high gas pressures ($p_{\mathrm{SF}_6} \geq 0.5\,\mathrm{MPa}$), the partial discharge current amplitudes remained almost constant. At a gas pressure of 0.1 MPa, an increased voltage led to an increase of the amplitude of the impulse current in one order of magnitude. If the voltage was increased further up to a voltage of $U \approx 2.6 \cdot U_{\mathrm{i}}$, no current impulses could be measured any longer, whereas the measurement with the transimpedance amplifier was still showing a pulseless current (cf. Section 5.3). Obviously, the partial discharge physics had changed. It could be assumed that a pulseless glow discharge built up at the protrusion. In the investigated pressure-voltage range, this behavior could only be observed under a gas pressure of 0.1 MPa and a negative polarity of the protrusion.

When increasing the voltage at a positive protrusion and a gas pressure of 0.5 MPa, a significant increase of the amplitude of the impulse current could be measured.

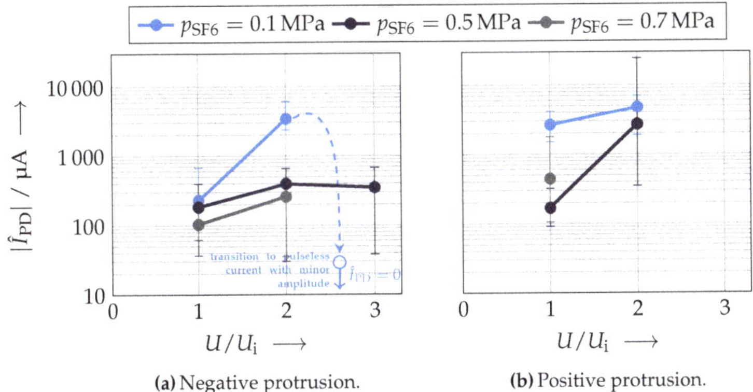

(a) Negative protrusion. **(b)** Positive protrusion.

Figure 9. Mean value and maximum/minimum of the amplitude of the impulse current in SF_6 depending on the voltage polarity and the gas pressure.

5.2.2. Time Difference between Subsequent Sequences

The observed time differences between subsequent sequences varied in a huge range between a few microseconds and several minutes (Figure 10). The scatter for almost every voltage-pressure constellation was rather high. Due to the breakdown danger, it was not possible to perform measurements at 0.7 MPa for a positive protrusion.

In general, the time differences between subsequent sequences at a positive protrusion were smaller than at a negative one, and a higher voltage led to a decreased time between subsequent sequences. At a gas pressure of 0.1 MPa, a higher voltage led to a higher time difference between subsequent impulses. The space charge in front of the protrusion became more stable, preventing the inception of further impulses. Above $2.6 \cdot U_i$, no impulses could be measured any longer at the negative protrusion. As known from the measurements under Section 5.3, only a pulseless current occurred. No statement can be given for a positive protrusion above $2 \cdot U_i$, since the voltage could not be increased further due to the danger of a breakdown. At a positive protrusion and a gas pressure of 0.5 MPa, a remarkable change of the PD behavior could be observed, when increasing the voltage from $1 \cdot U_i$ to $2 \cdot U_i$: the amplitude of the impulse current increased by one order of magnitude (cf. Figure 9), whereas the time between subsequent sequences decreased by approximately seven orders of magnitude (Figure 10). Considering the time dependent PD current (Figure 11) at inception voltage, only one small current impulse could be measured per sequence (Figure 11a); whereas at $2 \cdot U_i$, one high current impulse in the range of several milliamperes was followed by several small current impulses with an amplitude of several hundred microamperes (Figure 11b).

Every recorded current impulse was followed by smaller impulses. The time differences Δt_{imp} between the first high impulse and the subsequent smaller impulses was in the range of several hundred nanoseconds (Figure 11c); its minimum was defined according to Table 2. A higher gas pressure led to lower time differences. At a gas pressure of 0.7 MPa, these subsequent impulses already occurred at inception voltage, whereas at gas pressures of 0.1 MPa and 0.5 MPa, this behavior could only be observed at twice the inception voltage. At gas pressures of 0.1 MPa, only approximately 5% of the recorded sequences contained these subsequent impulses.

This transition from a streamer discharge to a streamer discharge with subsequent impulses was in accordance with the literature [6,32,33,38]. The subsequent impulses were probably generated in the discharge channel of the first discharge streamer. Due to the present space charge of the first discharge, the critical volume decreased, leading to a smaller amplitude of the subsequent impulses [32]. This behavior occurred mainly at a positive protrusion due to the different drift velocities of ions and electrons (cf. Table 1), resulting in a different space charge distribution compared to a negative protrusion.

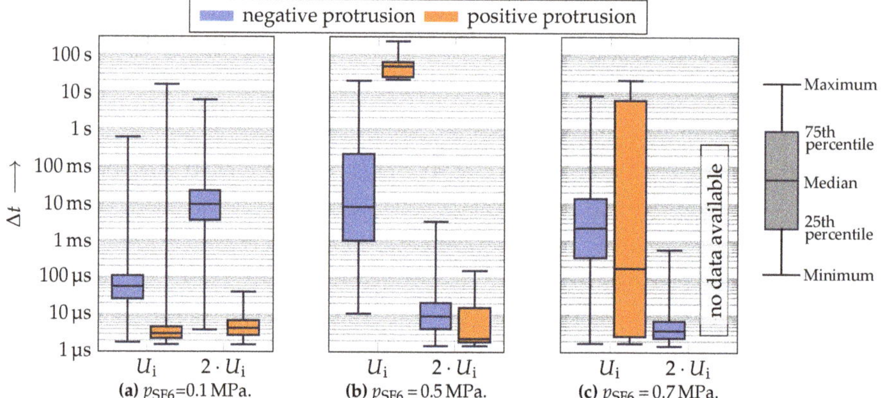

Figure 10. Time differences between subsequent sequences in SF_6 depending on the gas pressure and the applied voltage.

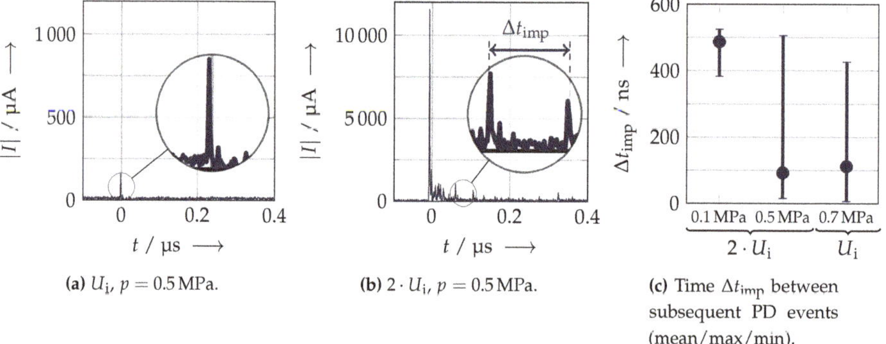

Figure 11. Comparison of time dependent partial discharge current and time differences Δt_{imp} between subsequent PD events at a positive protrusion in SF_6.

At a negative protrusion and gas pressures $p \geq 0.5\,\text{MPa}$, the time difference between subsequent sequences decreased by three orders of magnitude (Figure 10), and the amplitude of the impulses remained almost constant with increasing voltage (Figure 9). A behavior such as under positive voltage could only be observed for less than one percent of the recorded sequences at higher gas pressures ($p_{SF_6} \geq 0.5\,\text{MPa}$). The occurrence of these subsequent impulses at a negative protrusion became more probable with higher voltages.

5.3. Pulseless Partial Discharge Current

Measurements with the transimpedance amplifier showed the pulseless direct current share of the discharges (Figure 12).

This current was mainly related to the movement of slow positive and negative ions, which drifted along the electric field lines. The amplitude of the pulseless PD current was magnitudes below the impulse current amplitudes. As described in Section 5.1, the inception of the pulseless current was dependent on the polarity of the protrusion. At a negative protrusion, pulseless currents and impulse currents incepted at once at the same voltage during a VRT; whereas at a positive protrusion, the pulseless current incepted at almost twice the inception voltage of the impulse current. An increased voltage led to a higher pulseless PD current, regardless of the polarity of the protrusion and the gas pressure. A higher gas pressure led to a lower pulseless partial discharge current, but the differences at a negative protrusion were less pronounced compared to a positive protrusion.

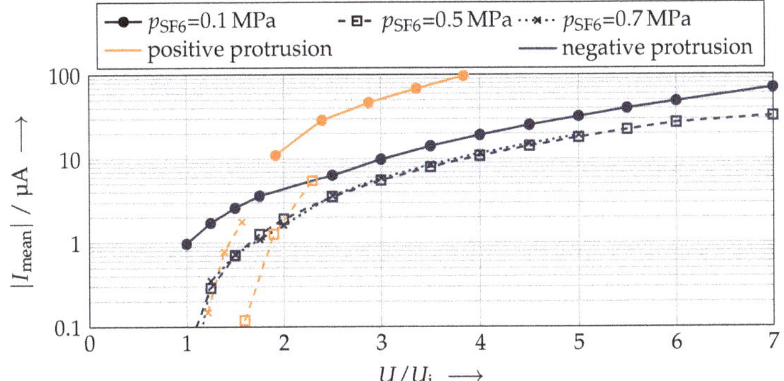

Figure 12. Pulseless partial discharge current dependent on the voltage, the polarity of the protrusion, and the gas pressure (limit of the measureable current: 100 µA).

6. Discussion

The results obtained by the analysis of the partial discharge impulse current amplitude, the time difference between subsequent discharge impulses (respectively sequences), and the pulseless partial discharge current allowed a classification of the PD types occurring in SF_6-insulated systems under DC voltage stress.

One main factor for the determination of the discharge types was the share between pulseless PD currents and current impulses. As expected from the literature, the pulseless currents (caused by the slowly moving ions) were always lower than the current impulses (caused by the fast moving electrons) (cf. Figure 2). The dependency on the applied voltage was more pronounced for the pulseless currents (Figure 13). At a positive protrusion, the pulseless current incepted at higher voltages than the streamer impulses.

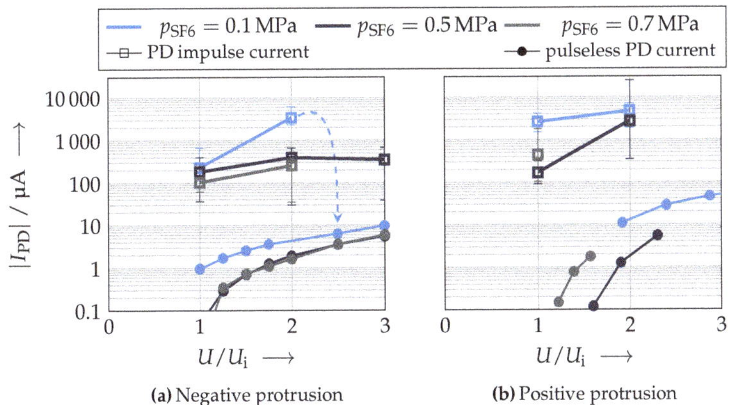

Figure 13. Comparison of pulseless PD current and the peak value of current impulses (values from Figures 9 and 12).

The pulseless current was already described by various authors for gas-insulated DC systems [10,11,21], but so far, no detailed description of the occurrence in dependence of the electric field strength, gas pressure, and voltage polarity in addition to the description of the impulse currents can be found. These pulseless discharges were caused by a permanent occurrence of avalanche discharges in front of the protrusion, which formed a stable space charge region. The parallel elapsing avalanche processes could not be measured separately, only the constant ion drift could be evaluated. If this space charge region became instable or an avalanche discharge turned over to a streamer discharge, additional current impulses could be measured. Therefore, these current impulses were superimposed on the pulseless current. Since more discharge avalanches could build up at higher voltages due to the increased critical volume, this space charge region and thus also the pulseless current increased.

According to the evaluation of the data, four different discharge types could be differentiated (Table 3).

The first type occurred at the inception voltage of the positive protrusion independently of the gas pressure; strong PD impulses could be measured, but no pulseless current was present. Hence, a streamer discharge generation took place at the protrusion. Due to the long time differences between subsequent impulses, no stable space charge region could be generated, and therefore, no pulseless discharge was superimposed on the streamer discharges.

The second one, PD current impulses superimposed on a pulseless PD current, could occur at various voltage-pressure-variations. This meant that the charges generated by the streamer impulses led to a stable space charge region in front of the protrusion, which affected the distribution of the electric field strength and therefore the size of the critical volume. In this region, continuous avalanche processes took place, and an equilibrium of charge generation and the drift of ions was reached. If this region became unstable, a streamer discharge could occur, superimposed on the glow discharge.

The third one was special for positive protrusions. A strong first discharge impulse changed the space charge region and therefore the electric field distribution significantly. Subsequent impulses with a minor amplitude took place in the first discharge channel as described in Section 2.

A PD behavior without any current impulses, so-called glow discharge, could only be determined under a gas pressure of 0.1 MPa at a negative protrusion in the investigated pressure-voltage range. This space charge region was very stable, and therefore, the generation and drift of ions were always in equilibrium. It could not be excluded that a pure glow discharge, as for low gas pressures, occurred close to the breakdown voltage even at higher gas pressures since the maximum applicable DC voltage was limited in this investigations.

Combining the described data, a classification of the determined PD types was possible, as well as the description of the transition between different types (Figure 14).

At a positive protrusion, a continuous transition between the PD types took places. This could be a transition between pure impulse currents to impulses superimposed on a pulseless current at lower gas pressures (Figure 14a) or a transition to impulses with subsequent impulses superimposed on a pulseless current for higher gas pressures (Figure 14b). At a negative protrusion in a low pressure environment, the transition to a pulseless PD current was abrupt. No change in the PD behavior could be observed at a negative protrusion at high gas pressures.

Probably a more precise distinction would be possible considering more data points, especially at gas pressures of 0.3 MPa. A defined statement for 0.7 MPa was not possible due to the fact that the breakdown occurred close to the PD inception, and therefore, no measurements were possible at higher voltages than inception voltage in order to protect the measurement devices. However, the results obtained so far allowed us to assume that it was similar to the results obtained at 0.5 MPa.

Even though the presented effective ionization coefficient SF_6 rose with enhanced electric field strength (Figure 1), the impulse amplitudes did not always rise in the same manner. Hence, the build-up of space charges influenced the formation of the streamer impulses due to a changing critical volume.

Looking at the gas pressures used in technical applications ($p_{SF_6} \approx 0.5\,\mathrm{MPa}$), it was evident that challenges arose during PD measurements under DC voltage stress. In principle, it seemed possible to identify a protrusion in gas-insulated DC systems with the well-known measurement principles based on IEC 60270 or by UHF measurements. it was difficult to distinguish between PD impulses and noise, e.g., due to the long time differences between subsequent impulses at a positive protrusion or low impulse amplitudes at a negative protrusion. Hence, a low noise environment must be established during PD measurements under DC, to achieve high sensitivity. Further challenges may arise in the precise evaluation of subsequent current impulses with low time differences [9]. It may be beneficial for a meaningful PD measurement to measure and evaluate the pulseless currents as well.

Table 3. Classification of the determined PD types and their occurrence in the investigated experimental parameter range depending on the polarity of the protrusion, the applied voltage, and the gas pressure.

	Description	Schematic Representation	Occurrence Polarity	Voltage	Gas Pressure
①	PD impulses	I_{PD}, t, ①	+	$\approx U_i$	(0.1 ... 0.7) MPa
②	Pulseless PD current with superimposed PD impulses	I_{PD}, t, ②	+/−	$\geq U_i$	(0.1 ... 0.7) MPa
③	Pulseless PD current with superimposed PD impulses and small subsequent impulses	I_{PD}, t, ③	+	$\geq 2 \cdot U_i$ $\geq U_i$	(0.1 ... 0.5) MPa 0.7 MPa
④	Pulseless PD current	I_{PD}, t, ④	−	$\gtrsim 2.5 \cdot U_i$	0.1 MPa *

* It could not be excluded that pulseless glow discharges occurred at higher pressures as well, because the measurement range (applied voltage) was limited in this investigation.

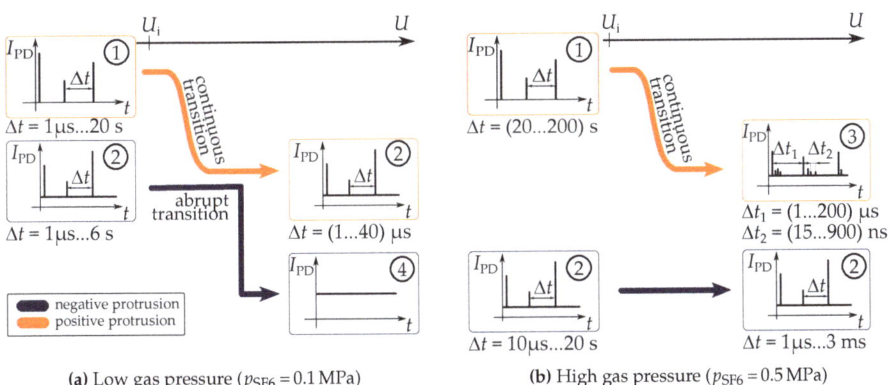

(a) Low gas pressure ($p_{SF6} = 0.1$ MPa) **(b)** High gas pressure ($p_{SF6} = 0.5$ MPa)

Figure 14. Transition between partial discharge types in dependence of the gas pressure, voltage polarity, and the applied voltage.

7. Outlook

The necessity to reduce the global greenhouse gas emissions leads to an intensified research and usage of alternative insulating gases with lower greenhouse potential [18,41]. One main challenge of the application of SF_6 alternatives will remain: the safe and reliable operation of gas-insulated systems requires a decent knowledge of the partial discharge behavior in order to achieve a reliable risk analysis out of PD measurements. Therefore, the partial discharge physics, such as pulse amplitude and pulse distance, was compared between SF_6 and one possible alternative for medium and high voltage equipment: synthetic air under higher gas pressures. The presented investigations were all carried out at a gas pressure of $p = 0.5\,\text{MPa}$. The aim of this section is to underline the applicability of the presented experimental procedure to other insulating gases and to give an outlook for future research. In contrast to the presented results of the impulse currents, these investigations will also include the pulseless currents.

7.1. Inception Voltage (Voltage Rising Test)

The inception voltages of current impulses determined with a VRT as described in Section 5.1 showed the differences expected from the comparison of the effective ionization coefficients between the two insulating gases (Figure 15). Due to the lower dielectric strength of synthetic air, its inception voltage is lower than in SF_6. The differences in between both polarities of the protrusion are more prominent in SF_6. This might be an effect of the differences in the electron affinity and the ionization energies. Since the starting electrons at a positive protrusion must be generated by detachment processes, a lower number of free electrons was present in the vicinity of the protrusion in SF_6 compared to synthetic air. This could be justified comparing the ionization energies of both gases [26] and the low lifetime of free electrons in SF_6 [21]. Hence, the voltage could increase further, and the differences in the discharge inception voltage at a positive protrusion were higher comparing both gases.

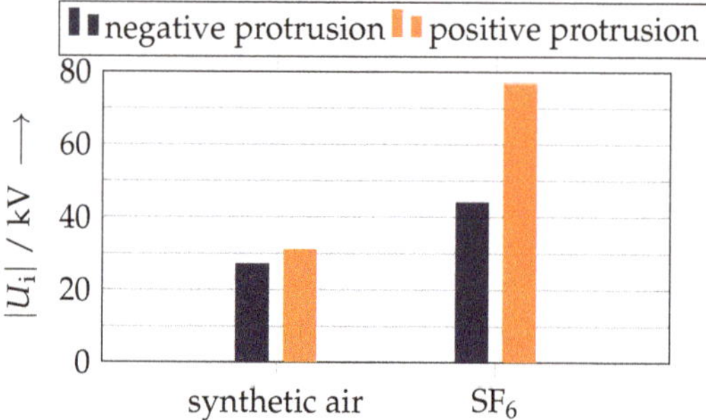

Figure 15. Inception of partial discharge impulse currents depending on the polarity of the protrusion in synthetic air and SF_6, $p = 0.5\,\text{MPa}$.

7.2. Amplitude of Partial Discharge Impulses

Comparing a single partial discharge impulse at inception voltage, it was evident that the PD behavior was significantly different (Figure 16a). Besides the higher PD amplitude at inception voltage, the time constants for the rise and decay of the impulses were different due to the significantly different effective ionization coefficients (Figure 16b).

This resulted in a higher converted charge for gases with a lower ionization coefficient. The comparison of Figure 16a,b may lead to the assumption that the rise times were dependent on the voltage polarity, especially for synthetic air as the insulating medium. This has to be evaluated in more detail during further investigations.

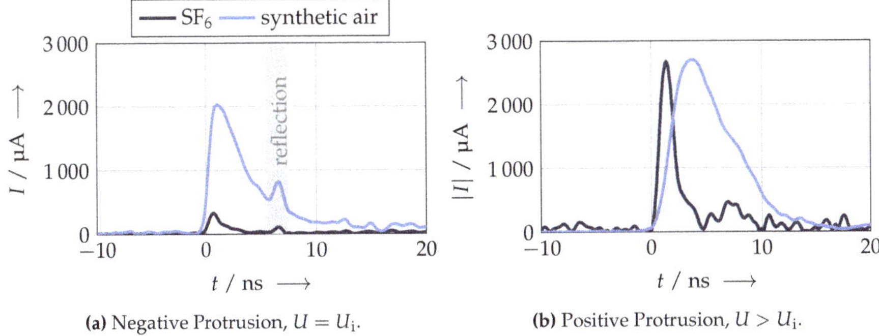

(a) Negative Protrusion, $U = U_i$.

(b) Positive Protrusion, $U > U_i$.

Figure 16. Comparison of single PD impulses in SF_6 and synthetic air $p = 0.5\,\text{MPa}$.

Besides the evaluation of a single PD impulse, the comparison of the peak values is of interest, because they can be an indicator for the detectability of PD with state-of-the-art PD measurement techniques [9]. PD impulses at a negative protrusion in synthetic air were approximately one order of magnitude higher than the impulses in SF_6, despite the applied voltage (Figure 17). Whereas the peak current amplitude for a negative protrusion in SF_6 was roughly constant, it increased in synthetic air with increasing electric field strength. At a positive protrusion, the peak currents increased significantly in SF_6 with increasing voltage, as already described in Section 5.2.1. At twice the inception voltage, there was no significant difference between the impulse amplitude in SF_6 and synthetic air. In contrast to SF_6, the PD impulses at a positive protrusion in synthetic air were in the same order of magnitude as the ones at a negative protrusion. An increasing amplitude with increasing electric stress as in SF_6 could not be observed. The described small subsequent current impulses in SF_6 (Figure 11b) could not be observed in synthetic air. Hence, the partial discharge physics seemed to be different, and no change of the type of PD could be observed with varying electric stress.

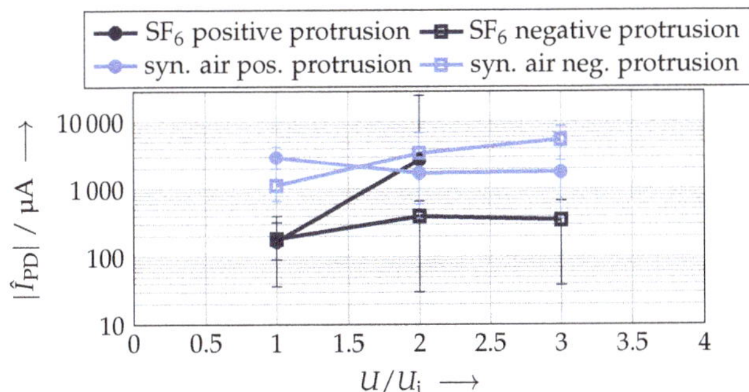

Figure 17. Mean value of partial discharge peak current and minimum/maximum in dependence of the applied voltage and polarity at a gas pressure $p = 0.5\,\text{MPa}$.

7.3. Time Difference between Subsequent Sequences

Besides the investigated variations in the impulse current amplitudes, the time differences between subsequent sequences were analyzed. At inception voltage, the time differences in synthetic air were lower than the ones observed for SF_6, independent of the voltage polarity (Figure 18).

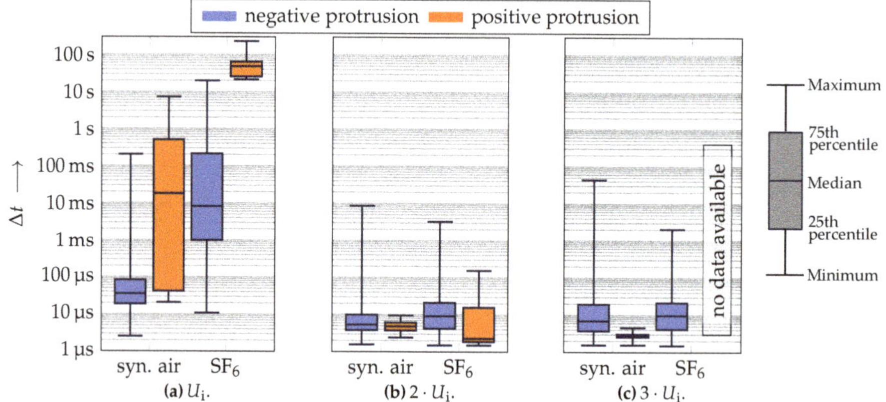

Figure 18. Comparison of time differences between subsequent sequences in synthetic air and SF_6 depending on the applied voltage and the polarity of the protrusion, $p = 0.5\,\text{MPa}$.

Increasing the voltage led to lower time differences between subsequent impulses; the differences between the two investigated gases were less pronounced and in the range of several microseconds. Considering several PD impulses within one recorded sequence ($\Delta t < 1\,\mu\text{s}$), a significant amount could only be observed in SF_6 for positive polarity (Figure 11). This behavior was only weakly developed in synthetic air and could only be observed for a few sequences at a negative protrusion at three times the inception voltage U_i.

The comparison of SF_6 with synthetic air at 0.5 MPa showed crucial differences in the discharge behavior. The different effective ionization coefficients, resulting from the different abilities of the gases to act as an electron scavenger, led to different PD current amplitudes, time differences between subsequent impulses, and time constants of the current impulses. This was explicitly shown by the comparison of single PD current impulse amplitudes (Figure 16).

It was confirmed that this type of investigation was not only applicable for SF_6-insulated systems, but also for alternative insulating gases.

8. Conclusions

Current measurements are a promising technique for the investigation of the partial discharge behavior in gas-insulated systems under DC voltage stress. The evaluation of the measured electron and ion current increased the understanding of discharge processes at one common PD source in gas-insulated systems, a protrusion. The analysis of the measurement data allowed classifying four PD types depending on the polarity of the protrusion, the electric field stress, and the pressure of the insulating gas used. The results improved the interpretation of the measurement data gained during tests and provided therefore a basis for a meaningful and robust PD analysis.

Challenges during partial discharge measurement in gas-insulated DC systems may arise as a result of varying time differences between subsequent impulses, PD current amplitudes changing over orders of magnitude, and pulseless PD currents. The time differences could be in the range from several tens of nanoseconds to several minutes and require therefore PD measurement over a sufficiently long time and with high bandwidth of the measurement system in order to obtain a detailed analysis. The precise measurement of partial discharge current impulses with a low amplitude could be beneficial for the interpretation (at the conductor or enclosure) of a defect, due to the different behavior of positive and negative protrusion. The measurement of the pulseless PD direct current in addition to the well-established measurements according to IEC 60270 and in the UHF range could be used advantageously during laboratory tests to increase the knowledge about the defect.

To obtain a meaningful PD analysis, the applied voltage during PD measurements should not be fixed to a certain value or polarity, due to the fact that glow discharges could occur at negative protrusions, which are not detectable with the state-of-the-art measurement techniques.

The studies complement the knowledge necessary for safe and reliable operation of gas-insulated DC systems, since they contributed to a meaningful PD measurement and analysis for one common PD source, a protrusion. Future research is necessary to facilitate the changeover from the currently used insulating gas SF_6 to more environmentally friendly gases like synthetic air or others.

Author Contributions: Conceptualization, T.G. and K.B.; methodology, T.G.; validation, T.G. and K.B.; formal analysis, T.G.; investigation, H.K. and T.G.; data curation, T.G. and H.K.; writing, original draft preparation, T.G.; writing, review and editing, T.G. and K.B.; visualization, T.G.; supervision, K.B.; funding acquisition, T.G. All authors read and agreed to the published version of the manuscript.

Funding: This research was funded by the Deutsche Forschungsgemeinschaft (DFG, German Research Foundation) Project Number 379542208. Open Access Funding by the Publication Fund of the TU Dresden.

Acknowledgments: The authors gratefully thank Joachim Speck, former leader of the high voltage group of the IEEH , for the valuable discussions during the funding acquisition, the construction of the experimental setup, and the interpretation of the results. Further, the authors would like to thank Maria Kosse for her continuous support and the discussions during the funding acquisition. The authors thank ABB Power Grids Switzerland Ltd. (Uwe Riechert) for providing the test vessel necessary for the execution of the experiments.

Conflicts of Interest: The authors declare no conflict of interest. The funders had no role in the design of the study; in the collection, analyses, or interpretation of data; in the writing of the manuscript; nor in the decision to publish the results.

Abbreviations

The following abbreviations are used in this manuscript:

AC	Alternating current
AIS	Air-insulated System
CIGRE	Conseil International des Grands Réseaux Électriques (International Council of Large Electric Systems)
DC	Direct current
GIS	Gas-insulated System
GWP	Global warming potential
HV	High voltage
PD	Partial discharge
PRPD	Phase-resolved partial discharge pattern
PTFE	Polytetrafluoroethylene
SF_6	Sulfur-hexafluoride
SMA	SubMiniature Version A
SNR	Signal-to-noise ratio
UHF	Ultra-high frequency
VRT	Voltage rising test

References

1. CIGRE. *GIS State of the Art 2008: TB 381*; CIGRE WG B3.17; CIGRE: Paris, France, 2009.
2. Gremaud, R.; Doiron, C.B.; Baur, M.; Simka, P.; Teppati, V.; Hering, M.; Speck, J.; Großmann, S.; Källstrand, B.; Johansson, K.; et al. *Solid-Gas Insulation in HVDC Gas-Insulated System: Measurement, Modeling and Experimental Validation for Reliable Operation: D1-363*; CIGRE Session: Paris, France, 2016.
3. Hering, M.; Speck, J.; Großmann, S.; Riechert, U. Field Transition in Gas-insulated HVDC Systems. *IEEE Trans. Dielectr. Electr. Insul.* **2017**, *24*, 1608–1616. [CrossRef]
4. Hauschild, W.; Lemke, E. *High-Voltage Test and Measuring Techniques*; Springer: Berlin/Heidelberg, Germany, 2014. [CrossRef]
5. Hering, M.; Speck, J.; Großmann, S.; Riechert, U.; Neuhold, S. *Detection of Particles on the Insulator Surface in Gas Insulated DC Systems*; Highvolt Kolloquium'15; Steiner, T: Dresden, Germany, 2015.

6. Takahashi, T.; Hayakawa, N.; Yuasa, S.; Okabe, S.; Okubo, H. Space charge behavior and corona stabilization effect in SF_6 gas viewed from sequential generation of a dc partial discharge. *J. Phys. D Appl. Phys.* **2001**, *34*, 1878–1884. [CrossRef]
7. Hering, M. Überschlagverhalten von Gas-Feststoff-Isoliersystemen unter Gleichspannungsbelastung. Ph.D. Thesis, Technische Universität Dresden, Dresden, Germany, 2016.
8. Götz, T.; Linde, T.; Simka, P.; Speck, J.; Backhaus, K.; Gabler, T.; Riechert, U.; Großmann, S. Surface discharges on dielectric coated electrodes in gas-insulated systems under DC voltage stress. In *Beiträge der Fachtagung VDE-Hochspannungstechnik*; VDE e. V.: Berlin, Germany, 2018.
9. Götz, T.; Wenger, P.; Beltle, M.; Backhaus, K.; Tenbohlen, S.; Riechert, U. *Partial Discharge Analysis in Gas-Insulated HVDC Systems Using Conventional and Non-Conventional Methods: D1-109*; CIGRE Session: Paris, France, 2020.
10. Pirker, A.; Schichler, U. HVDC GIS/GIL—PD Identification by NoDi* Pattern. In Proceedings of the International Symposium on High Voltage Engineering (ISH), Buenos Aires, Argentinia, 27 August–1 September 2017.
11. Ouss, E.; Zavattoni, L.; Beroual, A.; Girodet, A.; Vinson, P. Measurement and analysis of Partial Discharges in HVDC gas insulated substations. *CIGRÉ Sci. Eng.* **2018**, *11*, 62–68.
12. Ouss, E.; Beroual, A.; Girodet, A.; Ortiz, G.; Zavattoni, L.; Vu-Cong, T. Characterization of partial discharges from a protrusion in HVDC coaxial geometry. *IEEE Trans. Dielectr. Electr. Insul.* **2020**, *27*, 148–155. [CrossRef]
13. IEC 60270:2000. High-Voltage Test Techniques—Partial Discharge Measurements. Available online: https://webstore.iec.ch/publication/1247 (accessed on 14 June 2020).
14. Abbasi, A.; Castellon, J.; Cavallini, A.; Fritsche, R.; Geissler, M.; Götz, T.; Hochbrückner, B.; Kharezy, M.; Kosse, M.; Küchler, A.; et al. *Interim Report of WG D1.63: Progress on Partial Discharge Detection under DC Voltage Stress*; CIGRE Joint Colloquium SCA2/SCB2/SCD1: Janpath/New Delhi, India, 2019.
15. European Parliament and of the Council. Regulation (EU) No 517/2014 of the European Parliament and of the Council on Fluorinated Greenhouse Gases and Repealing Regulation (EC) No 842/2006. 2014. Available online: https://eur-lex.europa.eu/legal-content/EN/TXT/?uri=CELEX%3A32014R0517 (accessed on 14 June 2020).
16. Diggelmann, T.; Tehlar, D.; Mueller, P. *170 kV Pilot Installation with a Ketone Based Insulation Gas with First Experience from Operation in the Grid*; CIGRE Session: Paris, France, 2016.
17. Kuschel, M.; Brandler, C.; Bütüner, C.; Hansen, L.; Mortensen, A.S.B.; Gaard, J. *On-Site Experiences of 72.5 kV Clean-Air GIS for Wind-Turbine on- and Offshore Application: B3-115*; CIGRE Session: Paris, France, 2018.
18. CIGRE. *Dry air, N_2, CO_2 and N_2-SF_6 Mixtures for Gas-Insulated Systems: TB 730*; CIGRE WG D1.51; CIGRE: Paris, France, 2018.
19. Fowler, R.H.; Nordheim, L. Electron emission in intense electric fields. In *Proceedings of the Royal Society of London*; Royal Society: London, UK, 1928.
20. Kindersberger, J. The Statistical Time-Lag to Discharge Inception in SF_6. Ph.D. Thesis, TU München, München, Germany, 1987.
21. Mosch, W.; Hauschild, W. *Hochspannungsisolierungen Mit Schwefelhexafluorid*; VEB Verlag Technik: Berlin, Germany, 1979.
22. Rabie, M. A Systematic Approach to Identify and Quantify Gases for Electrical Insulation. Ph.D. Thesis, ETH Zurich, Zurich, Switzerland, 2017. [CrossRef]
23. Berger, S. Onset or breakdown voltage reduction by electrode surface roughness in air and SF_6. *IEEE Trans. Power Appar. Syst.* **1976**, *95*, 1073–1079. [CrossRef]
24. Naidu, M.S.; Prasad, A.N. Mobility and diffusion of negative ions in sulfur hexafluoride. *J. Phys. D Appl. Phys.* **1970**, *3*, 951–956. [CrossRef]
25. Teich, T.H.; Sangi, B. Discharge Parameters for some Electronegative Gases and Emission of Radiation from electron avalances. In Proceedings of the International Symposium on High Voltage Engineering (ISH), Munich, Germany, 9–14 March 1972; pp. 391–395.
26. Beyer, M. *Hochspannungstechnik: Theoretische und Praktische Grundlagen für die Anwendung*; Springer: Berlin/Heidelberg, Germany, 1992.
27. Townsend, J.S. *Electricity in Gases*; Clarendon Press: Oxford, UK, 1915.
28. Sirotinski, L.I. *Hochspannungstechnik: Band 1: Teil 1: Gasentladungen*; VEB Verlag Technik: Berlin, Germany, 1955.

29. Kuffel, E.; Zaengl, W.S.; Kuffel, J. *High Voltage Engineering: Fundamentals*, 2nd ed.; Newnes Elsevier: Amsterdam, The Netherlands, 2008.
30. Raether, H. *Electron Avalanches and Breakdown in Gases*; Butterworths: London, UK, 1964.
31. Meek, J.M.; Craggs, J.D. *Electrical Breakdown of Gases*; Wiley: Chichester, UK, 1978.
32. Hayakawa, N.; Hatta, K.; Okabe, S.; Okubo, H. Streamer and leader discharge propagation characteristics leading to breakdown in electronegative gases. *IEEE Trans. Dielectr. Electr. Insul.* **2006**, *13*, 842–849. [CrossRef]
33. Wanninger, G. Ultrahochfrequente Teilentladungssignale in gasisolierten Schaltanlagen (GIS). Ph.D. Thesis, TU München, München, Germany, 1998.
34. Herbst, I.; Pietsch, R. The fast and slow signal components of partial discharges in SF_6 measurements of the electron and ion contributions to PD-signal. In Proceedings of the International Symposium on Electrical Insulation, Pittsburgh, PA, USA, 5–8 June 1994; pp. 283–287. [CrossRef]
35. Dezenzo, T.; Betz, T.; Schwarzbacher, A. The different stages of PRPD pattern for positive point to plane corona driven by a DC voltage containing ripple. *IEEE Trans. Dielectr. Electr. Insul.* **2018**, *25*, 30–37. [CrossRef]
36. Reid, A.J.; Judd, M.D. Ultra-wide bandwidth measurement of partial discharge current pulses in SF_6. *J. Phys. D Appl. Phys.* **2012**, *16*. [CrossRef]
37. Wetzer, J.M.; van der Laan, P. Prebreakdown currents: Basic interpretation and time-resolved measurements. *IEEE Trans. Electr. Insul.* **1989**, *24*, 297–308. [CrossRef]
38. Van Brunt, R.; Misakian, M. Mechanisms for Inception of DC and 60-Hz AC Corona in SF_6. *IEEE Trans. Electr. Insul.* **1982**, *EI-17*, 106–120. [CrossRef]
39. Gabler, T.; Backhaus, K.; Großmann, S.; Fritsche, R. Dielectric modeling of oil-paper insulation systems at high DC voltage stress using a charge-carrier-based approach. *IEEE Trans. Dielectr. Electr. Insul.* **2019**, *26*, 1549–1557. [CrossRef]
40. Pirker, A.; Schober, B.; Schichler, U. PD Monitoring in HVDC GIS/GIL. In Proceedings of the CIGRE Symposium, Chengdu, China, 20–26 September 2019.
41. Glaubitz, P.; Stangherlin, S.; Biasse, J.M.; Meyer, F.; Dallet, M.; Prüfert, M.; Kurte, R.; Saida, T.; Uehara, K.; Prieur, P.; et al. CIGRE Position Paper on the Application of SF_6 in Transmission and Distribution Networks. *Electra* **2014**, 274, 34–39.

Article

Enhancing the Protective Performance of Surge Arresters against Indirect Lightning Strikes via an Inductor-Based Filter

Mahdi Pourakbari-Kasmaei * and Matti Lehtonen

Department of Electrical Engineering and Automation, Aalto University, Maarintie 8, 02150 Espoo, Finland; Matti.Lehtonen@aalto.fi

* Correspondence: Mahdi.Pourakbari@aalto.fi; Tel.: +358-46-538-48-72

Received: 11 August 2020; Accepted: 8 September 2020; Published: 11 September 2020

Abstract: Preventing the medium voltage (MV) transformer fault by protecting transformers against indirect lightning strikes plays a crucial role in enhancing the continuous service to electricity consumers. Surge arresters, if selected properly, are efficient devices in providing adequate protection for MV transformers against transient overvoltage impulses while preventing unwanted service interruptions. However, compared to other protective devices such as the spark gap, their prices are relatively high. The higher the surge arrester rating and energy absorption capacity are, the higher the prices go. This paper proposes an inductor-based filter to limit the energy pushed into the surge arrester, and consequently to prevent any unwanted failure. An energy-controlled switch is proposed to simulate the fault of the surge arrester. Surge arresters with different ratings, e.g., 12 kV, 18 kV, 24 kV, 30 kV, 36 kV, and 42 kV with two different classes of energy, namely, type a and type b, are tested under different indirect lightning impulses such as 100 kV, 125 kV, 150 kV, 175 kV, 200 kV, 250 kV, 300 kV, and 500 kV. Furthermore, these surge arresters are equipped with different filter sizes of 100 μH, 250 μH, 500 μH, and 1 mH. Results prove that equipping a surge arrester with a proper filter size enhances the performance of the surge arrester significantly such that a high rating and somewhat expensive surge arrester can be replaced by a low rating and cheap surge arrester while providing similar or even better protective performance for MV transformers. Therefore, such configurations not only enhance the protective capability of surge arrester, but also reduce the planning and operating costs of MV networks.

Keywords: double exponential function; indirect lightning; medium voltage transformer; spark gap; filtered surge arrester; energy-controlled switch

---

1. Introduction

Indirect lightning phenomena are more common than direct lightning and are considered as one of the primary sources of failures and damages in the medium voltage (MV) equipment by causing stress on their insulation system [1,2]. The MV transformers, among all the equipment, are more expensive, and thus providing adequate protection for them is of high importance. Proper protection of transformers not only enhances the system reliability as well as social welfare, but also, by controlling the transient overvoltage stress, prevents/defers extra expenses imposed by transformer failure or severe damages [3].

In practice, typically spark gaps are used to protect the transformers in MV networks against lightning impulses [4]. Although spark gaps are rather cheap protective devices, their operation yields a service interruption due to voltage chop and such voltage chopping imposes steep voltage stress across the transformer terminal [5,6]. Besides, transients may also occur due to the energization of the transformer [7] after the follow current interruption due to spark gap operation. On the other hand, surge arresters, by their charming nonlinear behavior as well as surge energy capabilities, can provide adequate protection against lightning impulses while preventing any unwanted service interruption [8]. However, utilizing a surge arrester as the protective device may not always protect the MV transformers against lightning [9,10]. A report by the Cigre Study Committee A2.37 on the failure rate of MV transformers lists the failure rate of surge arrester-protected transformers due to lightning as ~3% [11].

In the literature, there exist several works that either only conventional surge arresters are considered to protect MV transformers against lightning, or some external or internal modifications are made aiming at enhancing the protective performance of surge arresters. In internal modifications, the microstructure and electrical properties of surge arresters are optimized to withstand different overvoltage conditions [12–15]. In external modifications, which is the focus of our work, more often than not, only other protective devices are combined with surge arresters, or different sides of the transformer are chosen for installing the surge arresters aiming at enhancing the protection outcome [6]. When a conventional surge arrester is used for protecting MV transformers, several characteristics should be taken into account to prevent unwanted failures or returning to normal operating condition after absorbing surge energy [16]. Among all, the thermal energy absorption limit plays a crucial role in guaranteeing the healthy operation of the surge arrester while absorbing surge energy. Moreover, the authors of [17] showed that if the surge arrester is installed closer to the transformer, the overvoltage stress transferred to the LV side of transformer tends to increase; although, to keep the overvoltage stress as low as possible on the MV side of the transformer, the surge arrester needs to be installed as close as possible to the transformer. The authors of [18] provided adequate information on selecting the surge arresters for residential areas. In a distribution network, the placement of surge arresters is a crucial challenge [19], where installing too many surge arresters increases the probability of surge arrester failure, and consequently unwanted outages may occur [20]. One approach is to decrease the number of surge arresters in the network [20], while strengthening the surge arrester and enhancing its performance against lightning overvoltages can be another alternative. In [6], the authors concluded that instead of using a surge arrester with a rather high rating, a series connection of a spark gap and a surge arrester might provide a similar protection level, and a lower rating surge arrester can be considered. However, in this work, the thermal energy absorption limit was not thoroughly investigated. Several other combinations of the spark gap and surge arresters have been studied to enhance the protection quality. For example, in [21], the spark gaps were installed at the MV terminals of the transformer and the LV terminals were protected by low rating surge arrester, while in [22] surge arresters and spark gaps were installed at the MV and LV terminals of the transformer, respectively. In [23], experts advised installing surge arresters at both the MV and LV terminals, which will impose additional costs to the system operator. The authors of [24] proposed a novel idea in which the surge arrester wa replaced by an inexpensive LC filter, and it was shown that even by removing all surge arresters in power system, the lightning overvoltages were kept below the BIL.

According to the literature review above, and to the best of our knowledge, (a) there is a gap on modeling the failure of surge arrester due to absorbing excess energy; (b) no work, so far, considered a filter for controlling the energy pushed into the surge arrester; and (c) there is a gap in investigating the impact of series-connected spark gap on the energy absorbed by surge arresters.

Therefore, the primary contributions of this work are fourfold:

- Proposing a filtered surge arrester to mitigate lightning overvoltages by limiting the energy pushed into the surge arrester.
- Proposing an energy-controlled switch to model the fault behavior of surge arrester due to absorbing excess energy.
- Investigating the impact of the proposed filter on enhancing the protective performance of the surge arrester for protecting MV transformers against indirect lightning phenomena.
- Investigating the impact of the joint spark gap and proposed surge arrester on protecting an MV transformer against indirect lightning phenomena.

The remainder of this paper is organized as follows. Section 2 presents the simulation set-up for different components of the test. Case studies and simulation results are provided in Section 3. Section 4 presents the concluding remarks and prospects of future works.

2. Simulation Setup

This section provides detailed information on the simulation set-up for testing the proposed filtered surge arrester. The components to be modeled in EMTP-RV are indirect lightning impulse, transformer, spark gap, surge arrester, and energy-controlled switch.

2.1. Indirect Lightning

In this paper, the double exponential Equation (1) is used to generate standard 1.2/50 µs indirect lightning, where 1.2 µs stands for the front-time and 50 µs represents the time-to-half value. It is worth mentioning that this standard waveform is subject to a simplification since this representation assumes that the lightning current rises linearly until reaching the peak and neglects the influence of channel-base current [25,26]. In EMTP-RV, there exist a voltage surge device that takes advantage of a double exponential function to generate lightning impulses, see Figure 1.

$$V(t) = V_m(e^{\alpha t} - e^{\beta t}) \tag{1}$$

where V_m is the maximum voltage of the source, and α and β are the coefficients to adjust the front time (e.g., 1.2 µs ± 30%) and time-to-half value (e.g., 50 µs ± 20%), respectively [27]. It is worth mentioning that, in EMTP-RV, this device has two options for start time and stop time to have more control over generating the surge impulses. However, generating indirect lightning impulse in software by using a double exponential function is not always an easy task. Determining the sensitive parameters of α and β for the double exponential function (1) is a challenging issue. In [28], an innovative approach was used to define the parameters of α and β, while for the sake of facilitating the parameter adjustments, an A factor was also used. Then, all the parameters A, α, and β were obtained via trial and error method, and the results show some violations compared to the standard 1.2/50 µs lightnng impulse. Therefore, adjusting these sensitive parameters requires utilizing some optimization model. In this work, the optimization model developed by Pourakbari-Kasmaei et al. in [29] is modified as (2)–(6).

$$\begin{aligned} &\max \alpha + \beta \\ &\text{s.t.} \end{aligned} \tag{2}$$

$$V_m(e^{\alpha \cdot t} - e^{\beta \cdot t}) \leq V_p; \;\; \forall t < t_p \tag{3}$$

$$V_m(e^{\alpha \cdot t_p} - e^{\beta \cdot t_p}) = V_p \tag{4}$$

$$V_m(e^{\alpha \cdot t} - e^{\beta \cdot t}) \leq V_p; \quad \forall t > t_p \tag{5}$$

$$V_h \leq V_m(e^{\alpha \cdot t} - e^{\beta \cdot t}) \leq \overline{V_h}; \;\; \forall t = t_h \tag{6}$$

where V_p is the peak value of the lightning impulse, t_p is the frontier time, t_h is the time to half value, and V_h is the half value. In this paper, the A factor is melted into the V_m such that $V_m = A{\cdot}V_p$, and more often than not, to determine the desired voltage, V_m is set a bit higher than V_p, i.e., A is greater than or equal to 1.

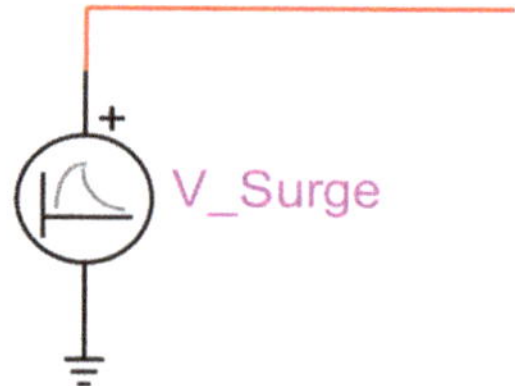

Figure 1. Voltage surge device to generate indirect lightning impulses.

To the best of our knowledge and according to practical experiences, to generate an indirect lightning impulse, parameters α and β take negative values. Therefore, maximizing the summation of these negative values guarantee to find the smallest values, which is extremely beneficial in enhancing the computational efficiency in transient studies via the EMTP-RV software [30]. It is noteworthy to mention that in order to support open access, the developed GAMS code used to obtain the parameters of the double exponential function for simulating an indirect standard 100 kV lightning impulse is available in [31].

2.2. Transformer

In this work, a Delta-Star ($\Delta - Y$) 22/0.4 kV transformer is modeled in the EMTP-RV environment, according to Cigre Guidelines [32]. The transformer comprises three basic building blocks as Figure 2 to present the 3-phase transformer model, see Figure 3. In high voltage studies, capacities on the MV side, LV side, and between MV and LV sides play a crucial role in obtaining more realistic results. The capacitor sizes are derived from the work in [33].

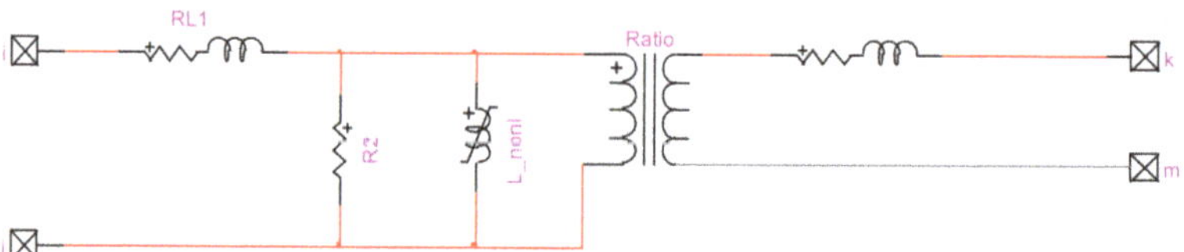

Figure 2. Basic building block for each phase of the transformer, EMTP-RV model.

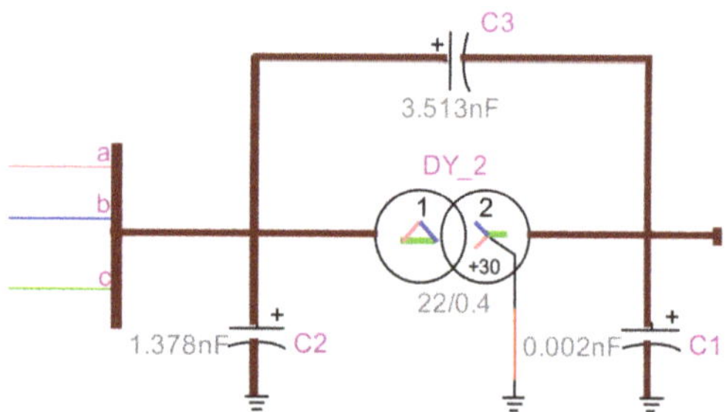

Figure 3. ΔY-connection model of the transformer with measured capacitances.

2.3. Spark Gap

A spark gap is the most commonly used device used to protect the transformers by limiting the overvoltage amplitude (chopping the voltage at a certain level). The spark gap is triggered when the overvoltage across its terminals exceeds the critical flashover voltage [34]. In most works, for the sake of simplicity, to model the behavior of the spark gap, a voltage-controlled switch is used [6,21], which cannot take into account the nature of different surge impulses. In this paper, to avoid such shortcomings and to simulate the behavior of the spark gap as close as the practical situation, the disruptive effect (DE) method (7) is used [35]. In this formulation, the flashover occurs when the integral part becomes greater than or equal to D. The integral function enables calculating the moment that the spark gap is ready to be triggered regardless of the waveform.

$$\int_{t_0}^{t} \left[\left| v_{gap}(t) \right| - v_0 \right]^k dt \geq D \tag{7}$$

where k and D are an empirical constant and the disruptive effect constant (kV μs), respectively; v_0 is the onset voltage of primary ionization (kV); and t_0 is the time corresponding to this primary ionization.

2.4. Surge Arrester

This subsection provides adequate information on the surge arrester and its fault model.

2.4.1. Frequency-Dependent Model

In order to simulate the behavior of a surge arrester, a frequency-dependent model of a metal oxide surge arrester, known as metal oxide varistor (MOV) surge arresters, is used. The modeling is done based on the guidelines provided by IEEE Working Group 3.4.11 [36]. The most commonly used frequency-dependent model is an RLC circuit with two nonlinear resistances Ao and A1 separated by an R-L filter, see Figure 4. The voltage-current (V-I) characteristics of the nonlinear resistances are derived from the work in [36], and the functionality of each type of surge arrester has been verified by performing a simulation-based test under a 10 kA current impulse (8/20 μs) and compared with the practical information obtained from [37].

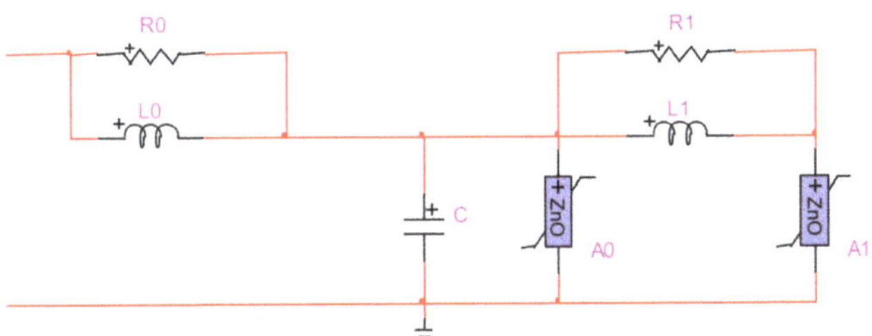

Figure 4. Frequency-dependent metal oxide surge arrester.

Practically, a transformer is under risk if the voltage at the MV side of the transformer is above 125 kV [38,39]. Therefore, in this paper, to have a safe margin from the risky overvoltage level, the goal is to keep the overvoltage amplitude below 100 kV, and to this end, only surge arresters with residual voltage below 100 kV are studied. Detailed information of the understudy surge arresters is presented in Table 1 [37], where "a" and "b" stand for the normal and high energy discharge capabilities of surge arresters, respectively.

Needless to say, the lower rating surge arresters and the normal energy class, due to lower costs, are of more interest to the system operators [6]. The nominal discharge current for the arresters is 10 kA, while the rated short circuit current is 50 kA (the maximum current flowing for a duration of 200 ms), and the peak value is ~125 kA, i.e., the currents for which the surge arrester withstand without sending the fractured part outside the circle with radius H (height of surge arrester) [37,40,41].

It is noteworthy to mention that to select a suitable surge arrester, several criteria should be taken into accounts, such as continuous operating voltage and rated voltage, nominal discharge current, protective level, energy class, and protective zone. More often than not, the protection capabilities of a surge arrester are evaluated according to its residual voltage while the nominal discharge current is flowing through it. A practical method to calculate a suitable residual voltage of the surge arrester is dividing the basic insulation level of the equipment to be protected by a factor of 1.4 as in (8) [37].

$$u_{pl} < {}^{BIL_{Tr}}/_{1.4} \tag{8}$$

where u_{pl} is the protective level of the surge arrester at a nominal current of 10 kA, 8/20 μs, and BIL is the basic insulation level of the equipment, i.e., an MV transformer in this work with a 150 kV BIL.

Table 1. Surge arresters' characteristics [37].

Surge Arrester Type	Rated Voltage (kV)	Maximum Continuous Operating Voltage (kV)	Residual Voltage at Nominal Discharge Current (kV)	Protective Level (kV)	Height (mm)	Thermal Energy Absorption Limit (kJ)
SA-12a	12	9.6	31.8	31.9	135	13.2
SA-12b	12	9.6	28.2	28.8	142	84
SA-18a	18	14.4	47.7	47.8	135	19.8
SA-18b	15	12	35.3	36.0	142	105
SA-24a	24	19.2	63.6	63.8	135	26.4
SA-24b	24	19.2	56.4	57.6	142	168
SA-30a	30	24.0	79.5	79.7	135	33.0
SA-30b	30	24.0	70.5	72.0	142	210.0
SA-36a	36	28.8	95.4	95.7	135	39.6
SA-36b	36	28.8	84.6	86.4	142	252
SA-42b	42	33.6	98.7	101	142	294

2.4.2. Surge Arrester Fault Model

More often than not, a failure in the arrester causes a complete short circuit. In most situations, dielectric breakdown is the source of such failure. Lightning surges, among others, are one of the most common circumstances against which an arrester faces an overvoltage beyond the withstand voltage and the dielectric breakdown results [42]. If the input energy to a surge arrester exceeds the thermal energy absorption limit, the arrester is pushed into a thermal runaway condition [43], and consequently the MOV blocks eventually become so conductive that these blocks cannot support even the maximum continuous operating voltage. More information on thermal runaway can be found in [44]. That is, the surge arrester will be completely short-circuited [42]. Therefore, to obtain more practical results, the fault behavior of surge arresters needs to be modeled. In this work, an energy-controlled switch is developed to simulate the behavior of surge arresters under fault conditions. The switch is closed once the absorbed energy of the surge arrester exceeds the thermal energy absorption limit, as in (9).

$$W = \int v(t)i(t)dt \geq \overline{W} \tag{9}$$

where W and $\overline{W}$ are the absorbed energy by surge arrester and its thermal energy absorption limit, respectively; $v(t)$ stands for instantaneous voltage across the surge arrester; and $i(t)$ represents the current flow through the surge arrester. The integral part of this equation stands for the area below the multiplication of the instantaneous voltage and current of the surge arrester, which is equal to energy. It is worth mentioning that the energy absorption capability of the surge arrester is not a constant value and highly depends on the microstructure (non-)uniformity and the stress conditions of the varistor [45]. However, more often than not, for the sake of simplicity, the factories provide a constant value, as presented in Table 1.

Figure 5 presents the proposed fault model of the surge arrester. An energy-controlled switch in the EMTP-RV environment is used to model the behavior of the surge arrester when excess energy is pushed into it. In this figure, for didactic purposes, alongside, below, or on top of each block, the measured signals and the utilized functions to calculate the absorbed energy of surge arrester have been presented.

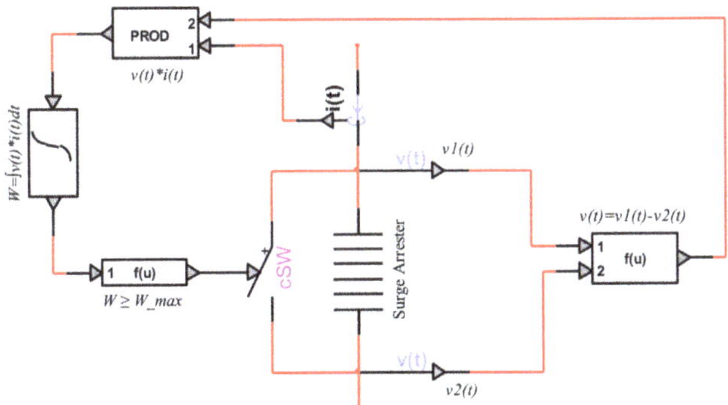

Figure 5. Modeling surge arrester fault via an energy-controlled switch.

2.5. Arrangement of Lightning Protection Devices

Figure 6 presents the arrangements of lightning protection devices for MV transformers. In Figure 6a, the transformer is protected by a sole surge arrester, while in Figure 6b, an inductor is used as the filter. The joint surge arrester and spark gap are presented in Figure 6c, and Figure 6d shows the filtered surge arrester and spark gap.

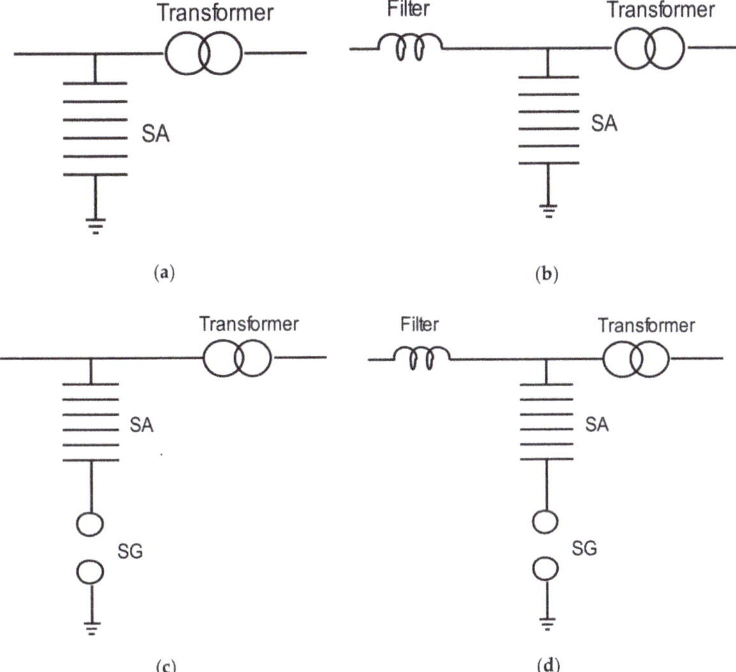

Figure 6. Lightning protection arrangement for MV transformers with (**a**) surge arrester, (**b**) filtered surge arrester, (**c**) joint surge arrester and spark gap, and (**d**) filtered surge arrester and spark gap.

Figure 7 resents the configuration of the transformer protected by surge arrester, spark gap, or a series connection of both devices in the EMTP-RV software. In this configuration, for the sake of simplicity, only the indirect lightning impulses are applied on one terminal of the MV transformer, namely, terminal a, while all the measurements are carried out on the connection point of the protective devices, see point (a) in Figure 7. All the arrangements presented in Figure 6 can be made in the EMTP-RV model in Figure 7. Wires (b) and (c) are used for adding/eliminating the surge arrester and spark gap by disconnecting/connecting the two oval points of these wires, respectively. In this figure, the R in R_L stands for the internal resistance of the L filter. In this paper, for the sake of simplicity, an 1Ω internal resistance has been considered for all inductors. In case that there is no filter, a 1Ω resistor will be considered to take into account the resistance of the connecting wire between the lightning impulse and the transformer. The switch across the surge arrester is an energy-controlled switch, modeled in Section 2.4.2. Rd is the damping resistor, and in this paper, it is set to 200 Ω.

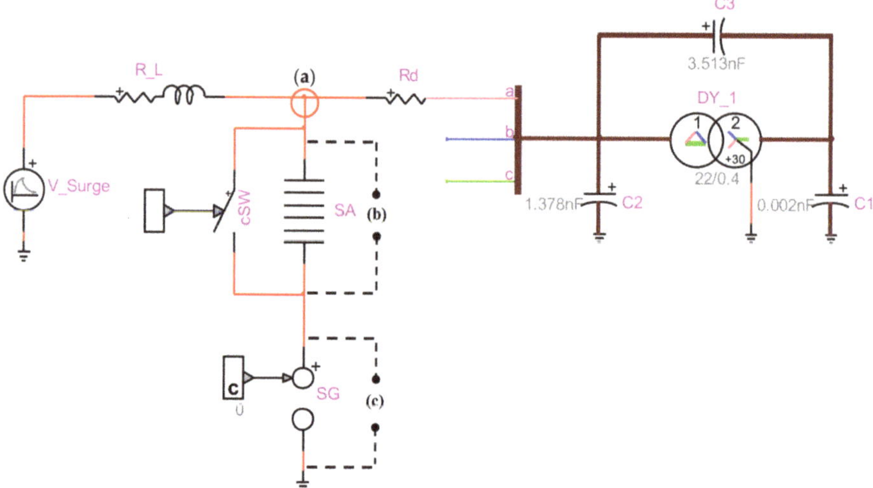

Figure 7. Series connection of surge arrester (SA) and spark gap (SG); (**a**) measurement point for overvoltage stress, (**b**) adding/eliminating the SA by disconnecting/connecting this wire, and (**c**) adding/eliminating the SG by disconnecting/connecting this wire.

3. Model Validation, Case Studies, and Simulation Results

In this section, first we validate the functionality of the proposed energy-controlled switch, which is used to model the surge arrester's fault condition due to absorbing energy higher than the thermal energy absorption limit. Then, several case studies are performed to see the potential of filtered surge arrester compared with the conventional surge arrester and series connection of surge arrester and spark gap. However, before validating the model and performing profound analyses, the lightning impulses need to be generated. Table 2 presents the optimal parameters of the double exponential model (1), obtained by solving (2)–(6). Most indirect lightning strikes have an overvoltage amplitude of less than 300 kV [46]. Therefore, in this work, we focus more on the overvoltages below 300 kV, though, a very high indirect overvoltage (500 kV) has also been introduced just to investigate the performance of protective devices under such high overvoltage stress. Table 2 can be used as a quick source for interested readers to generate the desired indirect lightning impulses.

Table 2. Optimal parameters of the double exponential model to generate indirect lightning impulses.

V_p (kV)	V_m (kV)	α	β
100	102	−14036.78	−4867830.80
125	127.7	−15100.47	−4790636.88
150	153	−14036.78	−4867830.71
175	178.5	−14027.23	−4864444.27
200	204	−14036.78	−4867830.80
250	255	−14036.78	−4867830.89
300	306	−14036.78	−4867831.15
500	510	−14099.63	−4889384.19

3.1. *Model Validation Validating the Surge Arrester Fault Model and the Spark Gap*

This subsection provides adequate information and comparisons to validate the models of protective devices.

3.1.1. Validating the Surge Arrester Fault Model

In order to validate the proposed surge arrester fault model, the surge arrester with the highest thermal energy absorption limit, i.e., SA-42b (see Table 1), which can resist better than the lower rating and energy class-a surge arresters against different lightning overvoltage levels, is used.

Figure 8a–h demonstrates the measured voltage at the MV side of the transformer (see point (a) in Figure 7), the absorbed energy by surge arrester, and the current flow through the surge arrester upon the ignition under different indirect lightning impulses. The time period in which the surge arrester is active can be verified by the current curve, i.e., whenever there is a current flow, the surge arrester is active, otherwise, it is an open circuit. As can be seen in Figure 8a, by applying a 100 kV lightning impulse, the surge arrester is active only for a short period of time while the current peak is about 5 kA. The energy absorbed by the surge arrester under this lightning impulse is ~4 kJ, which is well below its thermal energy absorption limit, 294 kJ; therefore, the surge arrester works properly, and no short circuit occurs. That is, under normal conditions, the switch remains open. Similarly, in Figure 8 b–e, the absorbed energy is below 294 kJ, and the surge arrester can effectively protect the transformer while not interrupting the service to the electricity consumers. In this figure, as the applied lightning impulse increases, the amplitude of current flow through the surge arrester increases, and consequently, the surge arrester absorbs a higher amount of energy. For example, in Figure 8b, compared with Figure 8a, increasing the lightning impulse by ~25 kV results in an ~12.5 kA increase in the current flow through the surge arrester while its absorbed energy faces a 21 kJ increase. Another observation from Figure 8a–e is that although the amplitude of the lightning impulse is increased by a step of 25 kV, the increases of the current and absorbed energy do not follow a linear pattern. This is mainly due to the nonlinear characteristic of metal oxide surge arresters [47]. As an instance, under the 200 kV lightning impulse, in Figure 8e, compared with the 175 kV lightning impulse, in Figure 8d, by increasing 12.5% in the lightning impulse, the absorbed energy is increased by ~56.8% (from 143.8 kJ to 228.31 kJ), while the operating duration and the peak current flow are increased by about 10 μs and 19.38 kA, respectively. The appropriate operation of the surge arrester continues until its absorbed energy exceeds the thermal absorption limit, as can be seen in Figure 8f–h where the surge arrester pushed into a failure. As can be seen from these figures, under such conditions a fault occurs, and the current flow through the surge arrester shows a considerable jump due to the proper functionality of the proposed energy-controlled switch under the faulty situation. When the surge arrester is short-circuited, the voltage at the transformer terminal reaches to zero. That is, continuous service to electricity consumers will be interrupted.

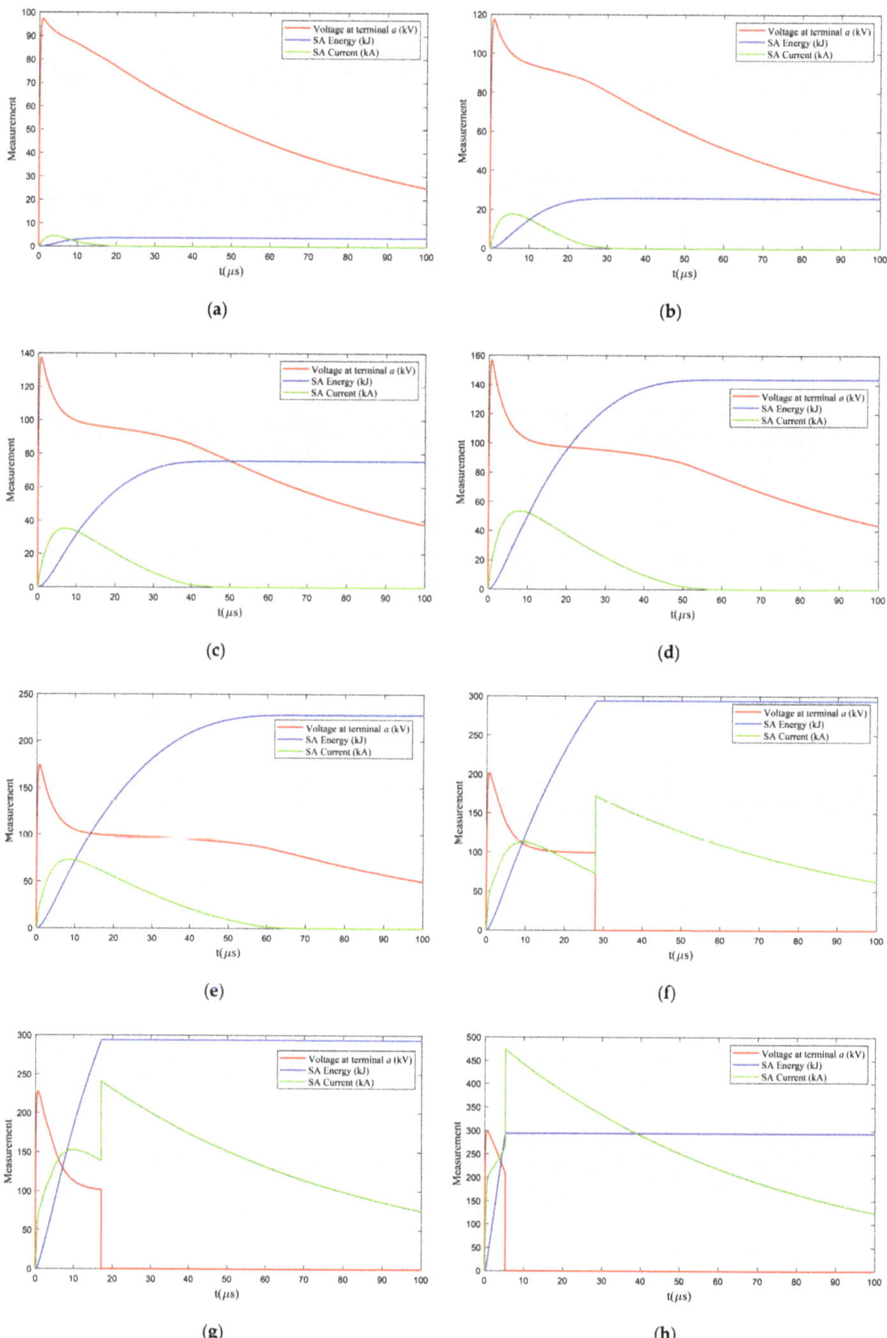

Figure 8. Performance of surge arrester SA-42b under lightning impulse voltage (**a**) 100 kV, (**b**) 125 kV, (**c**) 150, (**d**) 175 kV, (**e**) 200 kV, (**f**) 250 kV, (**g**) 300 kV, and (**h**) 500 kV.

Moreover, Figure 8f–h shows that the higher the lightning impulse is the faster the thermal absorption limit is reached and the surge arrester fault occurs such that under the 500 kV lightning impulse, Figure 8h, the surge arrester cannot even withstand to decrease the overvoltage amplitude below the risky overvoltage level, 125 kV [38,39].

Comparing the energy levels of the surge arrester under normal operating conditions, in Figure 8a–e, and the short circuit conditions, in Figure 8f–h, it can be deduced that the energy-controlled switch is working properly. That is, the proposed fault model (see Section 2.4.2) is validated. Please note that, besides a fault occurring due to pushing higher energy into the surge arrester, the current flow through it may cause failure as well. In this paper, for the sake of simplicity, after the failure, the impedance has been set to zero, while in practice, even when the varistor failure occurs, there is a non-zero impedance. Besides, in this work, the ground resistance of the transformer and the conductive coupling of the surge arrester have not been considered.

3.1.2. Validating the Spark Gap

Although more often than not some approximation methods are used to adjust the sensitive parameters of the DE method (7), i.e., k, v_0, and D, the best way to properly adjust these parameters is to have some experimental data. In this work, an 8 cm sphere–sphere spark gap, presented in Figure 9, is used [33].

Figure 9. Adjustable sphere–sphere spark gap [33].

The validation is done by applying a nonstandard 125 kV overvoltage impulse on the transformer terminal at the presence of a spark gap. In this work, first, the method proposed by the Cigre Working Group 01 [48] is used to estimate the V_0 ($V_0 \approx 0.9 \cdot \text{CFO} \approx 112$ kV), and according to the work in [35], if V_0 is large, then k is considered to be lower than or equal to 1, then a trial and error method is used to find the best values for k and D that yields to a better agreement with the laboratory results. By using this approach, the following parameter have been obtained; $V_0 = 112$ kV, k = 0.97, and D = 0.01. Figure 10a depicts the applied laboratory impulse (the green curve) and the simulated impulse in EMTP-RV (the blue curve). Figure 10b compares the performances of the spark gap under the laboratory test and via the DE method. As can be seen from this figure, there is a good agreement between the laboratory test and the results obtained in EMTP-RV software. Interested readers may refer to the work in [49] for obtaining a complete report on the laboratory setup, supporting figures, as well as a wide range of experimental results at the presence of spark gap, while more information regarding the applied impulses and the validation of the spark gap can be found in [33,50].

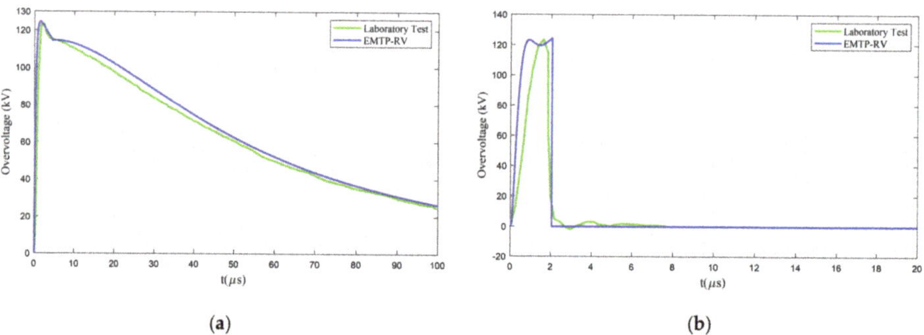

Figure 10. (**a**) Applied nonstandard 125 kV lightning impulse. (**b**) Measured overvoltage stress at the presence of spark gap [33].

3.2. Case Studies

To perform in-depth analyses on the protective performances of surge arresters against overvoltage surges, several case studies are conducted. The primary goals of the following case studies are (a) to investigate the degree of effectiveness of considering surge arrester in protecting the MV transformers against indirect lightning phenomena; (b) showing the pros and cons of considering a series connection of surge arrester and spark gap; and (c) investigating the effectiveness of considering an inductor-based filter, as the main contribution of this work, on enhancing the functionality of surge arrester in preventing unwanted service interruption while protecting the MV transformers against indirect lightning surges. In this work, and in order to achieve the aforementioned goals, the following case studies are conducted.

Case 1 Transformer protected by surge arrester. The main goal of this case is to reveal the protection capability of different surge arresters in protecting MV transformers against various indirect lightning overvoltages.

Case 2 Transformer protected by a series connection of surge arrester and spark gap. This case is used to investigate the impact of the combined protective devices on the overvoltage stress and protection of MV transformers against indirect lightning.

Case 3 Transformer protected by filtered surge arrester. The primary goal of this case is to investigate the impact of an inductor-based filter on the performance and operating margin of surge arresters.

Case 4 Transformer protected by a series connection of filtered surge arrester and spark gap. This case investigates the role of an inductor-based filter on the protective performance of a combined surge arrester and spark gap devices, presented in Case 2.

3.3. Simulation Results and Discussions

3.3.1. Case 1. Transformer Protected by Surge Arrester

This case investigates the protection capability of various surge arresters, listed in Table 1, in protecting MV transformers against different indirect lightning overvoltages, presented in Table 2. In this work, the protection capability is defined as the overvoltage level under which the surge arrester does not experience any failure. To perform the simulation setup, the two oval points of wire (b) in Figure 7 are disconnected to keep the surge arrester in the circuit, while the two oval points of wire (c) are connected to eliminate the spark gap.

As this case focuses on finding the proper functionality margin of the surge arresters, then, for the sake of unambiguity, only the performance of the surge arrester related to the lightning impulse with the lowest overvoltage amplitude against which a fault, i.e., short circuit, occurs is reported. In other words, for lightning impulses with overvoltage amplitude lower than this fault-causing lightning impulse, the surge arrester protects the MV transformer properly. Figures 11–15 present the failure occurrence of different surge arresters with two energy classes.

In Figure 11a, the surge arrester SA-12a is used, which has a lower thermal absorption capacity than the SA-12b; in Figure 11b, a 100 kV lightning impulse results in a fault within ~6.1 µs, while such failure for the SA-12b occurs under the 125 kV lightning impulse and within ~43.9 µs. Therefore, the surge arrester SA-12b can adequately protect the transformer against the 100 kV lightning impulses while guaranteeing the continuous service to the electricity consumers.

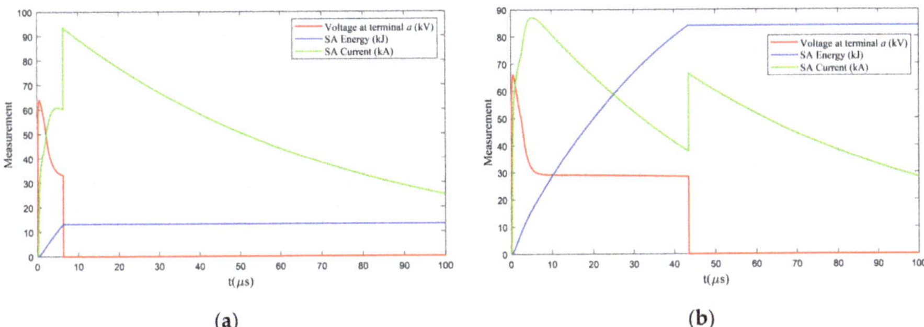

Figure 11. Failure of (**a**) SA-12a under 100 kV lightning impulse and (**b**) SA-12b under 125 kV lightning impulse.

Figure 12 shows that similar to the 12 kV surge arresters, surge arresters SA-18a and SA-18b experience failures under the 100 kV and 125 kV lightning impulses, respectively. However, compared with Figure 11, the times to failure are a bit longer, which is due to higher energy absorption levels of 18 kV rating surge arresters (see Table 1). Therefore, although the surge arresters SA-18a and SA-18b, compared with the surge arrester SA-12a and SA-12b, have higher energy absorption capacity, their performance against lightning overvoltages is similar. Both the class-a surge arresters failed against a 100 kV lightning impulse, while both the class-b surge arresters could not provide proper protection against a 125 kV lightning impulse.

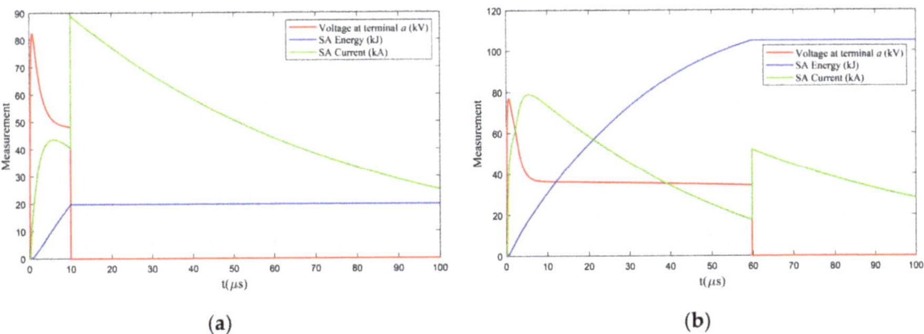

Figure 12. Failure of (**a**) SA-18a under 100 kV lightning impulse and (**b**) SA-18b under 125 kV lightning impulse.

Figure 13 presents the performance of the 24 kV rating surge arresters under failure condition. As can be seen in Figure 13b, surge arrester SA-24b provides proper protection for the MV transformers under the 100 kV, 125 kV, and 150 kV lightning impulses, and the failure occurs under the 175 kV lightning impulse. Therefore, its performance is much better than the performance of previous surge arresters in the same energy class, i.e., surge arresters SA-12b and SA-18b (see Figures 11b and 12b, respectively). However, the failure of surge arrester SA-24a, like previous type-a surge arresters (i.e., SA-12a and SA-18a in Figures 11a and 12a, respectively), occurs under the 100 kV lightning impulse. The fault for the surge arrester SA-24a occurs after ~17.9 μs, which takes 11.8 μs and 7.9 μs longer than the failure occurrence for surge arresters SA-12a and SA-18a, respectively. Although this shows a better instantaneous performance, in total, the protection performance matters, so, from the standpoint of providing continuous service under the 100 kV lightning impulse to electricity consumers, SA-24a does not have any advantage over the SA-12a and SA-18a.

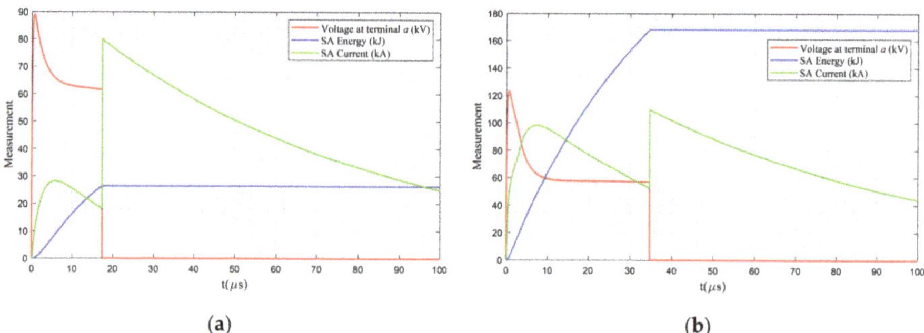

Figure 13. Failure of (**a**) SA-24a under 100 kV lightning impulse and (**b**) SA-24b under 175 kV lightning impulse.

Figure 14 presents the primary measurements of the 30 kV rating surge arresters experiencing failures. Figure 14a, compared with Figures 11a and 13a, shows that for providing proper protection against the 100 kV lightning impulses by energy class-a surge arresters, a surge arrester with a rating above 30 kV is required, while this goal can also be achieved by utilizing an energy class-b surge arrester with a lower rating, i.e., surge arrester SA-12b (see Figure 11b). Moreover, comparing the performances of surge arrester SA-30b in Figure 14b with performances of surge arrester SA-24b in Figure 13b shows that increasing the surge arrester rating by 6 kV although delays the failure time under the 175 kV lightning impulse by ~27.2 μs (due to the higher energy absorption capacity of SA-30b), still, it causes a service interruption to the electricity consumers and decreases their comfort level.

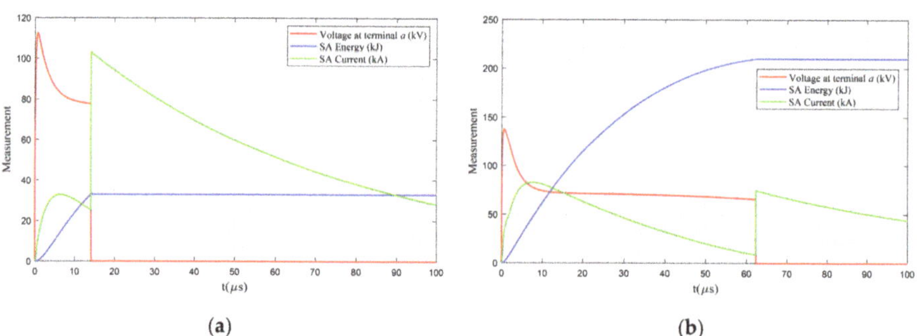

Figure 14. Failure of (**a**) SA-30a under 125 kV lightning impulse and (**b**) SA-30b under 175 kV lightning impulse.

Figure 15 presents the performance of the 36 kV rating surge arresters under failure condition. Figure 15a, compared with Figure 14a, shows that by increasing the rating of the type-a surge arrester from 30 kV to 36 kV, providing continuous service to consumers is guaranteed even for the 125 kV lightning overvoltage and the fault occurs under the 150 kV lightning impulse. A system operator, depending on the network condition and the available budget, may use an energy class-b 24 kV rating surge arrester, namely, SA-24b, instead of using a class-a 36 kV rating surge arrester for protecting against the 125 kV lightning impulses. On the other hand, Figure 15b shows that by using the surge arrester SA-36b, if indirect lightning with an overvoltage amplitude of 200 kV is applied, a failure will occur after ~50 µs and results in an interruption on the electricity supply to consumers. Therefore, it is deduced that an energy class-b 36 kV rating surge arrester (SA-36b) can properly protect the MV transformers against the 175 kV lightning impulses while guaranteeing to provide continuous service to electricity consumers.

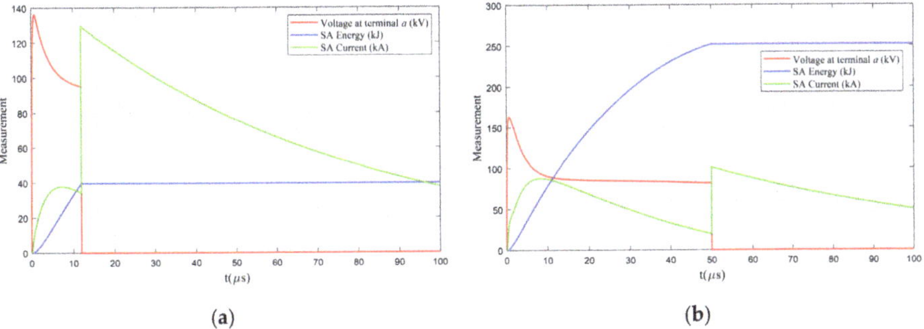

Figure 15. Failure of 36 kV surge arrester (**a**) for SA-36a under 150 kV lightning impulse and (**b**) for SA-36b under 200 kV lightning impulse.

The performance of the 42 kV rating surge arrester, namely, SA-42b, has been presented in Figure 8. As can be seen from this figure, the best performance among all the listed surge arresters in Table 1 belongs to this surge arrester, while even by applying the 200 kV lightning impulse, there is no failure, and consequently no service interruption occurs (see Figure 8e). However, when the 250 kV lightning impulse is applied, after ~28 µs, a failure occurs. Moreover, it can be seen from Figure 8 that by increasing the applied impulse, the time to fault is decreased while the overvoltage stress is increased.

The current flow through the surge arresters under the highest overvoltage stress against which these protective devices can properly protect the transformer (i.e., handleable overvoltage impulse) is depicted in Figure 16. From this figure, the range of di/dt for this case is between approximately 5.7 kA/µs and 11.0 kA/µs.

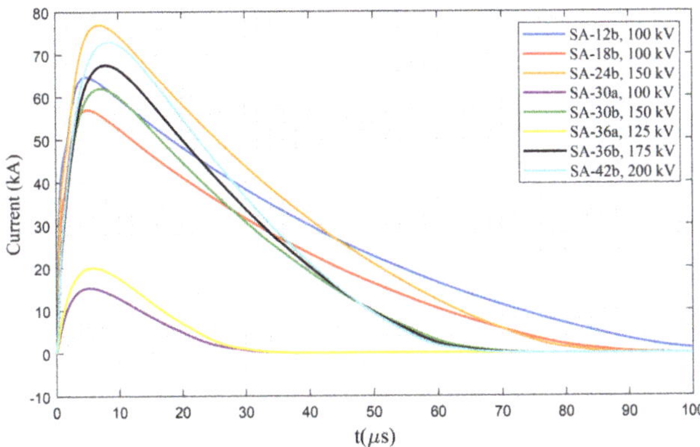

Figure 16. Current flow through surge arresters under their highest handleable overvoltage impulse, Case 1.

3.3.2. Case 2. Transformer Protected by a Series Connection of Surge Arrester and Spark Gap

This case investigates the mutual effects of two protective devices, connected in series, on the protection of MV transformers against indirect lightning. For systems protected only by a spark gap, when this device is triggered, it acts as a short circuit, and consequently the voltage at MV terminals of the transformer reaches to zero, i.e., the voltage is chopped, and the continuous service to electricity consumers is interrupted [33]. Figure 17 presents the performance of the spark gap under different indirect lightning impulses. As can be seen from Figure 17a, the spark gap is not triggered under the 100 kV lightning impulse, and the complete waveform is transferred to the MV terminal of the transformer. However, Figure 17b shows that for the higher impulse voltages, as the triggering criteria (7) are met, the voltage is chopped, and service interruption occurs. From Figure 17b, it is deduced that the higher the lightning impulse is, the faster the spark gap is triggered to prevent complete failure. Note that to show the functionality of the spark gap upon activation and for the sake of clarity, Figure 17b only depicts the overvoltage curves for the first 5 μs.

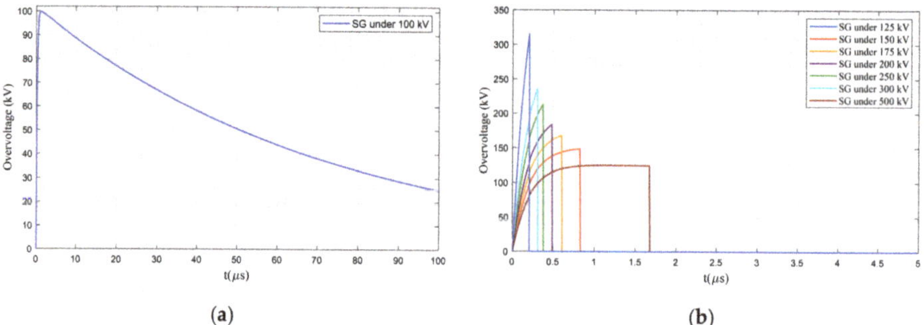

Figure 17. Overvoltage at the spark gap (SG) terminal for (**a**) not triggered condition and (**b**) triggered condition under different lightning impulses.

However, a series connection of spark gap and surge arrester, if both are selected properly, guarantees continuous service to consumers [6]. In a series connection of spark gap and surge arrester, different conditions may occur that can be explained via the following scenarios.

Scenario 1: *Spark Gap is not triggered*

In a series connection of spark gap and surge arrester, if the overvoltage at the terminal of one device is not strong enough to trigger it, the current flow through the protective devices is zero. That is, under this condition, they act as an open circuit and let the whole overvoltage pass toward the MV terminals of the transformer. This can be seen in Figure 18, in which the current through the protective devices (both the surge arrester SA-12a and spark gap) is zero during the 100 kV lightning impulse. This happens since the amplitude of lightning impulse is lower than the onset voltage of the spark gap (112 kV) and cannot trigger the spark gap. Therefore, the surge arrester does not work as well, and this can be seen by monitoring the energy absorbed by the surge arrester, which is zero during the lightning. This case shows that unlike Case 1 that the failure of surge arrester SA-12a causes a service interruption (see Figure 11a), a series combination of surge arrester and spark gap under the 100 kV lightning impulse prevents such an unwanted service interruption to the electricity consumers.

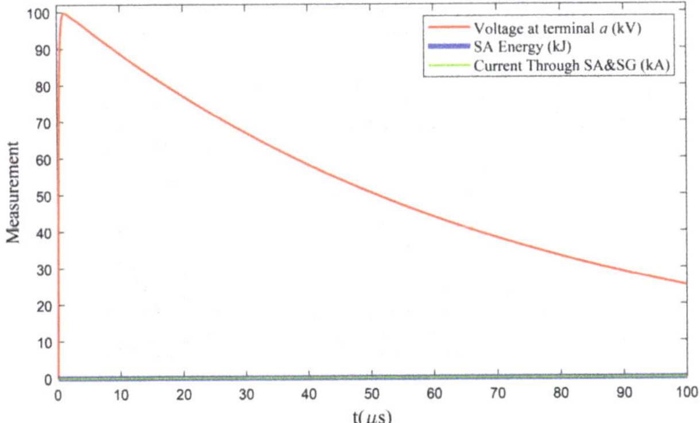

Figure 18. Performance of surge arrester and spark gap connected in series; scenario 1, SA-12a under 100 kV lightning impulse.

Scenario 2: *Spark Gap is triggered and Surge Arrester is pushed into Failure*

As in series connection of spark gap and surge arrester, the spark gap is triggered by a high voltage (above the onset voltage, which is 112 kV in this paper), a large current flows through the surge arrester and its protective role to attenuate the overvoltage stress is started. At this stage, the thermal energy absorption limit of the surge arrester is an important feature guaranteeing to provide continuous service to electricity consumers. If the failure occurs, the surge arrester will be short-circuited, and service interruption results. Point (1) in Figure 19 shows the time and corresponding overvoltage in which the spark gap is triggered, and the voltage dropped instantaneously to about 77 kV (see point (2)), where the current increased to ~48 kA. At this point, the surge arrester started its crucial role in preventing a short circuit condition. However, as can be seen in Figure 19, at point (3), the surge arrester has been pushed into a failure, and such a failure in the surge arrester caused a complete short circuit, and consequently, a service interruption resulted.

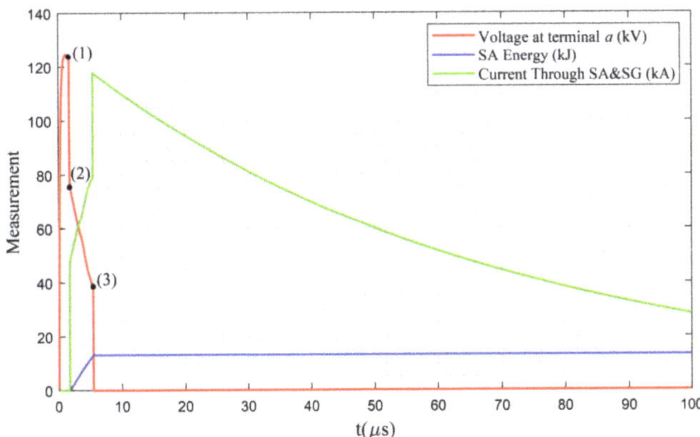

Figure 19. Performance of surge arrester and spark gap connected in series; scenario 2, SA-12a under 125 kV lightning impulse.

It is noteworthy to mention that in some works, the thermal energy absorption limit is neglected, and consequently, an imprecise conclusion is that the spark gap always enhances the protective performance of the surges arrester since the spark gap triggers to take away the overvoltage stress from the surge arrester and surge arrester can absorb the energy without any limit, and no interruption occurs. However, as has been shown in this case, each surge arrester has a thermal absorption limit, so any extra energy results in a failure and service interruption.

Scenario 3: *Spark gap is triggered and Surge Arrester works properly*

This condition is the most proper design for a series connection of a surge arrester and a spark gap, where the spark gap provides the primary protection, and after that, the surge arrester prevents any unwanted interruption. Figure 20 presents the proper protection of MV transformers under the 150 kV lightning impulse by connecting surge arresters SA-24b with the spark gap. The spark gap is triggered at point (1) when the voltage reached to about 149 kV in less than 0.82 µs. Then, the voltage drops rapidly to ~134 kV by discharging current through the protective devices. At point (2), the surge arrester starts its active role in protecting the transformer, and as can be seen, the absorbed energy (blue line) reaches to about 161 kJ, which is below the maximum absorption limit for this surge arrester (168 kJ). Consequently, no service interruption occurs, while in this case, the di/dt is about 9.8 kA/µs.

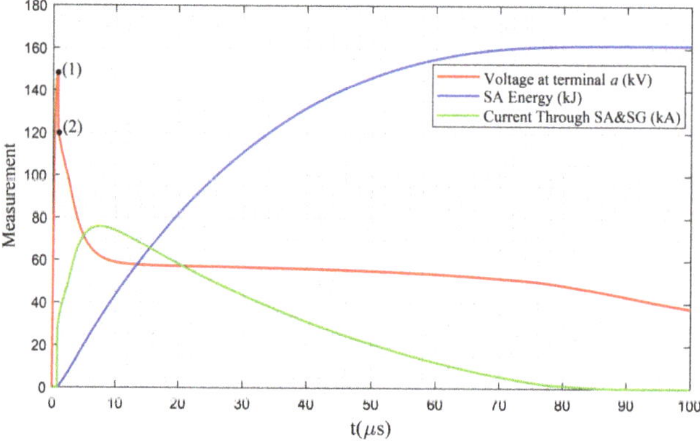

Figure 20. Performance of surge arrester and spark gap connected in series; scenario 3, SA-24b and SA-30b under 150 kV lightning impulse.

In order to compare the performance of joint surge arrester and spark gap protection against lightning overvoltages, the proper operation of such combination is investigated. After applying different lightning impulses, the protective performance of these combined protective devices is clustered into five different categories according to the highest amplitude of lightning impulses listed in Table 2 against which proper protection is provided. The first category provides proper protection for transformers only under the 100 kV lightning impulse. This category contains six surge arresters, namely, SA-12a, SA-12b, SA-18a, SA-18b, SA-24a, and SA-30a only, and the measurements show exactly similar output as depicted in Figure 18 as they act as an open circuit and let the whole overvoltage pass toward the MV terminals of the transformer.

The second category contains only the surge arrester SA-36a that when connected with the spark gap in series, the transformer is properly protected under 125 kV lightning impulses, as presented in Figure 21. Moreover, it is evident that similar to the case with only surge arresters (see Figure 15a, a series connection of surge arrester and spark gap provide protection for 125 kV lightning impulses. In this case, the di/dt is about 3.8 kA/µs.

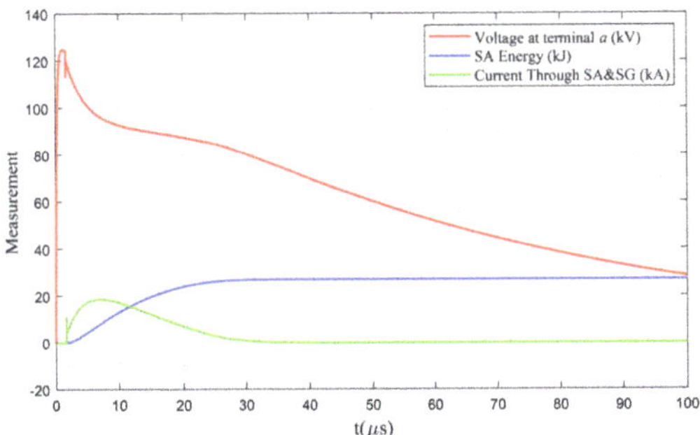

Figure 21. Performance of surge arrester and spark gap connected in series; scenario 3, SA-36a under 125 kV lightning impulse.

Surge arresters SA-24b and SA-30b provide proper protection for MV transformers against the 150 kV lightning impulses (see Figure 20). A comparison of Figure 20 with Figures 13b and 14b shows that the series combination does not provide better protective performance than utilizing only a surge arrester. Figures 22 and 23 present the other two categories that provide proper protection for MV transformers against the 175 kV and 200 kV lightning impulses, respectively, with SA-36b and SA-42b. By comparing Figure 23 with Figure 8e, it can be seen that by using the spark gap, the peak of overvoltage even goes higher than utilizing only the surge arrester, which is due to triggering the spark gap. The di/dt values in Figures 22 and 23 are approximately 8.7 kA/μs and 8.0 kA/μs, respectively.

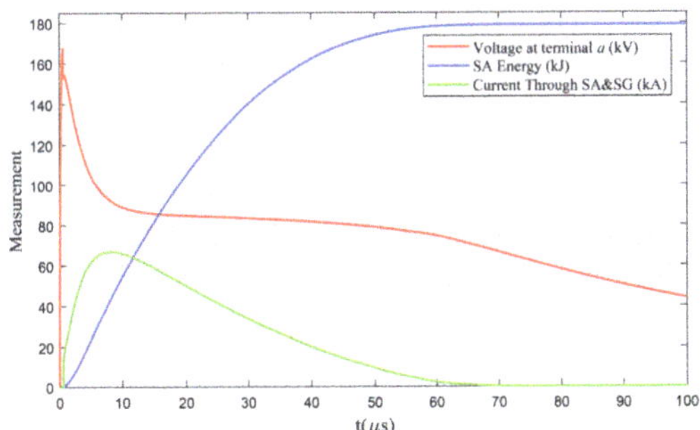

Figure 22. Performance of surge arrester and spark gap connected in series; scenario 3, SA-36b under 175 kV lightning impulse.

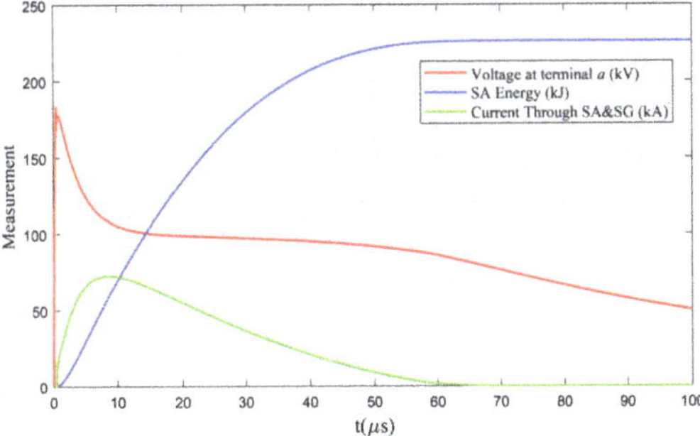

Figure 23. Performance of surge arrester and spark gap connected in series; scenario 3, SA-42b under 200 kV lightning impulse.

From the above results, it can be seen that connecting a spark gap in series with a surge arrester prevents triggering the surge arresters for the lightning impulses with an amplitude below the onset voltage of the spark gap and therefore prevents service interruptions. However, for higher overvoltage amplitudes, it does not have a positive effect on providing better protection for the transformer. Therefore, it is required to find a better approach to control the input energy pushed into a surge arrester and thereby prolong its lifetime.

3.3.3. Case 3. Transformer Protected by Filtered Surge Arrester

A surge arrester is modeled by an RLC circuit with two nonlinear resistances Ao and A1 separated by an R-L filter, see Section 2.4.1. Therefore, using an inductor before the surge arrester may control the input energy to the surge arrester. Case 3 investigates the viability and effectiveness of this idea and reveals its pros and cons. To do so, all the surge arresters listed in Table 1 are equipped with different inductor sizes, and the absorbed energy by surge arrester, current flow through it, as well as the overvoltage stress are measured. In this paper, the inductors to be used for test purposes are 100 μH, 250 μH, 500 μH, and 1 mH.

First, the surge arrester SA-12a, which is the lowest rating surge arrester with energy class a is considered. Figure 24 a–d presents the performance of filtered surge arrester SA-12a under the 100 kV indirect lightning impulse at the presence of the aforementioned inductors. By comparing Figure 24a with Figure 11a, it can be observed that by using a 100 μH inductor, the high initial overvoltage (~64 kV) presented in Figure 11a has been substantially decreased to about the surge arrester residual voltage level (~31.5 kV). However, as the energy pushed into the surge arrester is higher than the thermal absorption limit, a failure, and consequently, an interruption occurs. Moreover, by using the 100 μH inductor, it takes ~43 μs until the energy pushed into the surge arrester passes the thermal absorption limit (see Figure 24a), while for the case without the inductor, the time to failure is ~6.1 μs. Similarly, by increasing the inductor size in Figure 24b,c, the time to failure, compared with Figure 24a, is increased. Eventually, as depicted in Figure 24d, by using an appropriate filter, i.e., 1 mH, the surge arrester SA-12a properly protects the MV transformer against the 100 kV lightning impulse. That is, by increasing the filter (inductor) size, the absorbed energy is reduced until which the filtered surge arrester starts functioning properly and guarantees the continuous service to the electricity consumers. The energy absorbed by the surge arrester at the presence of 1 mH is ~10.87 kJ, which shows a safe margin for even protecting the lightning overvoltages above 100 kV; however, it fails in providing proper protection against the 125 kV lightning impulses. The di/dt of the surge arrester under the normal condition, Figure 24d, is ~0.03 kA/μs.

Table 3 presents the impacts of considering different filter sizes on the performance of surge arresters against indirect lightning overvoltages. This table contains the highest level of lightning impulses, among the impulses listed in Table 2, against which the surge arrester provides proper protection for MV transformers. From Table 3, it can be seen that by considering a 1 mH inductor to SA-12a, its performance is enhanced such that it provides similar protection as the non-filtered surge arresters SA-12b, SA-18b, and SA-30a. However, a significant improvement can be seen by considering SA-12b, where by installing only a 250 μH inductor, the energy pushed into the surge arrester is controlled such that it can be used instead of SA-24b and SA-30b, yet the protective performance is better than the energy class-a 36 kV rating surge arrester, i.e., SA-36a. By considering a 1 mH filter for SA-12b, a 100% enhancement in its protective performance is achieved. In other words, instead of using an expensive class-b surge arrester with an energy class-b 42 kV rating surge arrester, a much lower rating surge arrester with the same energy class (i.e., SA-12b) can be used.

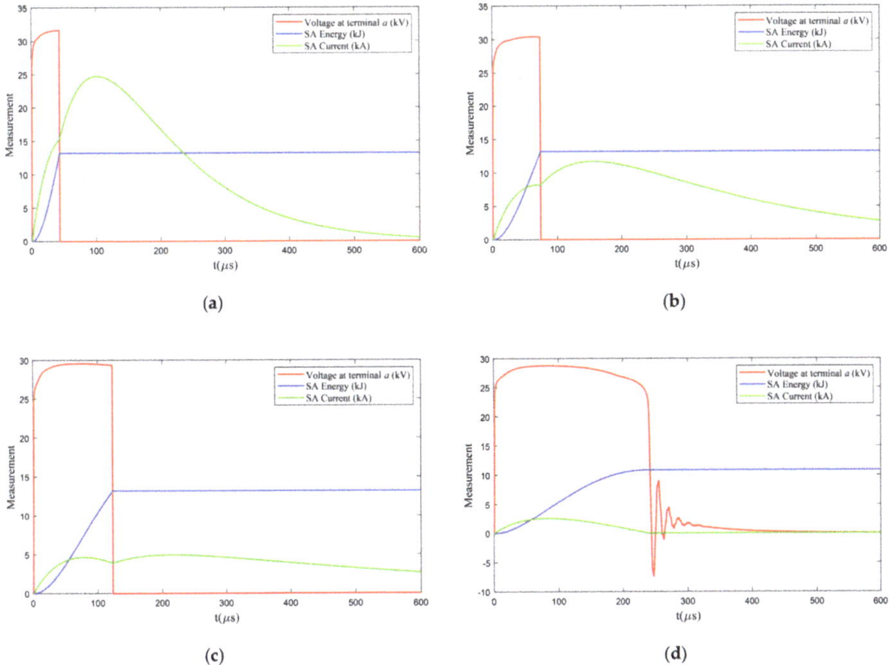

Figure 24. Performance of filtered surge arrester SA-12a under the 100 kV lightning impulse with inductor size (**a**) 100 μH, (**b**) 250 μH, (**c**) 500 μH, and (**d**) 1 mH.

Table 3. Performance of surge arrester and filtered surge arrester under indirect lightning impulses.

Surge Arrester		**No Filter**	**Filter Size**			
			100 μH	**250 μH**	**500 μH**	**1 mH**
Highest Lightning Impulse Level (kV)	SA-12a	-	-	-	-	100
	SA-12b	100	125	150	175	200
	SA-18a	-	-	100	100	125
	SA-18b	100	125	150	200	250
	SA-24a	-	100	125	125	150
	SA-24b	150	175	200	250	300
	SA-30a	100	125	125	150	175
	SA-30b	150	200	250	300	300
	SA-36a	125	150	150	175	200
	SA-36b	175	200	250	300	300
	SA-42b	200	250	300	300	300

In order to see the details of surge arrester performance enhancement, as an instance, the absorbed energy by surge arrester SA-12b under the 125 kV lightning impulse (as the risky overvoltage stress for transformers [38,39]) for different inductor sizes are compared in Figure 25. As can be seen from this figure, by using a conventional surge arrester (red curve), a voltage sag occurs at the beginning of lightning. After attenuating the voltage to ~30 kV, the surge arrester, due to absorbing more energy than its thermal absorption limit, is pushed into a failure and acts as a short circuit. Then, although even the overvoltage peak is much lower than 125 kV (~66 kV), all in all, the surge arrester is not suitable for guaranteeing a continuous service to consumers. On the other hand, it can be seen that when the surge arrester is equipped with an inductor, this newly formed filtered surge arrester provides proper protection against the 125 kV lightning surge. That is, first, the voltage sag is eliminated, and second, there will be no interruption in the supply service. Moreover, it can be deduced that by increasing the size of the filter, the peaks of the measured voltages are decreased, while the operating times of surge arrester are increased. For example, by increasing the inductor size from 100 µH to 250 µH, the peak voltage is decreased from 28.72 kV to 27.60 kV, while the operating time is increased from ~195 µs to ~243 µs.

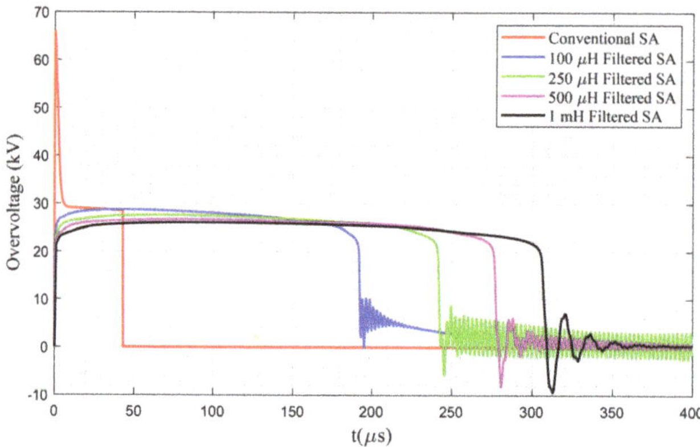

Figure 25. Comparison of the overvoltage stress of MV transformer protected by conventional surge arrester SA-12b (no filter) and filtered surge arrester with different filter sizes under the 125 kV lightning impulse.

Table 3 can be used as a lookup table for a system operator for choosing the right surge arrester as well as to adjust the inductor size aiming at obtaining the required protection for the MV transformers. For example, if the operator decides to protect the MV transformer against a 200 kV lightning impulse, Table 3 shows several appropriate designs. The most expensive way is to install a conventional surge arrester SA-42b; however, some filtered surge arresters can also provide proper protection against lightning amplitudes similar to the 1 mH filtered SA-12b, 500 µH filtered SA-18b, 250 µH filtered SA-24b, 100 µH filtered SA-30b, and 100 µH filtered SA-36b. To see what is the difference among these different configurations, the overvoltage stress on the MV terminal of the transformer and the absorbed energy by surge arresters are compared.

Figure 26 shows a comparison among all the surge arresters (with or without filter) that can provide proper protection against the 200 kV lightning impulse. As can be seen from this figure, the surge arresters with lower ratings decrease the overvoltage amplitude to a greater extent than the higher rating surge arresters. For example, the 1 mH filtered SA-12b lowers down the overvoltage stress at the MV terminal of the transformer to about 26.9 kV (peak value), which is ~85.8% lower than the peak voltage level at the presence of 100 µH filtered SA-36b. However, by utilizing the lower rating filtered surge arresters, it takes more time to absorb the excess energy and attenuate the overvoltage. As an instance, in case of using the 100 µH filtered SA-36b, it only takes ~380 µs to eliminate the overvoltage stress, while for the 1 mH filtered SA-12b, it takes ~505 µs. Furthermore, this figure shows that the only conventional surge arrester (i.e., non-filtered surge arrester) that can protect the MV transformer against 200 kV lightning impulse is the SA-42b. This surge arrester, at the beginning of lightning, does not work properly, and a voltage sag with an amplitude of 174.4 kV is reached the MV terminal of the transformer. This drawback of surge arrester SA-42b can be addressed by using a 100 µH filter (see Figure 27), while, from Table 3, it can be seen that the proper protection against lightning amplitudes is enhanced by 50 kV (it provides proper protection against 250 kV lightning impulses). It is noteworthy to mention that for old transformers that such high overvoltage stress can be a source of failure, a filtered low rating surge arrester is even better than a high rating surge arrester since it offers a lower residual overvoltage. As an instance, the residual voltage of SA-42b is 98.7 kV, while a 500 µH filtered surge arrester SA-18b provides a residual voltage of 35.3 kV (see Table 1 and Figure 26).

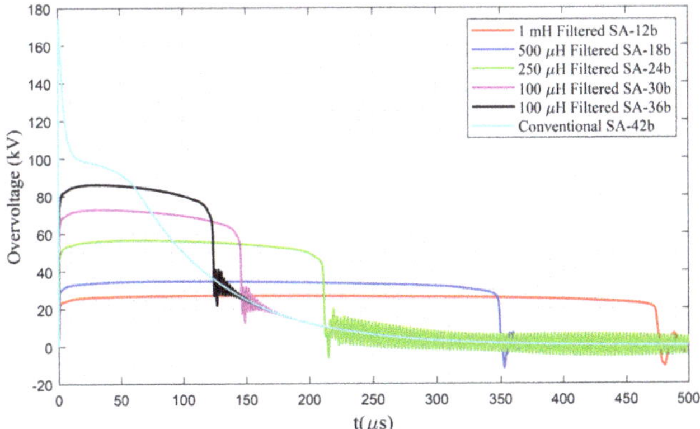

Figure 26. Comparison of the overvoltage stress of MV transformer protected by different configurations of surge arresters against the 200 kV lightning impulse.

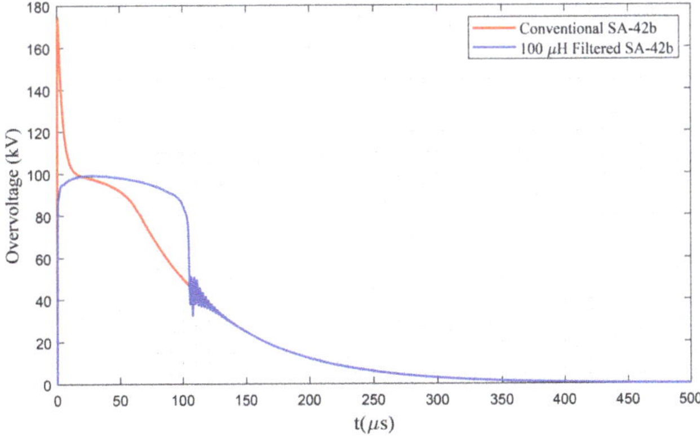

Figure 27. Comparison of the overvoltage stress of MV transformer protected by conventional SA-42b (red curve) and 100 µH filtered SA-42b (blue curve) under the 200 kV lightning impulse.

Figure 28 compares the absorbed energy by different configurations of surge arresters aims at providing proper protection for MV transformers against the 200 lightning impulse. The primary conclusion of this figure is that the proper sizes of filters can significantly enhance the performance of surge arresters by preventing any unwanted failure as a result of controlling the energy pushed into the surge arrester. Note that the surge arrester SA-42b has absorbed 228.1 kJ of energy to keep the overvoltage level below the desired voltage. That is, if the other surge arresters, presented in Figure 28, had a thermal energy absorption limit above 228.1 kJ, they could have been absorbed this energy and keep the overvoltage level below their desired residual voltage level, which is way below the residual voltage of SA-42b. Such capability has been awarded to these low thermal absorption-limit surge arresters by installing a filtering device and controlling their energy level such that even SA-12b with a 1 mH filter can provide proper protection, yet keeping the overvoltage stress below 24.6 kV (see Figure 26), which is well below the residual voltage of SA-42b, 98.7 kV. This is an incredible achievement by which the system operators can save their equipment against lightning overvoltages with a lower rating surge arresters, and consequently save a great amount of money.

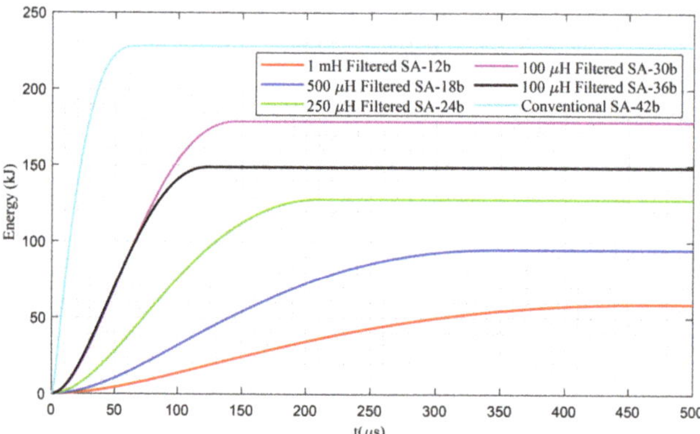

Figure 28. Comparison of the energy absorbed by different surge arresters against the 200 kV lightning impulse.

Figure 29 presents the current flowing through the filtered surge arresters under the highest handleable overvoltage impulse, see Table 3. The range of di/dt of the filtered surge arresters under such a condition is between about 0.04 kA/μs and 0.11 kA/μs. This range, compared with the reported di/dt in Case 1 and Case 2, shows a resounding performance of the filter in mitigating the steepness of the current, and consequently decreasing the stress over the surge arrester, which enhances the performance of this protective device.

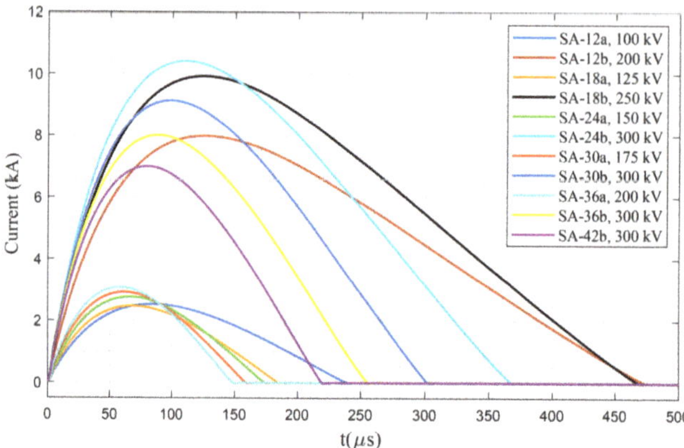

Figure 29. Comparison of the current flow through different surge arresters against the highest handleable lightning impulse.

In order to show the impact of different filters on di/dt of the surge arrester, SA-42b is tested under the 200 kV impulse (the impulse that can be handled via the conventional SA-42b, see Table 3), and results are depicted in Figure 30. As can be seen from this figure, the current flow through the surge arrester without any filter reached to about 73 kA with a di/dt of about 8 kA/μs, but after considering the 100 μH filter, the current has been limited to ~18 kA (i.e., an ~75.3% decrement) and the di/dt has been decreased to ~0.42 kA/µs. By applying the 250 µH, 500 µH, and 1 mH filters, the current decrease to about 9.8 kA, 5.8 kA, and 2.5 kA, respectively, with the di/dt of about 0.19 kA/µs, 0.09 kA/µs, and 0.04 kA/µs. Results confirm the superiority of using filtered surge arrester over the conventional one.

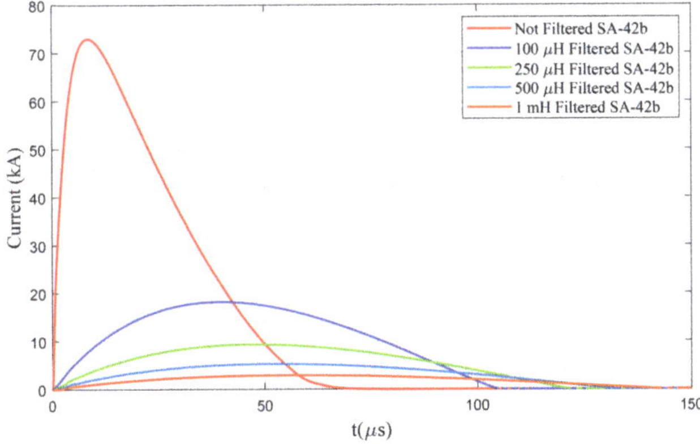

Figure 30. Comparison of the current flow through surge arresters SA-42b against the 200 kV lightning impulse under different filtering conditions.

It is noteworthy to mention that equipping the surge arresters with an inductor-based filter does not alter the residual voltage of surge arresters. This can be seen by comparing the residual voltage presented in Figure 26 with the residual voltage of each surge arrester in Table 1. The filtering devices, by controlling the energy pushed into the surge arrester, enhance their performances against lightning surges with higher overvoltage amplitudes, and in some cases, they eliminate the unwanted voltage sags, besides, they highly decrease the steepness of the current flow through the surge arrester.

3.3.4. Case 4. Transformer Protected by the Series Connection of Filtered Surge Arrester and Spark Gap

This case investigates the impact of a joint filtered surge arrester and spark gap on the protection against indirect lightning impulses. It is worth mentioning that the same scenarios presented in Case 2 (see Section 3.3.2) are also valid for this case. However, the focus of this case is to show the impact of the spark gap on enhancing the overvoltage stress and absorbed energy by surge arrester, compared to Case 3 (see Section 3.3.3). To this end, the performances of surge arresters presented in Figure 26 (which provided proper protection against the 200 kV lightning impulse without considering the spark gap) are studied at the presence of a joint filtered surge arrester and spark gap.

Figure 31a–f provides a comparison among the protection mechanisms of the filtered surge arrester and joint filtered surge arrester and spark gap against a 200 kV lightning impulse. In this figure, the solid curves present the measurements made at the presence of filtered surge arresters, while the dashed curves stand for the measurements made at the presence of a filtered surge arrester and spark gap. Note that Figure 31f presents the protection mechanisms of protecting devices with conventional (non-filtered) surge arrester SA-42b. As can be seen from the configurations in which a filter has been used (Figure 31a–e), the main difference between the protection mechanisms of filtered surge arrester (solid curve) and the joint filtered surge arrester and spark gap (dashed curve) is the voltage sag resulted at the beginning of lightning due to triggering the spark gap, while the surge arrester alone does not cause such voltage sags. However, the spark has a small impact on decreasing the absorbed energy by surge arrester that increases the functionality margin against higher lightning impulses, although negligible. As an instance, in Figure 31a, the energy pushed into the surge arrester when no spark gap is considered is ~59.33 kJ, while for the case with the spark gap, the energy is decreased to ~57.65 kJ, which shows 2.83% enhancement. Moreover, for lower rating surge arresters connected with spark gaps, the operating time is a bit less than the time required by using only surge arresters (see Figure 31a–c). Such a situation is not valid for higher rating surge arresters (see Figure 31d–f).

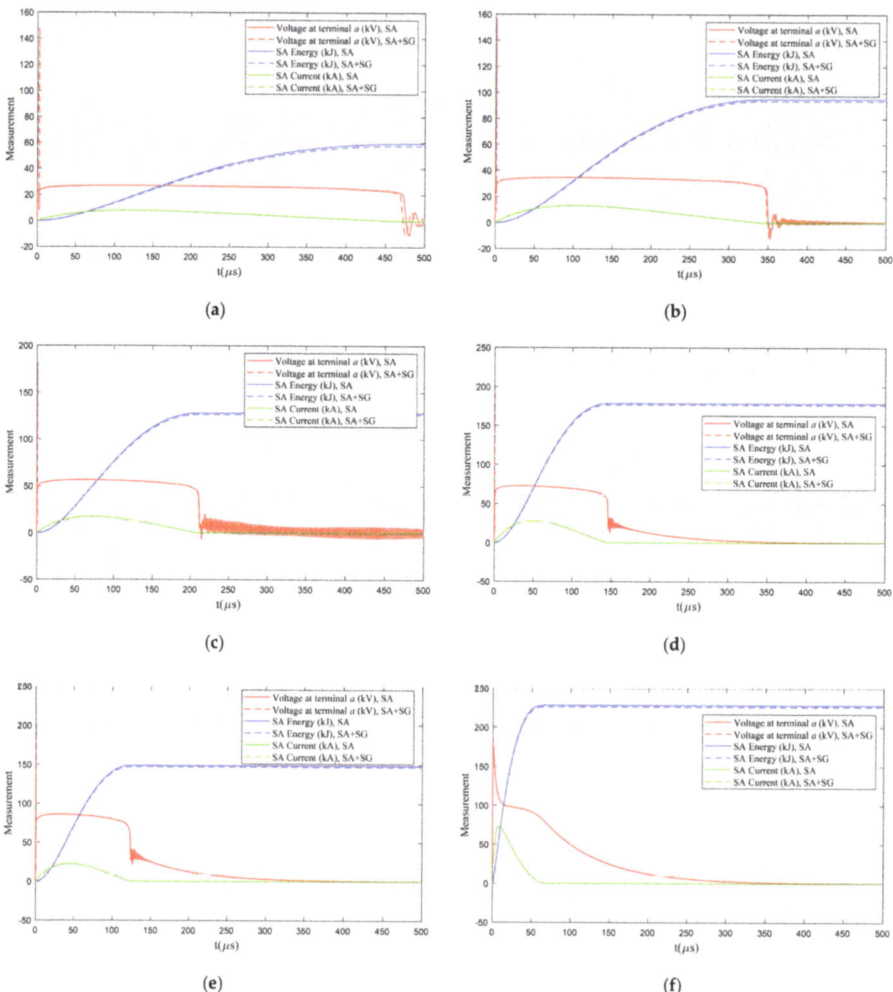

Figure 31. Comparison of the energy absorbed by different surge arresters against the 200 kV lightning impulse: (**a**) 1 mH filtered SA-12b, (**b**) 500 μH filtered SA-18b, (**c**) 250 μH filtered SA-24b, (**d**) 100 μH filtered SA-30b, (**e**) 100 μH filtered SA-36b, and (**f**) non-filtered SA-42b.

Figure 31f shows the mechanism of the non-filtered surge arrester and spark gap against the 200 kV lightning impulse. Unlike Figure 31a–e, in which at the presence of the filtered surge arresters (solid lines), no voltage sag was occurring (the voltage sag was happening only at the presence of joint filtered surge arrester and spark gap), in Figure 31f, and at the presence of non-filtered surge arrester SA-42b, voltage sag is observed. To see this phenomenon more clearly, Figure 32 presents the performance of SA-42b for the first 5 μs of lightning impulse. As can be seen from this figure, the spark gap is triggered at ~0.5 μs after the lightning strike, and this results in even sharper voltage sag than the case with only surge arrester. Although triggering the spark gap decreases the energy pushed into the surge arrester (blue dashed curve), the overvoltage trends for both protection approaches (i.e., surge arrester and joint surge arrester and spark gap) after ~14 µs are similar (see solid and dashed red curves in Figures 31f and 32).

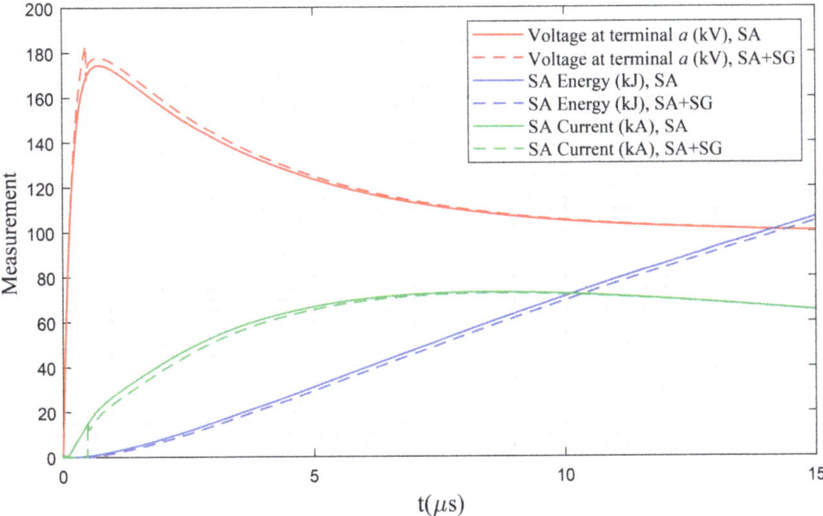

Figure 32. Comparison of the energy absorbed by different surge arresters against the 200 kV lightning impulse.

Figure 33 presents the current flowing through the filtered surge arresters under the highest handleable overvoltage impulse, see Table 3. As can be seen, this figure resembles very much Figure 29 and just a negligible enhancement has been obtained. Therefore, in this case, the range of the di/dt is almost similar to Case 3.

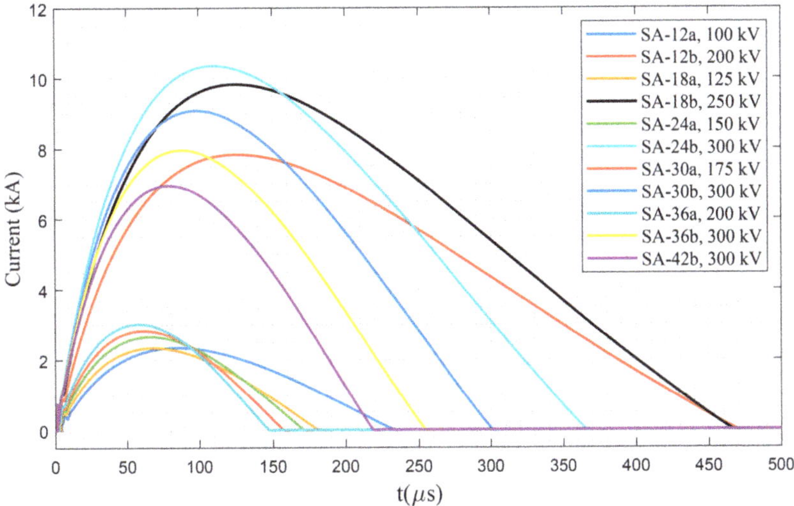

Figure 33. Comparison of the current flow through different surge arresters against the highest handleable lightning impulse.

All in all, a limited enhancement can be achieved by considering a spark gap in series with surge arresters where the absorbed energy by the surge arrester or the current flow through them are negligibly decreased, and therefore better protection for higher lightning impulses results. It should be noted that the spark gap can play an important role during the failure of the surge arrester. Each surge arrester, if not deteriorated under overvoltage stress, needs approximately 45 to 60 min for cooling down [47], and practically it behaves as a short circuit. Therefore, the spark gap can protect the MV transformers against the lightning impulses during this cooling down period, although by interrupting the continuous service to the electricity consumers.

4. Conclusions

This work has proposed an innovative technical approach in enhancing the performance of surge arresters to protect the MV transformers against indirect lightning overvoltages. More often than not, surge arresters fail after experiencing a thermal runaway as a result of receiving higher energy than their thermal energy absorption limit. An easy way to prevent such failures is to utilize a higher rating surge arrester with a desired energy class, however, it imposes extra costs to the system operator. In this paper, an inductor has been used as a filtering device to limit the energy pushed into the surge arrester. By controlling the energy of a surge arrester, the failure is prevented, and a lower rating surge arrester can be used instead of the high rating expensive surge arresters. In order to provide in-depth analyses, the performance of eleven surge arresters with ratings from 12 kV to 42 kV with two energy classes are investigated under different lightning impulses such as 100 kV, 125 kV, 150 kV, 175 kV, 200 kV, 250 V, 30 kV, and 500 kV. The filters that have been used to control the energy levels are 100 μH, 250 μH, 500 μH, and 1 mH inductors installed before the surge arrester. Besides considering the performance of the proposed filtered surge arrester configuration, the impacts of the spark gap on the performance of this configuration have also been studied. An energy-controlled switch has been proposed to monitor the thermal energy of the surge arrester and to simulate the failure. Results show the effectiveness of equipping surge arresters with an inductor-based filter. For instance, equipping a 1 mH filter resulted in a considerable enhancement in the protective performance of a 12 kV rating surge arrester (i.e., SA-12b). A non-filtered surge arrester SA-12b could only protect the MV transformer against 100 kV lightning impulses, while a 1 mH filtered surge arrester SA-12b provides proper protection against 200 kV lightning impulses. The only non-filtered surge arrester, among the considered surge arresters in this paper, that provides proper protection against 200 kV lightning overvoltages is a 42 kV surge arrester (i.e., SA-42b), though, unlike a 1 mH filtered surge arrester SA-12b, it shows inappropriate protection at the beginning of 200 kV lightning overvoltages where a voltage sag is reached to the transformer terminal. Moreover, the surge arrester SA-42b can only decrease the overvoltage tension to 98.7 kV, while a 1 mH filtered surge arrester SA-12b keeps the overvoltage tension below 28.2 kV. It has also been shown that the spark gap installed in series with the surge arrester may help in decreasing the absorbed energy by the surge arrester only slightly. Furthermore, results confirm that by using the filtered surge arrester, the di/dt decreases significantly such that, in this work, without installing a filter, the range of di/dt used was between 5.7 kA/μs and 11.0 kA/μs, and after considering the filtering device the range was between 0.04 kA/μs and 0.11 kA/μs, which is yet another resounding outcome.

All in all, the results (a) show the importance of considering the thermal absorption limit of surge arrester for transient overvoltage studies, and (b) reveal that instead of using an expensive high rating surge arrester to provide proper protection for the MV transformers, a low rating surge arrester, if equipped with a proper filter, can be used. A proper filtered low rating surge arrester not only provides a similar or even better protective performance against the lightning overvoltages, but also proposes a lower residual overvoltage than the proper high rating surge arresters. Therefore, the proposed filter, by boosting the performance of surge arresters, prolongs its lifetime by limiting the energy pushed into the surge arrester and preventing any unwanted failure. Above all, installing low rating surge arresters instead of high rating surge arresters results in considerable savings for the system operator. It is noteworthy to mention that the filtering device that has been applied to modify the performance of the surge arrester might be its longitudinal withstand voltage level, limited applicability against direct lightning strikes, as well as the size. These limitations need to be studied deeply.

In future works we aim to investigate (1) the impact of the proposed filter on mitigating the effects of lead length [51] and (2) the impact of the earthing system (soil ionization) as well as the conductive coupling of the surge arresters on the performance of the proposed protective device while taking into account the aforementioned limitations.

Author Contributions: M.P.-K. proposed the main idea, developed the model, performed the simulations, and prepared the original research draft. M.L. proposed the primary modifications, validated the approach, and supervised the work. All authors have read and agreed to the published version of the manuscript.

Funding: The APC was funded by Aalto University.

Conflicts of Interest: The authors declare no conflict of interest.

References

1. Visacro Filho, S. *Atmospheric Discharge: An Engineering Approach (Descargas Atmosféricas: Uma abordagem de engenharia)*; Artliber Editora: São Paulo, Brazil, 2005; ISBN 8588098318.
2. Keitoue, S.; Murat, I.; Filipović-Grčić, B.; Župan, A.; Damjanović, I.; Pavić, I. Lightning caused overvoltages on power transformers recorded by on-line transient overvoltage monitoring system. *J. Energy Energ.* **2018**, *67*, 44–53.
3. Furgał, J. Influence of Lightning Current Model on Simulations of Overvoltages in High Voltage Overhead Transmission Systems. *Energies* **2020**, *13*, 296. [CrossRef]
4. Sabiha, N.A.; Lehtonen, M. Overvoltage spikes transmitted through distribution transformers due to MV spark-gap operation. In Proceedings of the 2010 Electric Power Quality and Supply Reliability Conference, Kuressaare, Estonia, 16–18 June 2010; pp. 207–214.
5. Heine, P.; Lehtonen, M.; Oikarinen, A. Overvoltage protection, faults and voltage sags. In Proceedings of the 2004 11th International Conference on Harmonics and Quality of Power (IEEE Cat. No.04EX951), Lake Placid, NY, USA, 12–15 September 2004; pp. 100–105.
6. Mahmood, F.; Rizk, M.E.M.; Lehtonen, M. Evaluation of Lightning Overvoltage Protection Schemes for Pole-Mounted Distribution Transformers. *Int. Rev. Electr. Eng.* **2015**, *10*, 616. [CrossRef]
7. Pazos, F.J.; Amantegui, J.; Ferrandis, F.; Barona, A. Overvoltages in low voltage systems due to normal medium voltage switching. In Proceedings of the 19th International Conference on Electricity Distribution (CIRED), Vienna, Austria, 21–24 May 2007; pp. 1–4.
8. Meng, P.; Gu, S.; Wang, J.; Hu, J.; He, J. Improving electrical properties of multiple dopant ZnO varistor by doping with indium and gallium. *Ceram. Int.* **2018**, *44*, 1168–1171. [CrossRef]
9. Opsahl, A.M.; Brookes, A.S.; Southgate, R.N. Lightning Protection for Distribution Transformers. *Trans. Am. Inst. Electr. Eng.* **1932**, *51*, 245–251. [CrossRef]
10. Mikropoulos, P.N.; Tsovilis, T.E.; Koutoula, S.G. Lightning Performance of Distribution Transformer Feeding GSM Base Station. *Ieee Trans. Power Deliv.* **2014**, *29*, 2570–2579. [CrossRef]
11. Cigre Working Group A2.37. *Transformer Reliability Survey*; Technical brochure no. 642; International Council on Large Electric Systems: Paris, France, 2015.
12. Meng, P.; Yuan, C.; Xu, H.; Wan, S.; Xie, Q.; He, J.; Zhao, H.; Hu, J.; He, J. Improving the protective effect of surge arresters by optimizing the electrical property of ZnO varistors. *Electr. Power Syst. Res.* **2020**, *178*, 106041. [CrossRef]
13. Meng, P.; Lyu, S.; Hu, J.; He, J. Tailoring low leakage current and high nonlinear coefficient of a Y-doped ZnO varistor by indium doping. *Mater. Lett.* **2017**, *188*, 77–79. [CrossRef]
14. Meng, P.; Zhou, Y.; Wu, J.; Hu, J.; He, J. Novel ZnO Varistors for Dramatically Improving Protective Effect of Surge Arresters. In Proceedings of the 2018 34th International Conference on Lightning Protection (ICLP), Rzeszow, Poland, 2–7 September 2018; pp. 1–6.
15. He, J. *Metal. Oxide Varistors: From Microstructure to Macro-Characteristics*; Wiley-VCH Verlag GmbH & Co. KGaA: Weinheim, Germany, 2019; ISBN 9783527684038.

16. Zi-Ming, H.; Xiu-Juan, C.; Wei-Jiang, C. The Power Loss Calculation Model of Metal Oxide Resistors at Continuous Operating Voltage. In Proceedings of the 2019 11th Asia-Pacific International Conference on Lightning (APL), Hong Kong, 12–14 June 2019; pp. 1–6.
17. Piantini, A.; de Carvalho, T.O.; Obase, P.F.; Janiszewski, J.M.; Santos, G.J.G.; Fagundes, D.R.; Uchoa, J.I.L.; Kunz, E.N. The effect of the distance between transformer and MV arresters on the surges transferred to the LV side. In Proceedings of the 2012 International Conference on High Voltage Engineering and Application, Shanghai, China, 17 September 2012; pp. 105–109.
18. Josephine, M.M.; Ikechukwu, G.A. Performance of Surge Arrester Installation to Enhance Protection. In Proceedings of the World Congress on Engineering and Computer Science 2016 Vol I, San Francisco, CA, USA, 19–21 October 2016; pp. 1–7.
19. Christodoulou, C.A.; Vita, V.; Maris, T.I. On the optimal placement of surge arresters for the efficient protection of medium voltage distribution networks against atmospheric overvoltages. In Proceedings of the 2019 54th International Universities Power Engineering Conference (UPEC), Bucharest, Romania, 3–6 September 2019; pp. 1–4.
20. Firouzjah, K.G. Distribution Network Expansion Based on the Optimized Protective Distance of Surge Arresters. *Ieee Trans. Power Deliv.* **2018**, *33*, 1735–1743. [CrossRef]
21. Sabiha, N.A.; Lehtonen, M. Lightning-Induced Overvoltages Transmitted Over Distribution Transformer With MV Spark-Gap Operation—Part II: Mitigation Using LV Surge Arrester. *IEEE Trans. Power Deliv.* **2010**, *25*, 2565–2573. [CrossRef]
22. Somogyi, A.; Vizi, L. Overvoltage protection of pole mounted distribution transformers. *Period. Polytech. Electr. Eng.* **1997**, *41*, 27–40.
23. Application Note 2.0— Metal-oxide surge arresters in medium-voltage systems. In *Overvoltage Protection of Transformers*; ABB Switzerland Ltd.: Baden, Switzerland, 2018.
24. Piasecki, W.; Kuczek, T.; Stosur, M.; Steiger, M.; Szewczyk, M. Investigation on new mitigation method for lightning overvoltages in high-voltage power substations. *Iet Gener. Transm. Distrib.* **2013**, *7*, 1055–1062. [CrossRef]
25. Visacro, S. Statistical analysis of lightning current parameters: Measurements at Morro do Cachimbo Station. *J. Geophys. Res.* **2004**, *109*, D01105. [CrossRef]
26. De Conti, A.; Perez, E.; Soto, E.; Silveira, F.H.; Visacro, S.; Torres, H. Calculation of Lightning-Induced Voltages on Overhead Distribution Lines Including Insulation Breakdown. *IEEE Trans. Power Deliv.* **2010**, *25*, 3078–3084. [CrossRef]
27. Kuffel, E.; Zaengl, W.S.; Kuffel, J. *High. Voltage Engineering Fundamentals*; Elsevier: Amsterdam, The Netherlands, 2000; ISBN 9780750636346.
28. Jia, W.; Xiaoqing, Z. Double-Exponential Expression of Lightning Current Waveforms. In Proceedings of the The 2006 4th Asia-Pacific Conference on Environmental Electromagnetics, Dalian, China, 1–4 August 2006; pp. 320–323.
29. Pourakbari-Kasmaei, M.; Mahmoud, F.; Krbal, M.; Pelikan, L.; Orságová, J.; Toman, P.; Lehtonen, M. Optimal Adjustment of Double Exponential Model Parameters to Reproduce the Laboratory Volt-Time Curve of Lightning Impulse. Available online: https://drive.google.com/file/d/15NgOIIbDuCZIiqTqXzC7QeMKND8O22Tj/view?usp=usp=sharing (accessed on 15 August 2020).
30. Electromagnetic Transients Program-Restructured Version (EMTP-RV). Available online: https://www.emtp-software.com/ (accessed on 15 August 2020).
31. Pourakbari-Kasmaei, M.; Lehtonen, M. GAMS Code for Obtaining Optimal Parameters of Double Exponential Function. Available online: https://drive.google.com/drive/folders/19IBsft1DEfb57IJ_frpmvnTRsuJn5FpN?usp=sharing (accessed on 15 August 2020).
32. Cigre Working Group 02 (Internal overvoltages) Of Study Committee 33. *Guidelines for Representation of Network Elements When Calculating Transients*; International Council on Large Electric Systems: Paris, France, 1990; 30p.
33. Pourakbari-Kasmaei, M.; Mahmood, F.; Krbal, M.; Pelikan, L.; Orságová, J.; Toman, P.; Lehtonen, M. Evaluation of Filtered Spark Gap on the Lightning Protection of Distribution Transformers: Experimental and Simulation Study. *Energies* **2020**, *13*, 3799. [CrossRef]

34. Savadamuthu, U.; Udayakumar, K.; Jayashankar, V. Modified disruptive effect method as a measure of insulation strength for non-standard lightning waveforms. *IEEE Trans. Power Deliv.* **2002**, *17*, 510–515. [CrossRef]
35. Darveniza, M. The generalized integration method for predicting impulse volt-time characteristics for non-standard wave shapes-a theoretical basis. *IEEE Trans. Electr. Insul.* **1988**, *23*, 373–381. [CrossRef]
36. IEEE Working Group. 3.4.11: Application of Surge Protective Devices Subcommittee- Surge Protective Device Committee. Modeling of metal oxide surge arresters. *IEEE Trans. Power Deliv.* **1992**, *7*, 302–309. [CrossRef]
37. Siemens Catalogue 31.1 HG Medium-Voltage Surge Arresters: Product Guide. Available online: https://assets.new.siemens.com/siemens/assets/api/uuid:a6c3362fa89bb8e6ba4c444b07ab680a87a4105a/version:0/medium-voltage-surge-arresters-catalog-hg-31-1-2017-low-resoluti.pdf (accessed on 15 August 2020).
38. IEEE Standards Board. 1313-1993-IEEE Standard for Power Systems - Insulation Coordination. *IEEE Std 1313-1993* **1993**. [CrossRef]
39. Technical Specification of Transformer. Available online: https://procurement-notices.undp.org/view_file.cfm?doc_id=104038 (accessed on 15 August 2020).
40. National Electrical Manufacturers Association Understanding the Arrester Datasheet. Available online: https://www.nemaarresters.org/understanding-arrester-datasheet/ (accessed on 15 August 2020).
41. Göhler, R.; Hinrichsen, V. Metal-Oxide Surge Arresters in High-Voltage Power Systems: Fundamentals. Available online: https://assets.new.siemens.com/siemens/assets/api/uuid:64a37729c7408df1e15b39ca8c9cde8b957a5463/e50001-g630-h197-x-4a00-ableiterhandbuch-teil-1-a4.pdf (accessed on 15 August 2020).
42. INMR Principal Failure Modes for Surge Arresters. Available online: https://www.inmr.com/principal-failure-modes-surge-arresters/#:~{}:text=Inmostscenarios%2Cfailureoccurs,orswitching)orlightningor (accessed on 15 August 2020).
43. Novizon, N.; Malek, Z.A.; Ahmad, M.H.; Waldi, E.P.; Aulia, S.; Laksono, H.D.; Riska, N. Power loss estimation of polymeric housing surge arrester using leakage current and temperature approach. *Iop Conf. Ser. Mater. Sci. Eng.* **2019**, *602*, 012008. [CrossRef]
44. Tsovilis, T.E.; Topcagic, Z. DC Overload Behavior of Low-Voltage Varistor-Based Surge Protective Devices. *Ieee Trans. Power Deliv.* **2020**, 1. [CrossRef]
45. Topcagic, Z.; Tsovilis, T.E. Varistor Electrical Properties: Microstructural Effects. In *Reference Module in Materials Science and Materials Engineering*; Elsevier: Amsterdam, The Netherlands, 2020.
46. IEEE. 1410 IEEE guide for improving the lightning performance of electric power overhead distribution lines. *IEEE Std 1410-2010 (Revision IEEE Std 1410-2004)* **2011**, 1–73. [CrossRef]
47. Richter, B. *Application Guidelines: Overvoltage Protection Metal-Oxide Surge Arresters in Medium-Voltage Systems*; ABB Switzerland Ltd.: Wettingen, Switzerland, 2018.
48. Cigre Working Group 01 (Lightning) of Study Committee 33 (Overvoltages and Insulation Co-ordination). *Guide to Procedures for Estimating the Lightning Performance of Transmission Lines*; Technical brochure no. 063; International Council on Large Electric Systems: Paris, France, 1991.
49. Pourakbari-Kasmaei, M.; Mahmoud, F.; Krbal, M.; Pelikan, L.; Orságová, J.; Toman, P.; Lehtonen, M. Transformer with serial inductors and protection spark gap. Available online: https://drive.google.com/file/d/1ztdvBW2QNNZx1t6XsAY6IMTzq-nxAlVc/view?usp=sharing (accessed on 15 August 2020).
50. Pourakbari-Kasmaei, M.; Mahmood, F.; Lehtonen, M. Optimized Protection of Pole-Mounted Distribution Transformers against Direct Lightning Strikes. *Energies* **2020**, *13*, 4372. [CrossRef]
51. Datsios, Z.G.; Panteleimon, M.; Politis, Z.; Kagiannas, A.G.; Tsovilis, T.E. Protection of distribution transformer against arising or transferred fast-front overvoltages: Effects of surge arrester connection conductors length. In Proceedings of the 18th International Symposium on High Voltage Engineering, Seoul, Korea, 25–30 August 2013.

MDPI
St. Alban-Anlage 66
4052 Basel
Switzerland
Tel. +41 61 683 77 34
Fax +41 61 302 89 18
www.mdpi.com

Energies Editorial Office
E-mail: energies@mdpi.com
www.mdpi.com/journal/energies